general topology and applications

LECTURE NOTES

IN PURE AND APPLIED MATHEMATICS

1. *N. Jacobson,* Exceptional Lie Algebras
2. *L. -Å. Lindahl and F. Poulsen,* Thin Sets in Harmonic Analysis
3. *I. Satake,* Classification Theory of Semi-Simple Algebraic Groups
4. *F. Hirzebruch, W. D. Newmann, and S. S. Koh,* Differentiable Manifolds and Quadratic Forms (out of print)
5. *I. Chavel,* Riemannian Symmetric Spaces of Rank One (out of print)
6. *R. B. Burckel,* Characterization of C(X) Among Its Subalgebras
7. *B. R. McDonald, A. R. Magid, and K. C. Smith,* Ring Theory: Proceedings of the Oklahoma Conference
8. *Y.-T. Siu,* Techniques of Extension on Analytic Objects
9. *S. R. Caradus, W. E. Pfaffenberger, and B. Yood,* Calkin Algebras and Algebras of Operators on Banach Spaces
10. *E. O. Roxin, P.-T. Liu, and R. L. Sternberg,* Differential Games and Control Theory
11. *M. Orzech and C. Small,* The Brauer Group of Commutative Rings
12. *S. Thomeier,* Topology and Its Applications
13. *J. M. Lopez and K. A. Ross,* Sidon Sets
14. *W. W. Comfort and S. Negrepontis,* Continuous Pseudometrics
15. *K. McKennon and J. M. Robertson,* Locally Convex Spaces
16. *M. Carmeli and S. Malin,* Representations of the Rotation and Lorentz Groups: An Introduction
17. *G. B. Seligman,* Rational Methods in Lie Algebras
18. *D. G. de Figueiredo,* Functional Analysis: Proceedings of the Brazilian Mathematical Society Symposium
19. *L. Cesari, R. Kannan, and J. D. Schuur,* Nonlinear Functional Analysis and Differential Equations: Proceedings of the Michigan State University Conference
20. *J. J. Schäffer,* Geometry of Spheres in Normed Spaces
21. *K. Yano and M. Kon,* Anti-Invariant Submanifolds
22. *W. V. Vasconcelos,* The Rings of Dimension Two
23. *R. E. Chandler,* Hausdorff Compactifications
24. *S. P. Franklin and B. V. S. Thomas,* Topology: Proceedings of the Memphis State University Conference
25. *S. K. Jain,* Ring Theory: Proceedings of the Ohio University Conference
26. *B. R. McDonald and R. A. Morris,* Ring Theory II: Proceedings of the Second Oklahoma Conference
27. *R. B. Mura and A. Rhemtulla,* Orderable Groups
28. *J. R. Graef,* Stability of Dynamical Systems: Theory and Applications
29. *H.-C. Wang,* Homogeneous Branch Algebras
30. *E. O. Roxin, P.-T. Liu, and R. L. Sternberg,* Differential Games and Control Theory II
31. *R. D. Porter,* Introduction to Fibre Bundles
32. *M. Altman,* Contractors and Contractor Directions Theory and Applications
33. *J. S. Golan,* Decomposition and Dimension in Module Categories
34. *G. Fairweather,* Finite Element Galerkin Methods for Differential Equations
35. *J. D. Sally,* Numbers of Generators of Ideals in Local Rings
36. *S. S. Miller,* Complex Analysis: Proceedings of the S.U.N.Y. Brockport Conference
37. *R. Gordon,* Representation Theory of Algebras: Proceedings of the Philadelphia Conference
38. *M. Goto and F. D. Grosshans,* Semisimple Lie Algebras
39. *A. I. Arruda, N. C. A. da Costa, and R. Chuaqui,* Mathematical Logic: Proceedings of the First Brazilian Conference

40. *F. Van Oystaeyen,* Ring Theory: Proceedings of the 1977 Antwerp Conference
41. *F. Van Oystaeyen and A. Verschoren,* Reflectors and Localization: Application to Sheaf Theory
42. *M. Satyanarayana,* Positively Ordered Semigroups
43. *D. L. Russell,* Mathematics of Finite-Dimensional Control Systems
44. *P.-T. Liu and E. Roxin,* Differential Games and Control Theory III: Proceedings of the Third Kingston Conference, Part A
45. *A. Geramita and J. Seberry,* Orthogonal Designs: Quadratic Forms and Hadamard Matrices
46. *J. Cigler, V. Losert, and P. Michor,* Banach Modules and Functors on Categories of Banach Spaces
47. *P.-T. Liu and J. G. Sutinen,* Control Theory in Mathematical Economics: Proceedings of the Third Kingston Conference, Part B
48. *C. Byrnes,* Partial Differential Equations and Geometry
49. *G. Klambauer,* Problems and Propositions in Analysis
50. *J. Knopfmacher,* Analytic Arithmetic of Algebraic Function Fields
51. *F. Van Oystaeyen,* Ring Theory: Proceedings of the 1978 Antwerp Conference
52. *B. Kedem,* Binary Time Series
53. *J. Barros-Neto and R. A. Artino,* Hypoelliptic Boundary-Value Problems
54. *R. L. Sternberg, A. J. Kalinowski, and J. S. Papadakis,* Nonlinear Partial Differential Equations in Engineering and Applied Science
55. *B. R. McDonald,* Ring Theory and Algebra III: Proceedings of the Third Oklahoma Conference
56. *J. S. Golan,* Structure Sheaves over a Noncommutative Ring
57. *T. V. Narayana, J. G. Williams, and R. M. Mathsen,* Combinatorics, Representation Theory and Statistical Methods in Groups: YOUNG DAY Proceedings
58. *T. A. Burton,* Modeling and Differential Equations in Biology
59. *K. H. Kim and F. W. Roush,* Introduction to Mathematical Consensus Theory
60. *J. Banas and K. Goebel,* Measures of Noncompactness in Banach Spaces
61. *O. A. Nielson,* Direct Integral Theory
62. *J. E. Smith, G. O. Kenny, and R. N. Ball,* Ordered Groups: Proceedings of the Boise State Conference
63. *J. Cronin,* Mathematics of Cell Electrophysiology
64. *J. W. Brewer,* Power Series Over Commutative Rings
65. *P. K. Kamthan and M. Gupta,* Sequence Spaces and Series
66. *T. G. McLaughlin,* Regressive Sets and the Theory of Isols
67. *T. L. Herdman, S. M. Rankin, III, and H. W. Stech,* Integral and Functional Differential Equations
68. *R. Draper,* Commutative Algebra: Analytic Methods
69. *W. G. McKay and J. Patera,* Tables of Dimensions, Indices, and Branching Rules for Representations of Simple Lie Algebras
70. *R. L. Devaney and Z. H. Nitecki,* Classical Mechanics and Dynamical Systems
71. *J. Van Geel,* Places and Valuations in Noncommutative Ring Theory
72. *C. Faith,* Injective Modules and Injective Quotient Rings
73. *A. Fiacco,* Mathematical Programming with Data Perturbations I
74. *P. Schultz, C. Praeger, and R. Sullivan,* Algebraic Structures and Applications Proceedings of the First Western Australian Conference on Algebra
75. *L. Bican, T. Kepka, and P. Nemec,* Rings, Modules, and Preradicals
76. *D. C. Kay and M. Breen,* Convexity and Related Combinatorial Geometry: Proceedings of the Second University of Oklahoma Conference
77. *P. Fletcher and W. F. Lindgren,* Quasi-Uniform Spaces
78. *C.-C. Yang,* Factorization Theory of Meromorphic Functions
79. *O. Taussky,* Ternary Quadratic Forms and Norms
80. *S. P. Singh and J. H. Burry,* Nonlinear Analysis and Applications
81. *K. B. Hannsgen, T. L. Herdman, H. W. Stech, and R. L. Wheeler,* Volterra and Functional Differential Equations

82. *N. L. Johnson, M. J. Kallaher, and C. T. Long,* Finite Geometries: Proceedings of a Conference in Honor of T. G. Ostrom
83. *G. I. Zapata,* Functional Analysis, Holomorphy, and Approximation Theory
84. *S. Greco and G. Valla,* Commutative Algebra: Proceedings of the Trento Conference
85. *A. V. Fiacco,* Mathematical Programming with Data Perturbations II
86. *J.-B. Hiriart-Urruty, W. Oettli, and J. Stoer,* Optimization: Theory and Algorithms
87. *A. Figa Talamanca and M. A. Picardello,* Harmonic Analysis on Free Groups
88. *M. Harada,* Factor Categories with Applications to Direct Decomposition of Modules
89. *V. I. Istrăţescu,* Strict Convexity and Complex Strict Convexity: Theory and Applications
90. *V. Lakshmikantham,* Trends in Theory and Practice of Nonlinear Differential Equations
91. *H. L. Manocha and J. B. Srivastava,* Algebra and Its Applications
92. *D. V. Chudnovsky and G. V. Chudnovsky,* Classical and Quantum Models and Arithmetic Problems
93. *J. W. Longley,* Least Squares Computations Using Orthogonalization Methods
94. *L. P. de Alcantara,* Mathematical Logic and Formal Systems
95. *C. E. Aull,* Rings of Continuous Functions
96. *R. Chuaqui,* Analysis, Geometry, and Probability
97. *L. Fuchs and L. Salce,* Modules Over Valuation Domains
98. *P. Fischer and W. R. Smith,* Chaos, Fractals, and Dynamics
99. *W. B. Powell and C. Tsinakis,* Ordered Algebraic Structures
100. *G. M. Rassias and T. M. Rassias,* Differential Geometry, Calculus of Variations, and Their Applications
101. *R.-E. Hoffmann and K. H. Hofmann,* Continuous Lattices and Their Applications
102. *J. H. Lightbourne, III, and S. M. Rankin, III,* Physical Mathematics and Nonlinear Partial Differential Equations
103. *C. A. Baker and L. M. Batten,* Finite Geometries
104. *J. W. Brewer, J. W. Bunce, and F. S. Van Vleck,* Linear Systems Over Commutative Rings
105. *C. McCrory and T. Shifrin,* Geometry and Topology: Manifolds, Varieties, and Knots
106. *D. W. Kueker, E. G. K. Lopez-Escobar, and C. H. Smith,* Mathematical Logic and Theoretical Computer Science
107. *B.-L. Lin and S. Simons,* Nonlinear and Convex Analysis: Proceedings in Honor of Ky Fan
108. *S. J. Lee,* Operator Methods for Optimal Control Problems
109. *V. Lakshmikantham,* Nonlinear Analysis and Applications
110. *S. F. McCormick,* Multigrid Methods: Theory, Applications, and Supercomputing
111. *M. C. Tangora,* Computers in Algebra
112. *D. V. Chudnovsky and G. V. Chudnovsky,* Search Theory: Some Recent Developments
113. *D. V. Chudnovsky and R. D. Jenks,* Computer Algebra
114. *M. C. Tangora,* Computers in Geometry and Topology
115. *P. Nelson, V. Faber, T. A. Manteuffel, D. L. Seth, and A. B. White, Jr.* Transport Theory, Invariant Imbedding, and Integral Equations: Proceedings in Honor of G. M. Wing's 65th Birthday
116. *P. Clément, S. Invernizzi, E. Mitidieri, and I. I. Vrabie,* Semigroup Theory and Applications
117. *J. Vinuesa,* Orthogonal Polynomials and Their Applications: Proceedings of the International Congress
118. *C. M. Dafermos, G. Ladas, and G. Papanicolaou,* Differential Equations: Proceedings of the EQUADIFF Conference
119. *E. O. Roxin,* Modern Optimal Control: A Conference in Honor of Solomon Lefschetz and Joseph P. Lasalle
120. *J. C. Díaz,* Mathematics for Large Scale Computing
121. *P. S. Milojević,* Nonlinear Functional Analysis

122. *C. Sadosky,* Analysis and Partial Differential Equations: A Collection of Papers Dedicated to Mischa Cotlar
123. *R. M. Shortt,* General Topology and Applications: Proceedings of the 1988 Northeast Conference
124. *R. Wong,* Asymptotic and Computational Analysis: Conference in Honor of Frank W. J. Olver's 65th Birthday
125. *D. V. Chudnovsky and R. D. Jenks,* Computers in Mathematics
126. *W. D. Wallis, H. Shen, W. Wei, and L. Zhu,* Combinatorial Designs and Applications
127. *S. Elaydi,* Differential Equations: Stability and Control
128. *G. Chen, E. B. Lee, W. Littman, and L. Markus,* Distributed Parameter Control Systems: New Trends and Applications
129. *W. N. Everitt,* Inequalities: Fifty Years On from Hardy, Littlewood and Pólya
130. *H. G. Kaper and M. Garbey,* Asymptotic Analysis and the Numerical Solution of Partial Differential Equations
131. *O. Arino, D.E. Axelrod, and M. Kimmel,* Mathematical Population Dynamics: Proceedings of the Second International Conference
132. *S. Coen,* Geometry and Complex Variables
133. *J. A. Goldstein, F. Kappel, and W. Schappacher,* Differential Equations with Applications in Biology, Physics, and Engineering
134. *S. J. Andima, R. Kopperman, P. R. Misra, J. Z. Reichman, and A. R. Todd,* General Topology and Applications

Other Volumes in Preparation

general topology and applications

fifth northeast conference

edited by

Susan J. Andima
C. W. Post College
Long Island University
Brookville, New York

Ralph Kopperman
The City College of New York
The City University of New York
New York, New York

Prabudh Ram Misra
The College of Staten Island
The City University of New York
Staten Island, New York

Jack Z. Reichman
Washington National Life Insurance Co. of New York
New York, New York

Aaron R. Todd
Baruch College
The City University of New York
New York, New York

Marcel Dekker, Inc. **New York • Basel • Hong Kong**

Library of Congress Cataloging–in--Publication Data

General topology and applications: fifth Northeast conference/
edited by Susan J. Andima...[et al.].
p. cm. -- -- (Lecture notes in pure and applied mathematics; v. 134)
"Proceedings of the Fifth Northeast Conference on General Topology and Applications, held June 15-17, 1989 at The College of Staten Island" -- --Pref.
Includes bibliographical references and index.
ISBN 0-8247-8552-5
1. Topology-- --Congresses. I. Andima, Susan J.
II. Northeast Conference on General Topology and Applications (5th: 1989: College of Staten Island) III. Series
QA611.A1G44 1991
514-- --dc20 91-15709
CIP

This book is printed on acid-free paper.

Marcel Dekker, Inc.
270 Madison Avenue, New York, New York 10016

Current printing (last digit):
10 9 8 7 6 5 4 3 2 1

PRINTED IN THE UNITED STATES OF AMERICA

Preface

This collection is the proceedings of the Fifth Northeast Conference on General Topology and Applications, held June 15-17, 1989 at The College of Staten Island - The City University of New York. It is the fourth of these conferences for which proceedings exist, the second such published through Marcel Dekker, Inc. It seems appropriate to reflect on the history and present status of this series of conferences.

In recent years general topology has grown sufficiently to support a second major annual conference in the United States, as demonstrated by the success of the Summer Conference on General Topology and Applications series. This series began at The City College of New York - CUNY in 1984 as the Independence Day Conference on Limits with 13 talks (9 contributed) and a little over 30 participants. It grew steadily to a 3-day conference with over 70 participants in 1988 when it was held at Wesleyan University as the Northeast Conference on General Topology and Applications. This conference was the first in the series that could truly be called major. It remained to be seen whether these conferences would continue to be strong with their return to New York, and this question was resolved by the 20% increase at the 1989 conference in attendance at The College of Staten Island. Its participants came from 11 different countries; 41 gave talks, of whom 6 were invited: Mel Henriksen, Neil Hindman, Karl H. Hofmann, Peter T. Johnstone, Ellen E. Reed and Mary Ellen Rudin. This success in its turn formed a strong basis for the success of the 1990 conference (now called the Summer Conference on General Topology and Applications), which was held at the C. W. Post Campus of Long Island University and had 117 participants. A seventh in the series, to honor Mary Ellen Rudin and her work, is planned at Madison, Wisconsin, June 26-29, 1991.

A quick glance at the contents of these proceedings will inform the reader of the basic philosophy of our series of conferences; we stress not only general topology but its applications. In addition to papers in general topology and set theoretic topology, we find here papers with motivations from topological groups and semigroups, convergence structures, functional analysis, topological algebra, category theory, Lie group theory, topological dynamics, computer science and several other disciplines. We hope that these articles will give some further insight into the relationship between general topology and other areas of mathematics.

Several factors led to the success of this conference at The College of Staten Island. There was strong administrative support at every level. President Edmond L. Volpe enthusiastically backed his Dean of Science, Richard Resch - in granting the conference sufficient resources. Professor Jane Coffee, Chair of Mathematics, was an instrumental member of the local organizing committee. Ms. Marie Anderson very efficiently and patiently typed all the abstracts and some of the manuscripts. Together, they assured a pleasant conference for the participants.

Some financial and administrative backing came from Baruch College and The City College, both of The City University of New York, the C. W. Post Campus of Long Island University, the New York Academy of Sciences, Notre Dame College of Saint John's University, and Wesleyan University. This and the tradition of the conference allowed us to offer reasonable support to the invited speakers, and encouraged other topologists to contribute talks. We wish to thank all the participants for the positive atmosphere that made the conference thoroughly enjoyable, and the authors for the 27 articles that are contained in these proceedings.

Finally, we wish to thank the anonymous referees, whose careful work and adherence to deadlines led to the quality and timely appearance of this volume.

Susan J. Andima
Ralph Kopperman
Prabudh Ram Misra
Jack Z. Reichman
Aaron R. Todd

Contents

Contributors

K. P. S. Bhaskara Rao Indian Statistical Institute, Bangalore, India

Denis Blackmore New Jersey Institute of Technology, Newark, New Jersey

Bruce S. Burdick Roger Williams College, Bristol, Rhode Island

W. W. Comfort Wesleyan University, Middletown, Connecticut

Charles Dorsett Louisiana Tech University, Ruston, Louisiana

Alan Dow York University, North York, Ontario, Canada

Raquel Ruiz de Eguino Washington State University, Pullman, Washington

Sherif El-Helaly The Catholic University of America, Washington, D.C.

M. Ganster Graz University of Technology, Graz, Austria

Eraldo Giuli University of L'Aquila, L'Aquila, Italy

Douglass L. Grant University College of Cape Breton, Sydney, Nova Scotia, Canada

F. Gressl Graz University of Technology, Graz, Austria

T. R. Hamlett East Central University, Ada, Oklahoma

Melvin Henriksen Harvey Mudd College, Claremont, California

Neil Hindman Howard University, Washington, D.C.

Karl H. Hofmann Darmstadt Institute of Technology, Darmstadt, Germany

Taqdir Husain McMaster University, Hamilton, Ontario, Canada

Gerald L. Itzkowitz Queens College, The City University of New York, Flushing, New York

Peter T. Johnstone University of Cambridge, Cambridge, England

Darrell C. Kent Washington State University, Pullman, Washington

Yuri Kochetkov* Stockholm University, Stockholm, Sweden

T. Y. Kong Queens College, The City University of New York, Flushing, New York

Ralph Kopperman The City College of New York, The City University of New York, New York, New York

Dimitry Leites Stockholm University, Stockholm, Sweden

Anthony J. Macula Westfield State College, Westfield, Massachusetts

Paul R. Meyer Lehman College, The City University of New York, Bronx, New York

Vitor Neves Universidade da Beira Interior, Covilhã, Portugal

Ellen E. Reed Trinity School at Greenlawn, South Bend, Indiana

Ivan L. Reilly University of Auckland, Auckland, New Zealand

Frank Rhodes University of Southampton, Southampton, England

David A. Rose East Central University, Ada, Oklahoma

Niel Shell The City College of New York, The City University of New York, New York, New York

Rae Michael Shortt Wesleyan University, Middletown, Connecticut

Christopher L. Thompson University of Southampton, Southampton, England

F. Javier Trigos-Arrieta Wesleyan University, Middletown, Connecticut

Arkady Weintrob Stockholm University, Stockholm, Sweden

Scott W. Williams State University of New York, Buffalo, New York

Ta-Sun Wu Case Western Reserve University, Cleveland, Ohio

Haoxuan Zhou+ State University of New York, Buffalo, New York

---

Current Affiliations: *University of Twente, Enschede, The Netherlands
+Sichuan University, Chengdu, People's Republic of China

Conference Participants

Susan J. Andima New York, New York
George Baloglou Oswego, New York
Imants Baruss London, Ontario, Canada
Ed Beckenstein Staten Island, New York
Denis Blackmore Newark, New Jersey
Bruce S. Burdick Bristol, Rhode Island
Szvetlana Buzasi Debrecen, Hungary
Bruce Chandler Staten Island, New York
Jane P. Coffee Staten Island, New York
Leslie Cohn Charleston, South Carolina
W. W. Comfort Middletown, Connecticut
Dennis E. Davenport Oxford, Ohio
George Day Meadville, Pennsylvania
Charles Dorsett Ruston, Louisiana
Alan Dow North York, Ontario, Canada
Sherif El-Helaly Washington, D.C.
G. Bert Estes New York, New York
Ben Fitzpatrick Auburn, Alabama
Luther Fuller Old Westbury, New York
Sarvador Garcia Middletown, Connecticut
Charles Giordina Staten Island, New York
Eraldo Giuli L'Aquila, Italy
Douglass L. Grant Sydney, Nova Scotia, Canada
Egbert Harzheim Cologne, Germany

Mariam S. Hastings Tarrytown, New York
Robert Heath Auburn, Alabama
Steve Hechler Queens, New York
Melvin Henriksen Claremont, California
Joseph Hertzlinger Jericho, New York
Neil Hindman Washington, D.C.
Karl H. Hofmann Darmstadt, Germany
Gerald L. Itzkowitz Flushing, New York
David Jacobson Staten Island, New York
Dragan Jankovic Ada, Oklahoma
Don Johnson Lenox, Massachusetts
Peter T. Johnstone Cambridge, England
Darrell C. Kent Pullman, Washington
Efim Khalimsky Staten Island, New York
Joseph Kist Las Cruces, New Mexico
Ann Kizanis Middletown, Connecticut
T. Y. Kong Flushing, New York
Ralph Kopperman New York, New York
Stephen Landry Middletown, Connecticut
Dimitry Leites Stockholm, Sweden
Randi Lerohl New York, New York
Alfred Levine Staten Island, New York
John E. Mack Lexington, Kentucky
Anthony J. Macula Westfield, Massachusetts
James J. Madden South Bend, Indiana
J. Matkowski Bielsko-Biala, Poland
Roz Merzer Staten Island, New York
Prabudh Ram Misra Staten Island, New York
Andrew Molitor Middletown, Connecticut
Shokry Nada Egypt
Vitor Neves Covilhã, Portugal
Charles W. Neville Middletown, Connecticut
Ercument Ozizmir Staten Island, New York
Dev Pandian Naperville, Illinois
Eric Patridge Mansfield, Ohio
Albain Rarivoson Cincinnati, Ohio
Ellen E. Reed South Bend, Indiana
Ivan L. Reilly Auckland, New Zealand
Richard Resch Staten Island, New York
Frank Rhodes Southampton, England
David A. Rose Ada, Oklahoma
Mary Ellen Rudin Madison, Wisconsin
Laurie J. Sawyer Durham, New Hampshire
Niel Shell New York, New York
Rae Michael Shortt Middletown, Connecticut
Robert M. Stephenson Columbia, South Carolina
Helen Strassberg Queens, New York
M. Suarez M. Tunja, Colombia
Shan Li Sun Ottawa, Canada

Angel Tamariz-Mascarua Middletown, Connecticut
Gloria Tashjian Worcester, Massachusetts
Aaron R. Todd New York, New York
F. Javier Trigos-Arrieta Middletown, Connecticut
Jerry Vaughan Greensboro, North Carolina
P. Venugopalan Stamford, Connecticut
Scott W. Williams Buffalo, New York
R. Grant Woods Winnipeg, Manitoba, Canada
Ta-Sun Wu Cleveland, Ohio

ORGANIZING COMMITTEE

Susan J. Andima C. W. Post College, Long Island University, Brookville, New York
W. W. Comfort Wesleyan University, Middletown, Connecticut
Anthony W. Hager Wesleyan University, Middletown, Connecticut
Ralph Kopperman The City College of New York, The City University of New York, New York, New York
Prabudh Ram Misra The College of Staten Island, The City University of New York, Staten Island, New York (Chairman)
Richard I. Resch The College of Staten Island, The City University of New York, Staten Island, New York
Rae Michael Shortt Wesleyan University, Middletown, Connecticut
Aaron R. Todd Baruch College, The City University of New York, New York, New York

Topological Characterization of Isolated Singularities of Infinite Dimensional Complex Hypersurfaces

Denis Blackmore New Jersey Institute of Technology, Newark, New Jersey

1. INTRODUCTION

Let f be a complex-valued analytic function defined on an open neighborhood U of the origin in a complex Hilbert space H, such that $f(0) = 0$. We study the topology of $V(f|U)$ or $V(f)$ for short, the zero set of f in U, and the nature of certain neighborhoods of the origin in H which are some sense adapted to $V(f)$. Several extensions of known results for isolated critical points in the finite-dimensional case are discussed and developed, including cone and fibration theorems, Milnor numbers, and invariance of topological type under certain types of homotopies. The stage is set for the study of isolated singularities by first giving a brief treatment of known results for regular points and nondegenerate critical points. For more general isolated singularities we prove a cone theorem, a neighborhood fibration theorem and some analogs of the Lê-Ramanujam theorem.

To be more precise, let H be a complex Hilbert space with inner product $\langle \cdot, \cdot \rangle$ and induced norm $||\cdot||$, and let $|\cdot|$ denote the modulus of a complex number. We shall consider bounded analytic functions

$$f : (U, 0) \to (\mathbb{C}, 0),$$

where U is a connected, bounded open neighborhood of the origin in H. The space $V(f|U) = U \cap f^{-1}(0)$ is an open subset of a hypersurface in H; more exactly, it is an analytic subvariety of U of (complex) codimension one. Research aimed at the characterization of such hypersurfaces in topological, geometric and algebraic terms has a long and distinguished history, and continues to be an area of considerable activity (see [2], [3], [6], [8], [9], [10] and [11]). In this paper we shall ask and answer a number of questions regarding the topological nature of $V(f)$ and neighborhoods which contain it, and show that some of the topological characteristics are invariant under certain types of deformations of f.

The following notation will be useful. Given a connected, bounded open neighborhood U of the origin in H and a nonnegative integer m, we let $\Gamma^m(U)$ denote the set of bounded functions $f : (U, 0) \to (\mathbb{C}, 0)$ which are continuous and have continuous (Fréchet) derivatives on U of all orders $k \leq m$. Define $\Gamma^\infty(U) = \cap\{\Gamma^m(U) : 0 \leq m\}$ and let $\Gamma^\omega(U)$ be the functions in $\Gamma(U) = \Gamma^0(U)$ which are analytic (or equivalently, holomorphic) on U. By analytic we mean that for each x in U there is an open neighborhood W of x contained in U and a sequence $\{\Phi_k(x)\}$ with $\Phi_k(x) \in L_s^k(H, \mathbb{C})$, $1 \leq k$, where $L_s^k(H, \mathbb{C})$ is the space of continuous, symmetric k-multilinear forms on H, such that

$$\lim_{n \to \infty} |f(x+h) - f(x) - \sum_{k=1}^{n} (k!)^{-1} \Phi_k(x) h^k| = 0$$

whenever $x + h \in W$. here h^k denotes the point in H^k with all coordinates equal to h. It is easy to verify that $f \in \Gamma^\omega(U)$ implies that $f \in \Gamma^\infty(U)$, and the derivatives are given by the formulas

$$f^{(k)}(x) = \Phi_k = \Phi_k(x),$$

for all $k \geq 1$ and $x \in U$. Moreover, just as in the finite-dimensional case, one can show that $f \in \Gamma^\omega(U)$ if $f \in \Gamma^1(U)$. A simple proof can be found in [1] and [4]. Note that $L_s^1(H, \mathbb{C}) = L(H, \mathbb{C}) = H^*$, the dual space of H. We shall say that f is a Γ^k*-function*, $0 \leq k \leq \omega$, if there exists a U such that $f \in \Gamma^k(U)$. If E is another complex Hilbert space, we use completely analogous notation for mappings $\phi : (U, 0) \to (E, 0)$; so the spaces $\Gamma^k(U, E)$, $0 \leq k \leq \omega$, are defined in the obvious way. We say that ϕ is a Γ^k*-morphism* when there exist U and E such that $\phi \in \Gamma^k(U, E)$, and that ϕ is a Γ^k*-isomorphism* when ϕ is a local homomorphism (i.e., a Γ-isomorphism) such that ϕ and its local inverse are both Γ^k morphisms.

In the sequel we shall concentrate on the case of H an infinite-dimensional complex Hilbert space, since the finite-dimensional versions of the results obtained are for the most part well-known. However, we shall rarely find it necessary to refer specifically to the dimensionality in the statements of the results, as they will in most instances also be valid for the finite-dimensional case.

2. REGULAR POINTS

A point x is a *regular point* of a Γ^ω-function f if $f'(x)$ is not the zero linear functional. When 0 is a regular point of f, it follows from continuity that the same is true of every point in a sufficiently small neighborhood of 0 which we can take to be the domain of f. In the regular point case the basic result on the topology of $V(f|U)$ for finite dimensions generalizes completely as seen by the following

THEOREM 1. *If 0 is a regular point of a Γ^w-function f, there exists an open neighborhood W of the origin such that $V(f|W) = W \cap f^{-1}(0)$ is an analytic submanifold of H of codimension 1.*

SKETCH OF PROOF. As $f'(0) \neq 0$, the kernel $\kappa = \ker f'(0)$ is a closed subspace of H of codimension one. The projection theorem then yields the orthogonal direct sum decomposition $H = \kappa \oplus \kappa^\perp$, where $\kappa^\perp$ is the orthogonal complement of κ–a 1-dimensional subspace. Thus we can write every vector in H uniquely in the form $x = y + z$, with $y \in \kappa$ and $z \in \kappa^\perp$. Since $V(f)$ is tangent to κ at 0, the implicit function theorem implies the existence of a Γ^ω-isomorphism

$$\phi : W \longrightarrow \kappa \times \kappa^\perp$$

of the form $\phi(y+z) = (y, z - \theta(y))$ with $\theta(0) = \theta'(0) = 0$, such that $\phi(V(f|W)) \subset \kappa \times 0$ (see [1] or [5] for details). The local chart (W, ϕ) therefore induces the required submanifold structure on $V(f|W)$.

Using Theorem 1 it is easy to construct a neighborhood of the origin which is a trivial fiber bundle over $V(f)$. All we need do is use a tubular neighborhood of the submanifold.

THEOREM 2. *Let f and W be as in Theorem 1. Then there is a closed neighborhood N of the origin which has the structure of an analytic, trivial, 2-disk bundle over a subset of $V(f|W)$.*

PROOF. Recall from the proof of Theorem 1 that there is a Γ^ω-isomorphism $\phi : W \longrightarrow \kappa \times \kappa^\perp$ with $\phi(W \cap f^{-1}(0)) \subset \kappa \times 0$. Choose $\epsilon > 0$ so small that $B_\epsilon = \{x \in H : ||x|| \leq \epsilon\}$ is contained in W. Then $F = \phi(V(f|B_\epsilon))$ is a closed set in $\kappa \times 0$, which naturally serves as the base space for a disk bundle as follows. Select $\delta > 0$ small enough so that

$$E = F \times \Delta = F \times \{z \in \kappa^\perp : |z| \leq \delta\}$$

is contained in $\phi(W)$. Define $p : E \longrightarrow F$ to be the projection onto the first factor of E, and set $N = \phi^{-1}(E)$. Clearly N is a closed neighborhood of the origin. We obtain the desired fiber bundle

$$\pi : N \longrightarrow V(f|B_\epsilon)$$

by setting $\pi = \phi^{-1} \circ p \circ \phi$, thus completing the proof.

Our last result for regular points concerns the cone structure adapted to $V(f)$ which characterizes a neighborhood of the origin. Given $\epsilon > 0$, the *sphere* $S_\epsilon = \{x \in H : ||x|| = \epsilon\}$ is a smooth ($= C^\infty$) submanifold of H of real codimension one which forms the boundary ∂B_ϵ of the *ball* B_ϵ. Define $V_\epsilon(f) = B_\epsilon \cap V(f)$ and $K_\epsilon(f) = S_\epsilon \cap V(f)$ when $B_\epsilon \subset U$. There is a natural *cone functor* C for pairs (A, B) with $B \subseteq A \subseteq H$, which we denote by $C(A, B)$. When A is such that each of its points lies on a unique ray from the origin, $C(A, B)$ can be defined in the usual geometric fashion so that its vertex is at 0.

THEOREM 3. *If 0 is a regular point of the Γ^ω-function f, there exists $\epsilon_0 > 0$ such that there is a homeomorphism of pairs*

$$(B_\epsilon, V_\epsilon(f)) \approx C(S_\epsilon, K_\epsilon(f))$$

for every $0 < \epsilon \leq \epsilon_0$.

SKETCH OF PROOF. Again we use the submanifold chart (W, ϕ) defined in the proof of Theorem 1. Select δ so small that $Q = \{(y, z) \in \kappa \times \kappa^\perp : ||y||^2 + |z|^2 = \delta^2\}$ is contained in $\phi(W)$. Define for each $(y, z) \in Q$ the ray segment $R(y, z) = \{t(y, z) : 0 \leq t \leq 1\}$. It is easy to see from the form of ϕ that there exists an $\epsilon_0 > 0$ such that $\phi^{-1}(R(y, z))$ intersects S_ϵ in a unique point for every $0 < \epsilon < q$ whenever $(y, z) \in Q$. Taking the segment of $\phi^{-1}(R(y, z))$ from 0 to the point where it intersects S_ϵ for all $(y, z) \in Q$, we obtain the rays for the cone structure. Moreover, we see that $K_\epsilon(f)$ is smoothly diffeomorphic with a sphere in κ. This completes the sketch of the proof.

3. NONDEGENERATE CRITICAL POINTS

The function f has $x \in U$ as a *critical point* if $f'(x) = 0$. In this section we shall assume that 0 is a critical point of f, so that $f'(0) = 0$; but we are going to make a further assumption which guarantees that the singularity is of the mildest possible kind. A critical point x is called a *nondegenerate critical point* of f if the continuous, symmetric bilinear form $f''(x)$ has the property that $f''(x)(h_1, h_2) = 0$ for all $h_1 \in H$ implies that $h_2 = 0$. This is tantamount to saying that the linear mapping $H \to H^*$ associated with $f''(x)$ is an isomorphism (in the category of topological vector spaces). If 0 is a nondegenerate critical point of f, it follows from the inverse function theorem that it is isolated. Hence we may assume that 0 is the critical point of f in its domain U.

We shall show that Theorem 1–3 for regular points have counterparts in the case of a nondegenerate critical point. It is easy to see in this case that $V(f)$ is not, in general, a submanifold. The function $f(z, w) = z^2 + w^2$ of two complex variables illustrates this in finite dimensions, while the following example supports the same conclusion in the infinite-dimensional case: Let $\ell_2(\mathbb{C})$ be the separable complex Hilbert space consisting of all square-summable complex sequences $x = (c_1, c_2, c_3, \ldots)$. Define $f(x) = \sum_{k=1}^{\infty} c_k^2$. One can readily show that f is analytic, that $f'(0) = 0$ and that $f''(0)(x, y) = 2 \sum_{k=1}^{\infty} c_k d_k$, where $y =$

$(d_1, d_2, d_3, \ldots)$. Hence 0 is a nondegenerate critical point. Moreover, $\sum_{k=1}^{\infty} c_k^2 = 0$ is not a submanifold of $\ell_2(\mathbb{C})$. The analogy with Theorem 1 must therefore be made from a different perspective. This can be accomplished by using the concept of normal forms (cf. [2]). An elementary modification of the proof of Theorem 1 yields the existence of a Γ^ω-isomorphism ψ such that locally

$$f \circ \psi(y+z) = z.$$

The following analog for the case at hand can be obtained via a slight alteration of the proof of the Morse-Palais lemma (cf. [1] and [5]).

THEOREM 4. *Suppose that 0 is a nondegenerate critical point of the Γ^ω-function f. Then there exists a Γ^ω-isomorphism ψ of the form $\psi(x) = x + \theta(x)$, with $\theta(0) = \theta'(0) = 0$ such that locally*

$$f \circ \psi(x) = 2^{-1} x^*(x),$$

where x^* is the functional defined by $x^*(y) = f''(0)(x, y)$.

PROOF. As $f(0) = f'(0) = 0$, the series expansion of f yields

$$f(h) = 2^{-1} f''(0)(h,h) + \sum_{k=3}^{\infty} (k!)^{-1} f^{(k)}(0) h^k.$$

On setting $h = \psi(x) = x + \theta(x)$, we observe that ψ has the desired property if

$$\Theta(x, \theta(x)) = 0$$

for all small x, where

$$\Theta(x,y) = f''(0)(x,y) + 2^{-1} f''(0) y^2 + \sum_{k=3}^{\infty} (k!)^{-1} f^{(k)}(0)(x+y)^k.$$

We compute that $\Theta(0,0) = 0$ and that $D_2\Theta(0,0)$ is the linear isomorphism defined by $D_2\Theta(0,0)h = h^*$. Hence an application of the implicit function theorem completes the proof by producing the desired solution θ.

Observe that it follows from this result that the example for ℓ_2 adduced to show that $V(f)$ is not a submanifold when 0 is a nondegenerate critical point is actually typical for separable Hilbert spaces (since they are isomorphic to ℓ_2). In addition, we can easily infer a cone theorem akin to Theorem 3 from Theorem 4; namely

THEOREM 5. *The conclusion of Theorem 3 also obtains for the case when 0 is a nondegenerate critical point of f.*

SKETCH OF PROOF. Let $x \neq 0$ be such that

$$f \circ \psi(x) = 0 = 2^{-1} x^*(x),$$

where ψ is the isomorphism of Theorem 5. Then it follows that

$$f \circ \psi(tx) = 2^{-1}(tx)^*(tx) = 2^{-1}t^2x^*(x) = 0$$

for all $t \geq 0$. Hence $f \circ \psi$ possesses a natural cone structure. As ψ is such that it can be made arbitrarily C^1 close to the identity on a sufficiently small neighborhood of the origin, it follows that the ψ-image of each short ray segment will have a unique intersection point with a sphere S_ϵ when $\epsilon > 0$ is small enough. Consequently, the cone structure for $f \circ \psi$ is mapped by ψ into the desired cone structure for f.

In order to produce a fiber bundle neighborhood theorem akin to Theorem 2, we must introduce a notion of a singular fiber bundle. We represent the standard 2-disk by

$$\Delta = \{z \in \mathbb{C} : |z| \leq 1\}.$$

A quadruple (π, X, Y, y_0) is a *locally trivial, 2-disk bundle with a singular point at* y_0 if $\pi : X \to Y$ is a continuous surjection between topological spaces X and Y, $y_0 \in Y$ and the following properties obtain: Every point y of Y except y_0 has open neighborhood U in Y for which there is a homeomorphism ϕ such that the diagram

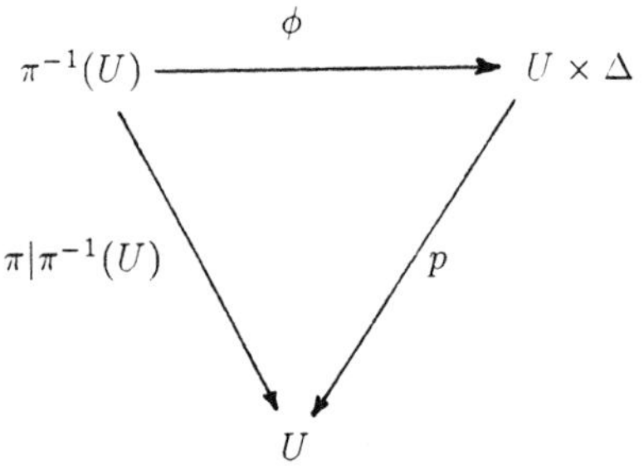

commutes. Here p is the projection on the first factor. X is called the *total space* of the bundle, Y the *base space* and π the *projection* of the bundle. For each $y \in Y$, the space $\pi^{-1}(y)$ is the *fiber over* y, and $\pi^{-1}(y_0)$ is the *singular fiber*. Observe that all nonsingular fibers are homeomorphic to Δ, while the singular fiber may be of an entirely different topological type. Our next result shows how these ideas come into play for nondegenerate hypersurface singularities. We leave the proof, which can be readily fashioned using Theorems 4 and 5, to the reader. It should be noted that the proof we give in the next section for a general isolated critical point is also applicable.

THEOREM 6. *If 0 is a nondegenerate critical point of a Γ^ω-function f, there exists a closed neighborhood N of the origin in H which has the structure of a locally trivial, 2-disk bundle with singular point at 0 over $V(f|N)$.*

4. ISOLATED CRITICAL POINTS

From now on we shall assume that the Γ^ω-function $f : (U, 0) \to (\mathbb{C}, 0)$ has an isolated critical point at the origin; that is, 0 is a critical point of f and every point of $U \setminus 0$ is regular. As we observed in the preceding section, this includes the case where 0 is a nondegenerate critical point. It would therefore be reasonable to pursue a course signaled by Theorem 4: one which leads to a succession of normal forms for increasingly more general singularities. However, we choose not to strike off in this direction; rather, we shall obtain local topological characterizations of isolated singularities satisfying the fewest possible additional properties.

We first obtain a description of $V(f)$ which somewhat resembles the cone theorem (Theorem 5) of the previous section. Our proof requires the introduction of some additional notation. H is a complex Hilbert space with complex inner product $\langle \cdot, \cdot \rangle$. But we can also consider H to be a real Hilbert space with inner product $\langle \cdot, \cdot \rangle_r$ defined to be the real part of the complex inner product, $Re\ \langle \cdot, \cdot \rangle$. Let $\langle \cdot, \cdot \rangle_i$ denote the imaginary part of the inner product, $Im\ \langle \cdot, \cdot \rangle$. When H is treated as a real space, the definitions of Section 1 have real analogs which we denote by adding a subscript r; for example, Γ_r^k-functions. Define $u = Re f$ and $v = Im\ f$ so that

$$f = u + iv,$$

where $u, v : (U, 0) \to (\mathbb{R}, 0)$ are Γ_r^ω-functions. Then

$$f'(x) = u'(x) + iv'(x)$$

for all $x \in U$. Note that $f'(x) \in L(H, \mathbb{C})$, while $u'(x),\ v'(x) \in L(H, \mathbb{R})$.

By the Riesz representation theorem (see, e.g. [1]) there exists for each $x \in U$ a unique vector $\nabla f(x)$ such that

$$f'(x)h = \langle h, \nabla f(x) \rangle$$

for all h in H. This defines a function $\nabla f : (U, 0) \to (H, 0)$ called the *gradient* of f. Clearly ∇f is a Γ_r^ω-morphism which vanishes only at the origin in U. Similarly, there are Γ_r^ω-morphisms $\nabla u, \nabla v : (U, 0) \to (H, 0)$ vanishing only at the origin such that

$$u'(x)h = \langle h, \nabla u(x) \rangle_r \quad \text{and} \quad v'(x)h = \langle h, \nabla v(x) \rangle_r$$

for all $x \in U$ and $h \in H$ which are called the *gradient* of u and v, respectively. These gradients are related. To see this, we note that

$$\begin{aligned} f'(x)h = \langle h, \nabla f(x) \rangle &= \langle h, \nabla f(x) \rangle_r + i \langle h, \nabla f(x) \rangle_i \\ &= u'(x)h + iv'(x)h = \langle h, \nabla u(x) \rangle_r + i \langle h, \nabla v(x) \rangle_r . \end{aligned}$$

Hence, by equating real and imaginary parts we find that $\nabla u = \nabla f$ and

$$\langle h, \nabla v(x) \rangle_r = \langle h, \nabla f(x) \rangle_i .$$

Similarly, we compute that

$$\langle h, i \nabla f(x) \rangle = -i \langle h, \nabla f(x) \rangle = \langle h, \nabla f(x) \rangle_i - i \langle h, \nabla f(x) \rangle_r .$$

Consequently, $\langle h, \nabla f(x)\rangle_r = \langle h, i\nabla f(x)\rangle_r$. Thus we have shown that

$$\nabla u = \nabla f \text{ and } \nabla v = i\nabla f \tag{1}$$

Using (1), we obtain

$$\begin{aligned} f'(x)\nabla u(x) = &\quad \langle \nabla u(x), \nabla f(x)\rangle = \|\nabla f(x)\|^2 = \|\nabla u(x)\|^2 \\ = &\quad u'(x)\nabla u(x) + iv'(x)\nabla u(x) \\ = &\quad \langle \nabla u(x), \nabla u(x)\rangle_r + i\,\langle \nabla u(x), \nabla v(x)\rangle_r \\ = &\quad \|\nabla u(x)\|^2 + i\langle \nabla u(x), \nabla v(x)\rangle_r \,. \end{aligned}$$

This implies that

$$\langle \nabla u, \nabla v\rangle_r = 0; \tag{2}$$

hence ∇u and ∇v are orthogonal with respect to $\langle\cdot,\cdot\rangle_r$. We denote this by $\nabla u \perp_r \nabla v$, while orthogonality in the complex sense is symbolized by $x \perp y$.

There is no loss of generality in assuming that the bounded open set U contains the ball B_3; therefore, we include this in our basic assumption:

(A) $f : (U, 0) \to (\mathbb{C}, 0)$ is a Γ^ω-function such that 0 is is the only critical point of f in U, and the connected, bounded open set U contains B_3.

We shall use int X and cl X, respectively, to denote the interior and closure of the set X in a topological space.

THEOREM 7. *Suppose that f satisfies (A) and the following assumption:*

(B) *There exist a nondecreasing function $\gamma : \{a \in \mathbb{R} : a \geq 0\} \to \mathbb{R}$ with $\gamma(0) = 0$ and $\gamma(a) > 0$ for $a > 0$, and an $a_0 > 0$ such that*

$$\|\nabla f(x)\| \geq \gamma(\epsilon) > 0$$

for every $\epsilon > 0$ and all $x \in U \cap \{x : \|x\| \geq \epsilon\} \cap f^{-1}(\{\alpha \in \mathbb{C} : |\alpha| \leq a_0\})$.

Then there exists $\delta_0 > 0$ such that $M_\delta = \text{int } B_2 \cap f^{-1}(\delta)$ is a codimension one, analytic submanifold of int B_2 for every $0 < \delta \leq \delta_0$, and for each such δ there is an open neighborhood W_δ of the origin such that $V(f|W_\delta)$ is homeomorphic to M_δ/C_δ, where C_δ is a closed subset of M_δ. Moreover, the homeomorphism is actually a Γ_r^∞-isomorphism except at the origin.

PROOF. As f is a nonconstant analytic function, $f(x) = z$ for some $x \in \text{int } B_2$ for each $|z| \leq \delta_1$ when δ_1 is sufficiently small. Fix $0 < \delta \leq \delta_1$. Every point of M_δ is regular by virtue of (A), so the submanifold structure follows from the proof of Theorem 1.

The smooth gradient vector field $\nabla u : U \to H$, which has the origin as its only stationary point, induces a local flow $X = x(t) = \eta(t, x_0)$ obtained by integrating

$$\frac{dx}{dt} = -\nabla u(x). \tag{3}$$

We compute that

$$\frac{d}{dt}u(x(t)) = \left\langle \frac{dx}{dt}, \nabla u \right\rangle_r = -||\nabla u(x(t))||^2 \tag{4}$$

and

$$\frac{d}{dt}v(x(t)) = \left\langle \frac{dx}{dt}, \nabla v \right\rangle_r = -\langle \nabla u(x(t)), \nabla v(x(t)) \rangle_r \, .$$

Hence it follows from (A) and (2) that u decreases and v is constant along a positive semiorbit $c_+(x_0) = \{\eta(t, x_0) : t \geq 0\}$ of ∇u when $x_0 \neq 0$. In particular, if $x_0 \in M_\delta$, then $u(x(t))$ decreases from δ at $t = 0$ and $v(x(t)) = 0$ as t increases. Hence $f(x(t))$ is real for all $t \geq 0$, $f(x(0)) = f(x_0) = \delta$ and $f(x(t))$ decreases with increasing t.

It follows from (1) and (3) that

$$\left| \frac{d}{dt}||x(t)|| \right| \leq ||\nabla f(x(t))||.$$

Integrating this inequality and using (B), we conclude that any positive semiorbit which starts at $x_0 \in M_\delta$ and leaves B_3 must do so for $t \geq \gamma(2)^{-1}$ For this semiorbit we find by integrating (4) and using (B) that

$$u(x_0) - u(x(t)) = \delta - u(x, t) \geq \gamma(2)^2 t;$$

so $c_+(x_0)$ hits $V(f)$ in a unique point for $t \leq \delta\gamma(2)^{-2}$. Hence if $\delta \leq \min(a_0, 2^{-1}\gamma(2))$, $c_+(x_0)$ hits $V(f)$ before exiting B_3. Similarly, when $x_0 \in B_1 \cap (V(f) \setminus 0)$, the negative semiorbit $c_-(x_0) = \{\eta(t, x_0) : t \leq 0\}$ hits M_δ at a unique point before leaving B_2 if $\delta \leq \min(a_0, 2^{-1}\gamma(1))$. Consequently if we set $\delta_0 = \min(\delta_1, a_0, 2^{-1}\gamma(1))$, then if $x_0 \in M_\delta$ with $0 < \delta \leq \delta_0$, $c_+(x_0)$ must hit $V(f)$ at a unique point before exiting B_3, and there exist x_0 such that $c_+(x_0)$ hits $V(f) \setminus 0$.

We now show that if $x_0 \in M_\delta$ with $0 < \delta \leq \delta_0$, then $c_+(x_0)$ intersects $V(f)$ even if it does not leave B_3. When $||\eta(t, x_0)|| \geq \epsilon > 0$ for all $t \geq T > 0$, (4) and (B) imply that

$$u(x_1) - u(\eta(t, x_1)) \geq \gamma(\epsilon)^2 t$$

for $t \geq 0$, where $x_1 = \eta(T, x_0)$. Hence $\eta(t, x_0)$ hits $V(f)$ for a unique $t \leq T + \delta\gamma(\epsilon)^{-2}$. Alternatively, 0 is an ω-limit point of x_0, which we denote by $0 \in \omega(x_0)$. Define

$$C_\delta = \{x_0 \in M_\epsilon : 0 \in \omega(x_0)\}.$$

Then for each point $x_0 \in M_\delta \setminus C_\delta$ there exists a unique $t(x_0) > 0$ such that $\eta(t(x_0), x_0) \in V(f)$. If $x_0 \in C_\delta$, we write $t(x_0) = \infty$.

Let $\phi : M_\delta \to V(f)$ be the mapping defined by $\phi(x_0) = \eta(t(x_0), x_0)$, and let ϕ_* be defined by passing to the quotient in the diagram

$$\begin{array}{ccc} M_\delta & \xrightarrow{\phi} & V(f) \\ \downarrow & & \downarrow \\ M_\epsilon / C_\delta & \xrightarrow{\phi_*} & V(f)/0 \approx V(f) \end{array}$$

It follows from standard results on flows that ϕ_* has the desired properties. This completes the proof.

This result is valid if (B) is replaced by the Palais-Smale condition (cf. [9]).

(PS) Any sequence $\{x_n\}$ for which $\{f(x_n)\}$ is bounded and $f'(x_n) \to 0$ has a convergent subsequence.

In the finite-dimensional case, where (B) follows from (A), this theorem is essentially known, at least as part of the folklore of singularity theory.

Our next major result is a true cone theorem, but first we need to prove

LEMMA 1. *If f satisfies (A), then there exists $\epsilon > 0$ such that x and $\nabla f(x)$ are $\mathbb{C}$-linearly independent ($\mathbb{C} - LI$) whenever $x \in V(f) \cap \{x : 0 < ||x|| < \epsilon\}$.*

PROOF. It follows from the Cauchy-Schwartz inequality that the set of points Y of $V(f)$ for which x and $\nabla f(x)$ are $\mathbb{C}$-linearly dependent is characterized by $f(x) = 0$ and $|\langle x, \nabla f\rangle|^2 = (||x||\,||\nabla f||)^2$. Hence Y is a real analytic subvariety of U. Suppose, on the contrary, that there is a sequence $\{x_n\} \subset Y$ such that $x_n \to 0$. Then the nature of Y guarantees the existence of a $\delta > 0$ and a nonconstant continuous curve $\sigma : [0, \delta) \to Y$ which is smooth on $(0, \delta)$ and satisfies $\sigma(0) = 0$. We compute that

$$\begin{aligned} 0 = & \frac{d}{dt} f(\sigma(t)) = \left\langle \frac{d\sigma}{dt}, \nabla f(\sigma(t))\right\rangle = \lambda(t)\left\langle \frac{d\sigma}{dt}, \sigma(t)\right\rangle \\ = & \lambda(t)\left[2^{-1}\frac{d}{dt}||\sigma(t)||^2 + i\left\langle \frac{d\sigma}{dt}, \sigma(t)\right\rangle_i\right]. \end{aligned}$$

Consequently, $\sigma(t) = 0$ for all $0 \le t < \delta$. As this contradicts the choice of σ, the proof is complete.

It follows from Lemma 1 that the *Milnor radius*, defined by

$$r_M(f) = \sup\{\epsilon \geq 0 : \text{ int } B_\epsilon \subset \text{ domain } (f) \text{ and } x \text{ and } \nabla f(x) \text{ are } \mathbb{C} - LI \text{ on } 0 < ||x|| \leq \epsilon\}$$

is positive for f satisfying (A). We now use this in the statement of the following result which is a generalization of the Milnor cone theorem proved in [8].

THEOREM 8. *Let f satisfy (A) and (PS) and suppose that H is separable. Then for every $0 < \epsilon < r_M(f)$ there is a homeomorphism of pairs $(V_\epsilon, V_\epsilon(f)) \approx C(S_\epsilon, K_\epsilon(f))$.*

SKETCH OF PROOF. Since H is separable, the existence of smooth partitions of unity is guaranteed by Theorem 2 of II.4 of [5]. Using this fact together with Lemma 1 and (PS), one obtains the desired result by making the obvious modifications in the proof of Theorem 2.10 of [8].

We now conclude this section with the following generalization of Theorem 6, which we call the neighborhood fibration theorem:

THEOREM 9. *If f satisfies (A) and (B), then there exists a closed neighborhood N of the origin which is the total space of a locally trivial, 2-disk bundle with singular point 0 having $V(f|N)$ as its base space.*

PROOF. Define $\delta_0 = \min(\delta_1, a_0, 2^{-1}\gamma(1))$, where δ_1 is as in the proof of Theorem 7 and a_0 and Y are as in (B), and set

$$\Delta(\delta) = \{z \in \mathbb{C} : |z| \leq \delta\}.$$

Fix $0 < \delta \leq \delta_0$ and define $N_1 = B_3 \cap f^{-1}(\Delta(\delta))$. Then N_1 is a closed neighborhood of the origin in H. Moreover, (A) implies that F is transverse to $\partial\Delta(\delta)$; hence $M = B_3 \cap f^{-1}(\partial\Delta(\delta))$ is a smooth submanifold of B_3 of real codimension 1. We shall show that a portion of N_1 has the desire bundle structure.

In order to construct the fibers, we introduce the family of smooth vector fields

$$\frac{dx}{dt} = X_\theta(x) = -e^{i\theta}\nabla u(x) \tag{5}$$

for $\theta \geq 0$ and denote the local flow of X_θ by $x = \eta_\theta(t, x_0)$. A simple computation using (1) yields

$$\frac{d}{dt}u(\eta_\theta(t, x_0)) = -\cos\theta||\nabla u(\eta_\theta(t, x_0))||^2 \tag{6}$$

and

$$\frac{d}{dt}v(\eta_\theta(t, x_0)) = -\sin\theta||\nabla u(\eta_\theta(t, x_0))||^2 \tag{7}$$

Hence, we can use virtually the same reasoning as in the proof of Theorem 7, together with (5)–(7), to reach the following conclusions for $x_0 \in V(f|B_1) \setminus 0$: (a) For $0 \leq \theta < \pi$ none of the orbits of X_θ through x_0 intersect except at x_0, and none of the orbits through $x_1 \neq x_0$ intersect any of the orbits through x_0. (b) The orbit of X_θ through x_0 is just the orbit of $X_{\theta+\pi}$ through x_0 with reversed orientation. (c) An orbit $c(x_0, \theta) = \{\eta_\theta(t, x_0) : t \in \mathbb{R}\}$ intersects M in exactly two points, and does so transversely. Furthermore, there exist $t_-(x_0, \theta) < 0 < t_+(x_0, \theta)$ corresponding to these points such that

$$f(\eta_\theta(t_-(x_0, \theta), x_0)) = \delta e^{i\theta}, \quad f(\eta_\theta(t_+(x_0, \theta), x_0)) = -\delta e^{i\theta},$$
$$|f(\eta_\theta(t, x_0))| < \delta \quad \text{for } t_-(x_0, \theta) < t < t_+(x_0, \theta)$$

and

$$|f(\eta_\theta(t, x_0))| > \delta \quad \text{for } t < t_-(x_0, \theta) \text{ or } t > t_+(x_0, \theta).$$

In addition, it follows from (A), (b), (5)–(7) and standard properties of flows (see [5]) that t_- and t_+ depending smoothly on $x_0 \neq 0$ and $\theta \geq 0$, and they are bounded from below and above, respectively, on any subset W of $V(f|B_1) \setminus 0$ such that $0 \notin \text{cl } W$.

For $x_0 \in V(f|B_1) \setminus 0$ we define the fiber

$$F(x_0) = \{\eta_\theta(t, x_0) : t_-(x_0, \theta \leq t \leq t_+(x_0, \theta), 0 \leq \theta < \pi\}$$

and set $\pi(F(x_0)) = x_0$. Let C be the subset of $M \cap B_2$ such that if $x_0 \in C$ with $f(x_0) = \delta e^{i\theta}$, then 0 is an ω-limit point of x_0 with respect to η_θ. We define the singular fiber to be

$$F(0) = \{\eta_\theta(t, x_0) : x_0 \in C, f(x_0) = \delta e^{i\theta}, t > 0, 0 \leq \theta < 2\pi\}$$

and set $\pi(F(0)) = 0$. Now define

$$N = \cup\{F(x_0) : x_0 \in V(f|B_1)\}.$$

In virtue of its construction N is a closed neighborhood of 0, and we note that $V(f|B_1) = V(f|N)$.

We claim that $\pi : N \to V(f|N)$ has the bundle structure we are seeking. The smoothness of π on $N \setminus F(0)$ follows from its construction, so it only remains to verify the triviality on a suitably chosen neighborhood of each $x_0 \in V(f|N) \setminus 0$. It follows from properties (a)–(c) that if we choose an open neighborhood W of x_0 in $V(f|N)$ whose closure does not include the origin, then the morphism

$$\pi^{-1}(W) \to W \times \Delta$$

defined by mapping $\eta_\theta(t, x_0)$ into $(x_0, \delta^{-1}f(\eta_\theta(t, x_0)))$ for $t_-(x_0, \theta) \leq t \leq t_+(x_0, \theta)$, $0 \leq \theta < \pi$ is a Γ_r^∞-isomorphism which yields the required trivialization. Our proof is therefore complete.

5. MILNOR NUMBERS

In this section we shall give a very brief description of a an infinite-dimensional generalization of the Milnor number and discuss a few of its properties. Suppose that H is finite-dimensional for the moment. Then H can be assumed to be the complex hermitian space $\mathbb{C}^n$, for some $n \geq 0$, which can be viewed as the euclidean space $\mathbb{R}^{2n}$ by making the usual identification. Let 0 be an isolated critical point of f such that f' is not zero on $B_\epsilon \setminus 0$. The Milnor number of f is defined to be

$$\mu(f) = d_0(2^{-1}(\nabla u + i\nabla v), 0, B_\epsilon),$$

where d_0 is the Brouwer degree. This is different from the usual definition (cf. [8]), but it can easily be shown to be equivalent by noting that (1) implies that $\nabla f = 2^{-1}(\nabla u - i\nabla v)$.

Motivated by the definition of the Milnor number in the finite-dimensional case, we define the *conjugate gradient* $\nabla_* f$ of an analytic function by $\nabla_* f = 2^{-1}(\nabla u + i\nabla v)$. Now let H be infinite-dimensional. The natural way to generalize μ is to use the Leray-Schauder degree in place of the Bruower degree. But his requires that we focus on a special class of mappings. Let E be a subset of H such that E is contained in the domain of a mapping ϕ having codomain H. We say that ϕ is LS on E is if its restriction to E can be written in the form $\phi|E = id + \chi$, where id is the identity and χ is compact (i.e., maps bounded sets onto relatively compact sets). Let f be a Γ^ω-function such that f' does not vanish on $B_\epsilon \setminus 0$, for some $\epsilon > 0$ and $\nabla_* f$ (which is also nonzero on $B_\epsilon \setminus 0$) is LS on B_ϵ. The *Milnor number* of f is defined to be

$$\mu(f) = d(\nabla_* f, 0, B_\epsilon),$$

where d is the Leray-Schauder degree. A nice treatment of the Leray-Schauder degree appears in [1]. We remark that $\nabla_* f$ is automatically LS in the finite dimensional case.

If f satisfies the properties in the definition of the Milnor number and 0 and is a regular point, then $\mu(f) = 0$ since $\nabla_* f$ does not vanish in B_ϵ. Perhaps the most basic property of μ is the following:

LEMMA 2. *Let f be a Γ^ω-function such that $\mu(f)$ is defined. Then $\mu(f) \geq 0$, and $\mu(f) > 0$ when 0 is a critical point.*

PROOF. It can be readily verified that $\nabla_* f$ is analytic. Hence the results follow from theorem 5.3.16 (v) of [1].

The Milnor number is, in a certain sense, a topological invariant for isolated singularities. Namely, suppose that $g = \psi \circ f \circ \phi$, where ϕ and ψ are Γ^1-isomorphisms in H and $\mathbb{C}$, respectively, such that ϕ and ψ are both orientation preserving or reversing and ϕ', f and g are LS on some ball. Then if 0 is an isolated critical point of f, the same is true of g, and it follows from standard properties of the Leray-Schauder degree that $\mu(g) = \mu(f)$. As one might expect, μ is not a complete topological invariant in this, or any, sense. To see this, consider the functions $f(z, w) = z^2 + w^5$ and $g(z, w) = z^3 + w^3$ in $\mathbb{C}^2$. Both of these functions have critical points only at the origin and have Milnor numbers equal to four. Yet $V(f)$ is a cone over a torus knot of type (2,5), while $V(g)$ is a cone over a torus knot of type (3,3), so $V(f)$ and $V(g)$ are not homeomorphic in any neighborhood of the origin (cf. [8]).

6. DEFORMATION OF SINGULARITIES

Suppose $F(x, t)$ determines a family of functions $\{f_t\}$, where the $f_t = F(\cdot, t)$ all have 0 as an isolated critical point for t ranging over a real interval. Such a family represents a deformation of each of the singularities f_t. We shall investigate conditions on F from which we can infer that the hypersurfaces $V(f_t)$ are all topologically equivalent in some sense.

Our first deformation theorem generalizes a result which is essentially known for the finite-dimensional case. For a given $\epsilon > 0$, a Γ^∞-function

$$F : (U \times (a, b), 0 \times t) \to (\mathbb{C}, 0),$$

where $a < 0$, $b > 1$ and U is an open neighborhood of 0 in H which contains B_ϵ, is an *$RT(\epsilon)$-homotopy* from f_0 to f_1 if it satisfies the following properties: (i) f_t is a Γ^ω-function on an open neighborhood W of S_ϵ for each $t \in I = [0, 1]$. (ii) There exists $\delta > 0$ such that

$$(\|x\| \, \|\nabla f_t(x)\|)^2 - |\langle x, \nabla f_t(x)\rangle|^2 > 0$$

for all points of $T(3\delta) = \{(x, t) \in H \times I : |(x, t)| = \epsilon, |F(x, t)| \leq 3\delta\}$.

It follows from the Cauchy-Schwarz inequality that (ii) implies that x and and $\nabla f_t(x)$ are $\mathbb{C}-LI$ at every point of $T(3\delta)$, and that in the finite-dimensional case (ii) can be inferred by requiring that x and $\nabla f_t(x)$ be $C-$LI on $K_\epsilon(f_t) = V(f_t) \cap S_\epsilon$, because of compactness.

We say that f and g have the same *topological type* if there exists a homeomorphism of pairs $(B_\epsilon, V_\epsilon(f)) \approx (B_\epsilon, V_\epsilon(g))$.

THEOREM 10. *If f_0 and f_1 are $RT(\epsilon)$-homotopic and f_0 and f_1 have the cone structure of Theorem 8, then f_0 and f_1 have the same topological type.*

PROOF. In virtue of the hypotheses, it suffices to prove that $(S_\epsilon, K_\epsilon(f_0)) \approx (S_\epsilon, K_\epsilon(f_1))$. The homeomorphism will be constructed from the flow of a vector field. We first note that (ii) implies that $F^{-1}(\alpha)$ has transverse intersection with $S_\epsilon \times I$ for each $|\alpha| \leq 3\delta$. Hence $M_\alpha = F^{-1}(\alpha) \cap (S_\epsilon \times I)$ and $Q_a = F^{-1}(\{\alpha : |\alpha| = a\}) \cap (S_\epsilon \times I)$ are smooth submanifolds of $S_\epsilon \times I$ when a, $|\alpha| \leq 3\delta$.

There exists a smooth function $\theta : \mathbb{R} \to I$ such that $\theta(t) = 1$ for $t \leq 0$ and $\theta(t) = 0$ for $t \geq 2\delta$. Now define $\Theta : S_\epsilon \times I \to I$ by setting $\Theta(x,t) = \theta(a)$ when $(x,t) \in T(3\delta)$ and $(x,t) \in Q_a$ and $\Theta(x,t) = 0$ for $(x,t) \in (S_\epsilon \times I) \setminus T(2\delta)$. Clearly Θ is a smooth function, $\Theta = 1$ on M_0 and $\Theta = 0$ on $(S_\epsilon \times I) \setminus T(2\delta)$.

Consider the vector field $X(x,t) = (\Theta(x,t)Y(x,t), 1)$ on $S_\epsilon \times I$, where

$$Y = D_2F\left[|\langle x, \nabla f_t\rangle|^2 - (||x||\,||\nabla f_t||)^2\right]^{-1}\left[||x||^2\nabla f_t - \langle x, \nabla f_t\rangle x\right].$$

It follows from the construction of Θ and the fact that F is a $RT(\epsilon)$-homotopy, that X is a smooth vector field. Extending $\langle\cdot,\cdot\rangle$ on H to $\langle\cdot,\cdot\rangle_\#$ on $H \times \mathbb{C}$ in the obvious way, namely $\langle(x,\alpha),(y,\beta)\rangle_\# = \langle x,y\rangle + \alpha\beta^*$, where $*$ denotes complex conjugate, we find that $\nabla F = (\nabla f_t, D_2F^*)$. X is tangent to $S_\epsilon \times I$, since a simple computation yields $\langle X,(x,0)\rangle_\# = 0$. Similarly, one computes that $\langle X, \nabla F\rangle_\# = 0$ on M_0, which shows that $X|M_0$ is tangent to M_0. Note that the t-component of X is $X^t = 1$.

The field X generates a smooth flow η on $S_\epsilon \times I$ such that M_0 is η-invariant. Define $\phi : S_\epsilon \to S_\epsilon$ by $\phi(x) = p \circ \eta(1,(x,0))$, where p is the projection on the first factor. It follows from the definition of ϕ and standard properties of flows that $\phi : (S_\epsilon, K_\epsilon(f_0)) \to (S_\epsilon, K_\epsilon(f_1))$ is a Γ_r^∞-isomorphism. Thus the proof is complete.

Our next, and last, result is an infinite-dimensional version of the Lê-Ramanujam theorem (theorem 2.1 of [6]; see also [3]). A Γ^∞-function

$$F : (U \times (a,b), 0 \times t) \to (\mathbb{C}, 0),$$

where $a < 0$ and $b > 1$ and U is a connected, bounded, open neighborhood of B_ϵ for some $\epsilon > 0$, is a *μ-homotopy* from f_0 to f_1 if the following properties obtain: (I) f_t is a Γ^ω-function having 0 as its only critical point in U for each $t \in I$. (II) F satisfies the following strong Palais-Smale condition: If $\{(x_n,t_n)\}$ is a sequence in $U \times I$ such that $\{F(x_n,t_n)\}$ is bounded and $\{\nabla f_{t_n}(x_n)\}$ is convergent, then $\{(x_n,t_n)\}$ has a convergent subsequence.

In the finite-dimensional case, (II) follows from (I) and, as shown in [6], condition (I) is equivalent to requiring that the Milnor number $\mu(f_t)$ be constant. Hence the μ-homotopy (or μ-equivalence). But why do we persist with this name even when the dimension is infinite? An answer is provided by observing that it can be readily verified that if each $\nabla_* f_t$ is LS on B_ϵ (so that the Milnor number $\mu(f_t)$ is defined according to section 5), then (I) implies that $\mu(f_t)$ is constant. Property (II) is needed to satisfy certain compactness requirements in the technical details of the proof of

THEOREM 11. *Let H be an infinite-dimensional Hilbert space. Then f_0 and f_1 have the same topological type if they are μ-homotopic and both have the cone structure of Theorem 8 for some $\epsilon > 0$.*

SKETCH OF PROOF. On a superficial level this theorem is a transparent corollary of some deep results of differential topology. To wit, we can choose ϵ, $\delta > 0$ so small that $M_\alpha = F^{-1}(\alpha) \cap (S_\epsilon \times I)$ is a smooth submanifold of $S_\epsilon \times I$ for every $|\alpha| \leq \delta$ and f_0 and f_1 have cone structure on B_ϵ. Consequently, M_0 is a smooth cobordism between $K_\epsilon(f_0)$ and $K_\epsilon(f_1)$. Moreover, $K_\epsilon(f_0)$ and $K_\epsilon(f_1)$ are simply connected and both are deformation retracts of M_0. Hence M_0 is an h-cobordism. As the dimension of M_0 is infinite, it is certainly greater than five, and so the h-cobordism theorem [7] implies that M_0 is Γ_r^∞-isomorphic to $K_\epsilon(f_0) \times I$. In fact, this isomorphism can be extended to a thickened M_0 to produce a smooth diffeomorphism of $T(\delta) = \cup\{M_\alpha : |\alpha| \leq \delta\}$ onto $[T(\delta) \cap (S_\epsilon \times 0)] \times I$, which can be extended to a homeomorphism $(S_\epsilon, K_\epsilon(f_0)) \approx (S_\epsilon, K_\epsilon(f_1))$. Then we obtain a homeomorphism $(B_\epsilon, V_\epsilon(f_0)) \approx (B_\epsilon, V_\epsilon(f_1))$ by virtue of the cone structure. This is essentially the argument used in [6] when the real dimension of H is finite and greater than or equal to eight.

Unfortunately this simple approach begs the question of whether or not the h-cobordism theorem is actually valid for the case at hand? The answer is yes, but the argument is a long and intricate one which involves demonstrating that the steps in the proof of the h-cobordism theorem given in [7] can be appropriately modified. Limitations of space militate against embarking on such a venture here, so we shall give a sketch of the proof which lies somewhere between the superficial and completely rigorous arguments, but which is decidedly closer to the former.

In view of (I) and the hypotheses, we can pick $\epsilon > 0$ so small that $V(F)$ is transverse to $S_\epsilon \times I$ and both f_0 and f_1 have cone structure on B_ϵ. Hence M_0 is a smooth submanifold of $S_\epsilon \times I$. It follows from (II) that there exists $\delta > 0$ such that the same is true for every M_α with $|\alpha| \leq \delta$. Define $\phi : T(\delta) \longrightarrow \mathbb{R}$ by $\phi(x,t) = t + u^2(x,t) + v^2(x,t)$ and set $\phi_\alpha = \phi|M_\alpha$. Observe that $\phi_0^{-1}(0) = K_\epsilon(f_0)$ and $\phi_0^{-1}(1) = K_\epsilon(f_1)$. Again using (II) we can mimic the finite-dimensional procedure to perturb each ϕ_α in the complement of the boundary of M_α to obtain a ψ_α having only isolated critical points. As in [7] we can associate a gradient-like vector field X_α with each ψ_α, and the infinite-dimensional space provides ample room to cancel the stationary points of the X_α's. Hence we may assume that each X_α has no stationary points on M_α.

It follows from the compactness condition (II) that by reducing the size of δ, if necessary, the gradient-like fields X_α can be combined to form a smooth gradient-like field X for ψ on $T(\delta)$ with the following properties: There exists $\sigma > 0$ such that $||X|| \geq \sigma$ on $T(\delta)$; $X|M_\alpha$ is tangent to M_α for all $|\alpha| \leq \delta$; and X generates a flow η on $T(\delta)$ such that the positive semiorbit starting at each $(x,0)$ of $T(\delta)$ hits $T(\delta) \cap (S_\epsilon \times 1)$ at a unique point and the orbit parameter intervals for the intersections are bounded. Therefore η can be used in the usual way to obtain a Γ_r^∞-isomorphism of $T(\delta)$ onto $[T(\delta) \cap (S_\epsilon \times 0)] \times I$. The sketch of the proof is completed by picking up the argument at the end of the superficial approach.

7. CONCLUDING REMARKS

We conjecture that the separability hypothesis in Theorem 8 can be dropped. In reference to the neighborhood fibration theorem (Theorem 9), observe that if two functions have the same topological type, then they have homeomorphic singular fibers. It seems quite likely that the converse is also true; namely, homeomorphic singular fibers imply identical topological type.

The reader familiar with hypersurface singularity theory will have taken note of the fact that we did not discuss generalizations of the Milnor fibration theorem [8] for

$$\phi : S_\epsilon \setminus K_\epsilon(f) \to S_1,$$

where $\phi(x) = f(x)/|f(x)|$. We believe that a significant portion of the well known properties obtain also for infinite dimensions if f satisfies a Palais-Smale condition. Finally, we remark that the real analogs of many of our results appear to be valid for C^2 functions.

ACKNOWLEDGEMENTS

The inspiration for this paper came as the result of my participation in a workshop on infinite-dimensional topology and analysis at the University of Alberta. My thanks to Mel Berger and Jim Timourian for inviting me to that workshop and to Vittorio Cafagna for several enlightening conversations. I completed this research while I was a Visiting Member at the Courant Institute, and I wish to express my appreciation for the warm hospitality and intellectually stimulating environment which I enjoyed during my visit. Lastly, I thank the organizers of the Northeast Topology Conference for giving me the opportunity to present this paper.

REFERENCES

1. Berger, M.S. (1977). *Nonlinearity and Functional Analysis*, Academic Press, New York.
2. Berger, M.S., Church, P.T., and Timourian, J.G. (1985). Folds and cusps in Banach spaces, with applications to nonlinear partial differential equations I, *Indiana Univ. Math. J.*, 34, pp. 1–19.
3. Blackmore, D. (1990). μ-Equivalence implies equality of topological type. To appear.
4. Hervé, M. (1989). *Analyticity in Infinite Dimensional Space*, de Gruyter, Berlin-New York.
5. Lang, S. (1972). *Differential Manifolds*, Addison-Wesley, Reading.
6. Lê Dung Tráng and Ramanujam, C.P. (1976). The invariance of Milnor's number implies the invariance of topological type, *Amer. J. Math.*, 98, pp. 67-78.
7. Milnor, J. (1965). *Lectures on the h-Cobordism Theorem*, Princeton Univ. Press, Princeton.

8. Milnor, J. (1968). *Singular Points of Complex Hypersurfaces*, Princeton Univ. Press, Princeton.

9. Nirenberg, L. (1981). Variational and topological methods in nonlinear problems, *Bull. Amer. Math. Soc.*, 4, pp. 267–302.

10. Randel, R., ed. (1989). *Proceedings of the Singularities Conference-University of Iowa*, 1986, Contemp. Math. 90, Amer. Math. Soc., Providence.

11. Zariski, O. (1932). On the topology of algebroid singularities, *Amer. J. Math.* 54, pp. 433–465.

A Note on Completeness of Hyperspaces

Bruce S. Burdick Roger Williams College, Bristol, Rhode Island

The hyperspace of a uniform space, as defined by Bourbaki (1951), satisfies certain properties if and only if the base space does. Examples of such properties would be metrizability and compactness. Completeness, however, is not one of these properties. Isbell (1964) gave necessary and sufficient conditions for a uniform space to have a complete hyperspace, and one of these conditions made use of functions, defined on partially ordered sets, and taking values in the space. Here we give a characterization of spaces with complete hyperspaces which is stated solely in terms of nets into the space. Our condition is closely related to the property of cofinal completeness, due to Howes (1971).

Given $(X,\mathcal{U})$ a uniform space, with $\mathcal{U}$ a collection of entourages, define 2^X to be the set of non-empty closed subsets of X. For each $U \in \mathcal{U}$ let $H(U) = \{(A,B) \in 2^X \times 2^X \mid A \subseteq U[B] \text{ and } B \subseteq U[A]\}$. Define $2^{\mathcal{U}}$ to be the uniformity on 2^X generated by the basis $\{H(U) \mid U \in \mathcal{U}\}$. Then the uniform space $(2^X, 2^{\mathcal{U}})$ is the *hyperspace* of $(X,\mathcal{U})$.

In a partial order $(P,\geq)$, a set $A \subseteq P$ is *residual* if for all $x \in A$ and all $y \in P$, if $y \geq x$ then $y \in A$. Note that the residual sets are the open sets of the partial order topology (see Rudin (1975), chap. 3) and that the cofinal sets in P are the dense sets for this topology.

If $(P,\geq)$ is a partial order, $(X,\mathcal{U})$ is a uniform space, and $f : P \to X$, we will say that f is *partially Cauchy* if for all $U \in \mathcal{U}$ there is a cofinal set $C \subseteq P$ which is a union $\cup R_\alpha$ of residual sets R_α each satisfying $f[R_\alpha] \times f[R_\alpha] \subseteq U$. We will say f is *partially convergent* to $x \in X$ if for every neighborhood O of x, $f^{-1}[O]$ contains a non-empty residual set. If $(P,\geq)$ is directed then partially Cauchy and partially convergent are the same as Cauchy and convergent, respectively. This terminology differs somewhat from that used by Isbell (1964).

If $(D,\geq)$ is a directed set, then we will say that a net $\nu : D \to X$ is *almost Cauchy* if for every $U \in \mathcal{U}$ there is a $d \in D$ and a collection C_α of cofinal subsets of D such that each $d' \geq d$ is in C_α for some α and each C_α satisfies $\nu[C_\alpha] \times \nu[C_\alpha] \subseteq U$. We will say (following Howes (1971)) that ν is *cofinally Cauchy* if for each $U \in \mathcal{U}$ there is a cofinal set $C \subseteq D$ such that $\nu[C] \times \nu[C] \subseteq U$. It is clear that Cauchy implies almost Cauchy which implies cofinally Cauchy.

If every cofinally Cauchy net in $(X,\mathcal{U})$ has a cluster point then $(X,\mathcal{U})$ is *cofinally complete*. If $(2^X, 2^\mathcal{U})$ is complete then $(X,\mathcal{U})$ is *hypercomplete* (which is the same as Isbell's (1964) *supercomplete*).

If $\nu : D \to 2^X$ is a net into the hyperspace, define $\overline{\lim}\ \nu$ to be the set of points $x \in X$ such that for any neighborhood O of x and any $d \in D$ there is a $d' \geq d$ such that $O \cap \nu(d') \neq \phi$ (this is a generalization to nets of Hausdorff's (1935) *closed upper limit*). It is easily shown that $\overline{\lim}\ \nu$ is closed and that if $A \in 2^X$ is a cluster point of ν then $A \subseteq \overline{\lim}\ \nu$.

If A is a set then cA will denote the closure of A. Any terms not defined herein are consistent with Kelley (1955).

Isbell (1964) proved the following:

THEOREM 1. For a uniform space $(X,\mathcal{U})$ the following are equivalent:

a. $(X,\mathcal{U})$ is hypercomplete.

b. Any partially Cauchy $f : P \to X$ is partially convergent.

The following is a characterization of hypercompleteness using nets, and to make the proof go smoothy we include (c.) below which is (superficially) a weakening of (b.) above.

THEOREM 2. For a uniform space $(X,\mathcal{U})$ the following are equivalent:

a. $(X,\mathcal{U})$ is hypercomplete.

b. Any almost Cauchy $\nu : D \to X$ has a cluster point.

c. For any partially Cauchy $f : P \to X$ there is a point $x \in X$ such that for any neighborhood O of x, $f^{-1}[O]$ has non-empty intersection with every residual cofinal (i.e., open dense in the partial order topology) subset of P.

Proof: a. implies b.: Assume 2^X is complete. Given an almost Cauchy net $\nu : D \to X$ define a net $\mu : \mathcal{U} \to 2^X$, where $\mathcal{U}$ is regarded as directed by reverse inclusion, by $\mu(U) = c\{x \in X|\ \nu \text{ is frequently in } U[x]\}$. Since ν is almost Cauchy, for each $U \in \mathcal{U}$ there is a $d \in D$ such that for every $d' \geq d$ we have $\nu(d') \in \mu(U)$. So for any $U, V \in \mathcal{U}$ and any $x \in \mu(V)$ we have ν is frequently in $V \circ V[x]$ and so $V \circ V[x] \cap \mu(U) \neq \phi$. Hence μ is Cauchy: given $U \in \mathcal{U}$, take $V \in \mathcal{U}$ with $V = V^{-1}$ and $V \circ V \subseteq U$; then if $W_1, W_2 \subseteq V$ we have $(\mu(W_1), \mu(W_2)) \in H(U)$. So μ converges in 2^X, consequently $\overline{\lim}\ \mu \neq \phi$.

Take $x \in \overline{\lim}\ \mu$. We claim x is a cluster point of ν. Given a neighborhood O of x take an open $U \in \mathcal{U}$ such that $U \circ U[x] \subseteq O$. $U[x]$ frequently intersects μ, so take $V \subseteq U$ such that $U[x] \cap \mu(V) \neq \phi$. Take $y \in U[x]$ such that ν is frequently in $V[y]$. Then ν is frequently in $U[y] \subseteq U \circ U[x] \subseteq O$.

b. implies c.: Given $f : P \to X$, partially Cauchy, let $D = \{(R, x)|\ x \in R \subseteq P$ and R is cofinal and residual in $P\}$. We make D a directed set by saying $(R_1, x_1) \geq (R_2, x_2)$ iff $R_1 \subseteq R_2$ (note that the open dense sets of any topological space are directed by reverse inclusion). Define $g : D \to P$ by $g(R, x) = x$.

Then $f \circ g$ is almost Cauchy: Given $U \in \mathcal{U}$, choose a family $\{R_\alpha\}$ of residual subsets of P such that $\cup R_\alpha$ is cofinal and for each α, $f[R_\alpha] \times f[R_\alpha] \subseteq U$. Let $R = \cup R_\alpha$ and choose $x \in R$. For each α let $C_\alpha = g^{-1}[R_\alpha]$. Then each C_α is cofinal in D. Moreover, if $(R', x') \geq (R, x)$ then x' must be in R_α for some α, and so $(R', x') \in C_\alpha$ for that α. Finally, $f \circ g[C_\alpha] \times f \circ g[C_\alpha] = f[R_\alpha] \times f[R_\alpha] \subseteq U$ for each α.

Since $f \circ g$ is almost Cauchy, let x be a cluster point of $f \circ g$. If O is a neighborhood of x then $(f \circ g)^{-1}[O]$ is cofinal in D, so every residual cofinal set in P meets $f^{-1}[O]$.

c. implies a.: This is essentially the same as Isbell's (1964) proof that (b.) implies (a.) in Theorem 1. We include a proof here in the interest of self-containment and because our terminology differs from Isbell's in several ways.

Given a Cauchy net $\nu : D \to 2^X$, we will show that ν converges to $\overline{\lim}\ \nu$. It is easy to show (see, e.g., Caulfield (1967)) that, by virtue of the fact that ν is Cauchy, we have for each $U \in \mathcal{U}$ a $d \in D$ such that for any $d' \geq d$, $\overline{\lim}\ \nu \subseteq U[\nu(d')]$. So it suffices to show that for each $U \in \mathcal{U}$ there is a $d \in D$ such that for any $d' \geq d$, $\nu(d') \subseteq U[\overline{\lim}\ \nu]$.

Suppose not. Take a symmetric $U \in \mathcal{U}$ such that for any $d \in D$ there is a $d' \geq d$ with $\nu(d') - U \circ U[\overline{\lim}\ \nu] \neq \phi$. Let $P = \{(A, x) |\ x \in A \subseteq X,\ A \cap U[\overline{\lim}\ \nu] = \phi$, and there is some $V \in \mathcal{U}$ such that for all $d \in D$ there is a $d' \geq d$ and a $y \in \nu(d')$ such that $V[y] \subseteq A\}$. $P \neq \phi$ since we could have $A = U[(\cup_{d \in D} \nu(d)) - U \circ U[\overline{\lim}\ \nu]]$. We make P a partial order by defining $(A_1, x_1) \geq (A_2, x_2)$ iff $A_1 \subseteq A_2$. Let $f : P \to X$ be defined by $f(A, x) = x$.

We show that f is partially Cauchy. It suffices to show that given $U^* \in \mathcal{U}$ and $(A, x) \in P$ there is a $(B, y) \geq (A, x)$ with $B \times B \subseteq U^*$. Take $V \in \mathcal{U}$ such that for all $d \in D$ there is a $d' \geq d$ and an $x' \in \nu(d')$ such that $V[x'] \subseteq A$. Take a symmetric $W \in \mathcal{U}$ such that $W \circ W \subseteq V$ and $W \circ W \circ W \circ W \subseteq U^*$. Take $d \in D$ such that $d_1, d_2 \geq d$ implies $(\nu(d_1), \nu(d_2)) \in H(W)$. Take $d' \geq d$ and $y \in \nu(d')$ such that $V[y] \subseteq A$. Let $B = W \circ W[y] \subseteq V[y] \subseteq A$. Since $y \in \nu(d') \subseteq W[\nu(d^*)]$ for any $d^* \geq d$, we have that $W[y]$ frequently intersects ν, and hence $(B, y) \in P$. Finally, $B \times B = W \circ W[y] \times W \circ W[y] \subseteq W \circ W \circ W \circ W \subseteq U^*$.

So, by (c.), there must be a point $x \in X$ such that for every neighborhood O of x, $f^{-1}[O]$ meets every residual cofinal set in P. We cannot have $x \in \overline{\lim}\ \nu$ since then $f^{-1}[U[x]] = \phi$. So take a neighborhood O of x and a $d \in D$ such that for all $d' \geq d$ we have $\nu(d') \cap O = \phi$, and take $V \in \mathcal{U}$ with $V^{-1} \circ V[x] \subseteq O$. Then $\{(B, y) \in P |\ B \subseteq V[X - O]\}$ is residual and cofinal in P, yet $f^{-1}[V[x]]$ does not meet it, which is a contradiction.

COROLLARY 1. Cofinal completeness implies hypercompleteness, which implies completeness.

It is customary to restate theorems about nets in terms of filters. The filter property which is analogous to "almost Cauchy net" was introduced by Isbell (1964): a filter $\mathcal{F}$ is *stable* if for any $U \in \mathcal{U}$ there is an $F \in \mathcal{F}$ such that for any $A \in \mathcal{F}$, $F \subseteq U[A]$. The filter property analogous to "cofinally Cauchy net" was introduced by Corson (1958): a filter $\mathcal{F}$ is *weakly Cauchy* if for any $U \in \mathcal{U}$ there is a set $A \subseteq X$ with $A \times A \subseteq U$, such that $A \cap F \neq \phi$ for each $F \in \mathcal{F}$. For filters, Cauchy implies stable, which implies weakly Cauchy.

COROLLARY 2. For a uniform space $(X, \mathcal{U})$, the following are equivalent:

a. $(X, \mathcal{U})$ is hypercomplete.
b. Every stable filter in X has a cluster point.
c. Every stable filter in X satisfies the property that for every $U \in \mathcal{U}$ there is an $F \in \mathcal{F}$ such that $F \subseteq U[C]$, where C is the set of cluster points of $\mathcal{F}$.

Proof. The equivalence of (a.) and (b.) follows from Theorem 2 and the discussion above. That (a.) implies (c.) follows from the following easily verified statements: if one defines a net $\nu : \mathcal{F} \to 2^X$ by $\nu(F) = cF$, with $\mathcal{F}$ regarded as a directed set via reverse inclusion, then, as Isbell (1964) noted, $\mathcal{F}$ stable implies that ν is Cauchy. So, as in the proof of Theorem 2, ν converges to $\overline{\lim}\ \nu$. But $\overline{\lim}\ \nu = C$, which shows that the property mentioned in (c.) is satisfied. Finally, that (c.) implies (b.) is obvious.

The next result would not be obvious without Theorem 2 (or Theorem 1) since even when $(X,\mathcal{U}_1)$ and $(X,\mathcal{U}_2)$ have the same topology, it frequently happens that $2^{\mathcal{U}_1}$ and $2^{\mathcal{U}_2}$ generate different topologies on 2^X.

COROLLARY 3. If $\mathcal{U}_1$ and $\mathcal{U}_2$ are uniformities for X which generate the same topology, such that $\mathcal{U}_1 \subseteq \mathcal{U}_2$, then $(X,\mathcal{U}_1)$ hypercomplete implies $(X,\mathcal{U}_2)$ hypercomplete; in fact, if $(X,\mathcal{U})$ and $(Y,\mathcal{V})$ are uniform spaces and $f : X \to Y$ is uniformly continuous, closed, and perfect, then $(Y,\mathcal{V})$ hypercomplete implies $(X,\mathcal{U})$ hypercomplete.

Theorem 2 makes it possible to give new proofs for the next four results, all of which are known (they are all explicitly stated in Caulfield (1967), and they all are implicit in Isbell (1964), Chapter VII).

COROLLARY 4. If $(X,\mathcal{U})$ is paracompact and fine, then it is hypercomplete.

Proof. In fact, Corson (1958) points out that paracompact and fine implies cofinally complete, and so by Corollary 1 this implies hypercomplete. Corson makes uses of the fact that in a paracompact uniformizable space every neighborhood of the diagonal in $X \times X$ is a member of the fine uniformity, together with Kelley's (1955) criterion, namely, a regular space $(X,\mathcal{T})$ is paracompact iff for every open cover $\mathcal{C}$ of X there is a neighborhood N of the diagonal in $X \times X$ such that $\{N(x) | \ x \in X\}$ is a refinement of $\mathcal{C}$.

COROLLARY 5. If $(X,\mathcal{U})$ is uniformly locally compact, then it is hypercomplete.

Proof. Again, Corson (1958) mentions that uniformly locally compact implies cofinally complete.

COROLLARY 6. If $(X,\mathcal{U})$ is metrizable and complete then it is hypercomplete.

Proof. Take the standard proof used to show that, in metric spaces, completeness in the sense of Cauchy sequences implies completeness in the sense of Cauchy nets, and adapt it to almost Cauchy nets.

We note that Corollary 6 has achieved notoriety lately in the study of fractals. See Barnsley (1988), section 2.7, where it is called "Completeness of the Space of Fractals." Kuratowski (1933) gives a proof of this result and credits it to Hahn (1932).

COROLLARY 7. If $(X,\mathcal{U})$ is hypercomplete then it is paracompact.

Proof. Assume $(X,\mathcal{U})$ satisfies (c.) of Theorem 2. Suppose $\mathcal{C}$ is an open cover of X which has no open locally finite refinement. Let P be the set of pairs (A,a) such that $a \in A \subseteq X$ and for every open cover $\mathcal{C}'$ of A refining $\mathcal{C}$, there is a point $x \in X$ every neighborhood of which intersects infinitely many members of $\mathcal{C}'$. Define $(A,a) \geq (B,b)$ iff $A \subseteq B$. Let $f : P \to X$ be given by $f(A,a) = a$.

We assert that f is partially Cauchy. Given $U \in \mathcal{U}$ let $\mathcal{S}_U = \{S \subseteq X | \ S \times S \subseteq U\}$ and for each $S \in \mathcal{S}_U$ let $R_S = \{(A,a) \in P | \ A \subseteq S\}$. Each R_S is residual and satisfies $f[R_S] \times f[R_S] \subseteq U$, so it remains to show that $C = \cup_{S \in \mathcal{S}_U} R_S$ is cofinal in P. Suppose not. Then there is $(A,a) \in P$ such that for any $B \subseteq A$ with $B \times B \subseteq U$ there is a locally finite refinement $\mathcal{C}'$ of $\mathcal{C}$ covering B. There is a pseudometric ρ on X such that $\{(x,y) \in X \times X | \ \rho(x,y) < 1\} \subseteq U$, and which generates a uniformity coarser than $\mathcal{U}$. Since any pseudometric space is paracompact, we can choose an open locally finite refinement $\mathcal{S}'$ of $\mathcal{S}_U$ covering A. Then for each $O \in \mathcal{S}'$ there is, by our assumption about A, an open locally finite refinement $\mathcal{C}_O$ of $\mathcal{C}$ covering $O \cap A$, and we may assume that $\cup\mathcal{C}_O \subseteq O$. Then $\mathcal{C}' = \cup_{O \in \mathcal{S}'}\mathcal{C}_O$ is an open locally finite refinement of $\mathcal{C}$ covering A, contradicting $(A, a) \in P$.

So there is some $x \in X$ such that for each neighborhood O of x, $f^{-1}[O]$ meets every residual cofinal set in P. But this is a contradiction, since some $O \in \mathcal{C}$ contains x, and since $O \in \mathcal{C}$ the clearly residual set $\{(A, a) \in P |\ A \cap O = \phi\}$ is also cofinal in P.

We conclude with a list of examples.

EXAMPLE 1. Howes (1988) shows that the Hilbert space H of square summable sequences is complete but not cofinally complete. Since H is a complete metric space it is hypercomplete. In fact, one could iterate the hyperspace construction any finite number of times on H and still have a complete metric space, suggesting the properties hyperhypercomplete, hyperhyperhypercomplete, and so on. However, this example shows that none of these properties is strong enough to imply cofinal completeness.

EXAMPLE 2. Let $\mathbf{R}$ be the reals with the half open interval (Sorgenfrey) topology and let $\mathcal{U}$ be the fine uniformity for this topology. Then, following Caulfield (1967), we note that $(\mathbf{R},\mathcal{U})$ is hypercomplete, by our Corollary 4. But $(2^{\mathbf{R}}, 2^{\mathcal{U}})$ is not even normal, since it contains a closed subset homeomorphic to the subset $\{(x, y)|\ x \leq y\}$ of the Sorgenfrey plane. So $(\mathbf{R},\mathcal{U})$ is paracompact and fine but not hyperhypercomplete, and its hyperspace is complete but not hypercomplete. This example also shows that the product of two hypercomplete spaces need not be hypercomplete.

EXAMPLE 3. Let I be the unit interval with the discrete topology, and let $\mathcal{U}$ be the uniformity on I generated by all countable coverings (in the sense of a covering uniformity). Isbell (1964) gives a short proof that this space is complete but not hypercomplete. In contrast to $(2^{\mathbf{R}}, 2^{\mathcal{U}})$ of Example 2, this space is paracompact (in fact, its topology is metrizable even though its uniformity isn't).

EXAMPLE 4. Let K be one of the examples given by Williams (1984) of a compact Hausdorff space such that the box product $\square^{\omega} K$ is not normal (or not paracompact). Let X be the disjoint union of a countably infinite family $\{K_i|\ i \in \omega\}$ of copies of K, and let $\mathcal{U}$ be the uniformity on X consisting of entourages $U \subseteq X \times X$ such that for each $i \in \omega$, $U \cap (K_i \times K_i)$ is in the unique uniformity on K_i. Then since $\cup_{i\in\omega}(K_i \times K_i) \in \mathcal{U}$, $(X,\mathcal{U})$ is uniformly locally compact, hence hypercomplete. But $(2^X, 2^{\mathcal{U}})$ has a closed subset homeomorphic to $\square^{\omega} K$, and so $(X,\mathcal{U})$ is not hyperhypercomplete.

REFERENCES

1. Barnsley, Michael (1988). *Fractals Everywhere*. Academic Press, San Diego.
2. Bourbaki, N. (1951). *Eléments de Mathématique, Topologie Générale*, 2nd ed., Chaps.

I and II. Herman, Paris.
3. Caulfield, Patrick Joseph (1967). *Multi-valued Functions in Uniform Spaces.* Thesis, The Ohio State University.
4. Corson, H.H. (1958). The Determination of Paracompactness by Uniformities, *American Journal of Mathematics*, 80: 185-190.
5. Hahn, H. (1932). *Reele Funktionen.* Akademische Verlagsgesellschaft, Leipzig.
6. Hausdorff, F. (1935). *Mengenlehre*, 3rd ed. Walter de Gruyter, Berlin.
7. Howes, Norman R. (1971). On Completeness, *Pacific Journal of Mathematics*, 38: 431-440.
8. — (1988). On Finite Refinements and the Cofinal Completion of Uniform Spaces, preprint.
9. Isbell, J. R. (1964). *Uniform Spaces.* American Mathematical Society, Providence.
10. Kelley, John L. (1955). *General Topology.* Van Nostrand, New York.
11. Kuratowski, C. (1933). *Topologie* I, *Espaces Métrisables, Espaces Complets.* Monografje Matematyczne, Warsaw.
12. Rudin, Mary Ellen (1975). *Lectures on Set Theoretic Topology.* American Mathematical Society, Providence.
13. Williams, Scott W. (1984). Box Products, *Handbook of Set-Theoretic Topology* (K. Kunen and J. E. Vaughn, eds.), Elsevier, Amsterdam.

Remarks on a Theorem of Glicksberg

W. W. Comfort and F. Javier Trigos-Arrieta Wesleyan University, Middletown, Connecticut

ABSTRACT

It is a theorem of Glicksberg that every locally compact Abelian topological group **G** = < **G**, t > *respects compactness* in the sense that a subset **A** of **G** is t-compact if and only if **A** is compact in the topology which **G** inherits from its Bohr compactification. We give a simple proof of this result for discrete Abelian groups, and we show further that in the obvious sense such groups also respect countable compactness and pseudocompactness. We show also that every discrete Abelian group becomes zero-dimensional in the topology inherited from its Bohr compactification.

AMS classification numbers: 54A05, 20K45, 22B99.

Key words and phrases: locally compact Abelian group, totally bounded Abelian group, discrete group, compact space, countably compact space, pseudocompact space, zero-dimensional space, Bohr compactification.

§0. Introduction and Background.

0.1. The symbols **Z**, **Q**, **R**, **T** and $\mathbf{Z}(p^\infty)$ have their usual meaning. The identity of **T** and its subgroups is written 1; for other Abelian groups we use additive notation and denote the identity by e. We write $\mathbf{U} = \{\zeta \in \mathbf{T} : \mathrm{Re}(\zeta) < 0\}$.

The class of Abelian groups is denoted by **AG**, and for $\mathbf{G} \in \mathbf{AG}$ we denote by **H**(**G**) the group Hom(**G**, **T**) of all homomorphisms from **G** into **T**. The set of subgroups of **G** is denoted by **S**(**G**), and t**G** is the torsion subgroup of **G**. If **A** is a nonempty subset of **G**, we denote by $\langle \mathbf{A} \rangle$ the subgroup of **G** generated by **A**, and for $a \in \mathbf{G}$ we write $\langle a \rangle$ in place of $\langle \{a\} \rangle$.

For every Abelian topological group **G** we denote by $\hat{\mathbf{G}}$ the set of continuous homomorphisms from **G** into **T**; thus $\hat{\mathbf{G}} = \mathbf{H}(\mathbf{G})$ when **G** is discrete. For a locally compact Abelian (hereafter: LCA) group **G** the function $i : \mathbf{G} \to \mathbf{T}^{\hat{\mathbf{G}}} = K$ given by $(ix)_h = h(x)$ is an isomorphism. Suppressing mention of the function i we consider $\mathbf{G} \subseteq K$ and we denote by b**G**, the so-called *Bohr compactification* of **G**, the closure of **G** in K. For **G** discrete the set **G** with the topological group topology inherited from b**G** is denoted $\mathbf{G}^\#$. This symbol, used in this context by Eric van Douwen [vD2], was suggested to one of us in letters of 1986 and 1987 [vD1].

The following elementary facts will be useful:

0.2. Proposition (see [CS]). Let $\mathbf{G} \in \mathbf{AG}$. Then

(a) The topology of $\mathbf{G}^\#$ is the topology induced on **G** by **H**(**G**).

(b) Every $\mathbf{A} \in \mathbf{S}(\mathbf{G})$ is closed in $\mathbf{G}^\#$.

(c) If $\mathbf{A} \in \mathbf{S}(\mathbf{G})$ then the topology of $\mathbf{A}^\#$ is the topology on **A** inherited from $\mathbf{G}^\#$.

(d) If **H** is a compact Abelian topological group then every $h \in \mathrm{Hom}(\mathbf{G}, \mathbf{H})$ is continuous from $\mathbf{G}^\#$ into **H**; in particular, $\mathbf{H}(\mathbf{G}) = (\mathbf{G}^\#)\hat{}$.

Proof. (a) restates the definition of $\mathbf{G}^\#$. (b) follows from the fact that for $\mathbf{A} \in \mathbf{S}(\mathbf{G})$ and $x \in \mathbf{G} \setminus \mathbf{A}$ there is $h \in \mathbf{H}(\mathbf{G})$ such that $h \equiv 1$ on **A** and $h(x) \neq 1$, so that $\mathbf{A} = \cap \{\ker h : h \in \mathbf{H}(\mathbf{G}), \mathbf{A} \subseteq \ker h\}$. From the fact that **T** is divisible it follows that for every $h \in \mathbf{H}(\mathbf{A})$ there is $\overline{h} \in \mathbf{H}(\mathbf{G})$ such that $h \subseteq \overline{h}$; statement (c) is immediate from this. We prove (d). As is well known (see for example [HR](24.3)), the inclusion map $i : \mathbf{H} \to \mathbf{T}^{\hat{\mathbf{H}}}$ of 0.1 is a topological isomorphism between $\mathbf{H}$ and its image, *i.e.*, $\mathbf{H}$ is a topological subgroup of the group $\mathbf{T}^{\hat{\mathbf{H}}}$. Now for $h \in \mathrm{Hom}(\mathbf{G},\mathbf{H})$ we have $\pi_\psi \circ h \in \mathbf{H}(\mathbf{G})$ for every $\psi \in \hat{\mathbf{H}}$, so $\pi_\psi \circ h$ is continuous from $\mathbf{G}^\#$ to $\mathbf{T}$. It follows that h itself must be continuous as a function from $\mathbf{G}^\#$ into the product space $\mathbf{T}^{\hat{\mathbf{H}}}$, as required. □

0.3. Remarks. (a) It follows from 0.2(d) that if $\mathbf{H}$ is a totally bounded Abelian group then every $h \in \mathrm{Hom}(\mathbf{G}, \mathbf{H})$ is continuous as a function from $\mathbf{G}^\#$. To see this, consider h as an element of $\mathrm{Hom}(\mathbf{G}, \overline{\mathbf{H}})$ with $\overline{\mathbf{H}}$ the *Weil completion* of $\mathbf{H}$ (defined as in [C](1.13), for example), and apply 0.2(d).

(b) The Theorem of Glicksberg cited in our abstract appears in [G]. In subsequent work [T2] and [T3], to which the present paper is preliminary, the second-listed author gives a simple and direct proof of Glicksberg's result, and extends it to show that LCA groups respect not only compactness but also pseudocompactness, the Lindelöf property, and the functionally bounded property. (Glicksberg's proof rests on work of Grothendieck concerning C*-algebras and Radon measures. The arguments of [T2] and [T3] require only Pontrjagin duality and an examination of quotient topologies on groups of the form G/H.) The present paper, which together with [T1] is essential to [T2] and [T3], is offered at the suggestion of colleagues in order to assure that the full arguments required for [T2] and [T3] are available in the literature.

(c) We announced the principal results of this paper in [CT].

(d) We are pleased to acknowledge helpful correspondence and conversations concerning this paper with Professors Eric K. van Douwen, Karl H. Hofmann, Lewis C. Robertson, and Dmitrii B. Shakhmatov.

§1. Compactness-type Properties of $\mathbf{G}^\#$.

1.1. Lemma. Let $\mathbf{G} \in \mathbf{AG}$ and let $\langle x_n \rangle_{n<\omega}$ be a sequence in $\mathbf{G}^\#$ converging to a point $p \in b\mathbf{G}$. Then (a) the sequence $\langle x_n \rangle_{n<\omega}$ is eventually constant, and (b) $p \in \mathbf{G}^\#$.

Proof. It is enough to prove (a). If $\mathbf{G}$ is a counterexample, we assume without loss of generality (replacing if necessary x_n by $x_n - x_{n+1}$) that $x_n \to e$. We assume further that the sequence $<x_n>_{n<\omega}$ is faithfully indexed. We will find a (necessarily continuous) homomorphism $h : \mathbf{G}^\# \to \mathbf{T}$ such that $h(x_n)$ does not converge to $h(e) = 1$, thus achieving a contradiction.

Case 1: $\mathbf{G} = \mathbf{Z}$. It is enough to show:

(*) for any sequence as above in $\mathbf{Z}$ there is $h \in \mathbf{H}(\mathbf{Z})$ such that the sequence $<h(x_n)>_{n<\omega}$ has at least two distinct cluster points in $\mathbf{T}$.

In [T1] a much stronger result is proved but the following direct argument, suggested by a referee of [T1], is sufficient for (*). First assume, replacing x_n by $- x_n$ if necessary and passing to a subsequence if necessary, that $x_{n+1} > 2^n \cdot x_n > 0$. Define $p_n = \Pi_{k \leq n} x_k$ for $n < \omega$, set $r_0 = 0$ and recursively, if r_k has been defined for $k < n$, choose $r_n \in [0 , p_{n-1}]$ so that

$$x_n(\Sigma_{k=0}^{n-1} r_k/p_k) + r_n/p_{n-1} = 0 \text{ (mod 1) if n is even}$$
$$= \tfrac{1}{2} \text{ (mod 1) if n is odd.}$$

Now define $\alpha = \Sigma_{k=0}^{\infty} r_k/p_k$, $h(m) = e^{2\pi i \alpha m}$ for $m \in \mathbf{Z}$, and $f(n) = x_n \cdot \Sigma_{k=n+1}^{\infty} r_k/p_k$. Then $h \in \mathbf{H}(\mathbf{Z})$ and $f(n) \to 0$, so from

$$x_n \cdot \alpha = 0 + f(n) \text{ (mod 1) if n is even}$$
$$= \tfrac{1}{2} + f(n) \text{ (mod 1) if n is odd}$$

it follows that

$$h(x_{2m}) = e^{2\pi i \alpha x_{2m}} = e^{2\pi i f(2m)} \to 1 \text{ and}$$
$$h(x_{2m+1}) = e^{2\pi i \alpha x_{2m+1}} = e^{2\pi i [\frac{1}{2} + f(2m+1)]} \to -1,$$

as required.

Case 2: $\mathbf{G} = \mathrm{t}\mathbf{T}$. Denote by $\mathbf{A}(n)$ the subgroup of $\mathbf{T}$ generated by $\{x_k: k \leq n\}$ and note that $\mathbf{A}(n)$ is finite. We assume without loss of generality, passing to a subsequence if necessary, that $x_{n+1} \notin \mathbf{A}(n)$. We suppose further that $x_0 \neq e$. Let $h_0 \in \mathbf{H}(\mathbf{A}(0))$ satisfy $h_0(x_0) \in \mathbf{U}$ and suppose inductively that $h_k \in \mathbf{H}(\mathbf{A}(k))$ has been defined for $k \leq n$ with $h_k(x_k) \in \mathbf{U}$ and with $h_k|\mathbf{A}(j) = h_j$ for $0 \leq j < k \leq n$. Let m be the least positive integer such that $m \cdot x_{n+1} \in \mathbf{A}(n)$; suppose that $h_n(m \cdot x_{n+1}) = \zeta \in \mathbf{T}$, choose $\xi \in \mathbf{U}$ so that $\xi^m = \zeta$, and define $h_{n+1}: \mathbf{A}(n+1) \to \mathbf{T}$ by the rule $h_{n+1}(x + k \cdot x_{n+1}) = h_n(x) \cdot \xi^k$ for $x \in \mathbf{A}(n)$, $k \in \mathbf{Z}$. It is a routine exercise to check that h_{n+1} is a well defined homomorphism on $\mathbf{A}(n+1)$ (that is, $h_{n+1} \in \mathbf{H}(\mathbf{A}(n+1))$), that $h_{n+1}(x_{n+1}) \in \mathbf{U}$, and that $h_k|\mathbf{A}(j) = h_j$ for $0 \leq j < k \leq n+1$. Define $h = \cup_{n<\omega} h_n$ and let $\overline{h} \in \mathbf{H}(\mathbf{G})$ satisfy $h \subseteq \overline{h}$ [HR](A.7). From $\overline{h}(x_n) = h(x_n) \in \mathbf{U}$ it follows that $\overline{h}(x_n) \nrightarrow 1$, as required.

Case 3. $\mathbf{G} = \mathbf{Q}$. We assume without loss of generality, passing to a subsequence if necessary, that either $x_n \in \mathbf{Z}$ for all n or $x_n \in \mathbf{Q} \setminus \mathbf{Z}$ for all n. In the former case, according to 0.2(c), from $x_n \to 0$ in $\mathbf{Q}^{\#}$ follows $x_n \to 0$ in $\mathbf{Z}^{\#}$; this contradicts case 1. In the latter case we have, writing $h(q) = e^{2\pi iq}$ for $q \in \mathbf{Q}$, that $h : Q^{\#} \to (t\mathbf{T})^{\#}$ is continuous (see Remark 0.3(a)) and that $1 \neq h(x_n) \to 1$ in $(t\mathbf{T})^{\#}$; this contradicts case 2.

Case 4. $\mathbf{G}$ arbitrary. There is a divisible group $\mathbf{D} \in \mathbf{AG}$ such that $\mathbf{G} \subseteq \mathbf{D}$ [HR](A.15). In view of 0.2(c) above it is enough to treat the case $\mathbf{G} = \mathbf{D}$, *i.e.* to assume that $\mathbf{G}$ itself is divisible. Standard structure theory (as in [Fu](23.1) or [HR](A.14)) furnishes the decomposition $\mathbf{G} = \oplus_{i\in I} \mathbf{G}_i$ where each of the groups $\mathbf{G}_i$ is equal either to $\mathbf{Q}$ or to one of the groups $\mathbf{Z}(p^\infty)$. For $n < \omega$ with $m \neq n$ we write

$S(n) = \{i \in I: (x_n)_i \neq e_i\}$ (the *support* of x_n)

and for m, $n < \omega$ we write

$\{m, n\} \in P_0$ if $S(m) \cap S(n) \neq \emptyset$

$\{m, n\} \in P_1$ if $S(m) \cap S(n) = \emptyset$.

The set $[\omega]^2$ of 2-element subsets of ω is thus expressed in the form $[\omega]^2 = P_0 \cup P_1$; according to the Erdős-Rado partition relation $\omega \to (\omega)^2_2$ (see for example [Je](Lemma 29.1) or [Ra]) there is an infinite set $\mathbf{S} \subseteq \omega$ such that $[S]^2 \subseteq P_0$ or $[S]^2 \subseteq P_1$ .We treat these cases separately, assuming for notational simplicity that $\mathbf{S} = \omega$.

If $[\omega]^2 \subseteq P_0$ there is $i_0 \in I$ such that $i_0 \in S(n)$ for infinitely many $n < \omega$. For each such n we have $(x_n)_{i_0} \neq e_{i_0}$ and from $x_n \to e$ in $\mathbf{G}^{\#}$ follows $(x_n)_{i_0} \to e_{i_0}$ in $\mathbf{G}_{i_0}{}^{\#}$. This contradicts case 2 if $\mathbf{G}_{i_0} = \mathbf{Z}(p^\infty)$ and it contradicts case 3 if $\mathbf{G}_{i_0} = \mathbf{Q}$.

If $[\omega]^2 \subseteq P_1$ then it is enough for $n < \omega$ to choose $h_n \in \mathbf{H}(\oplus_{i\in S(n)} \mathbf{G}_i)$ so that $h_n(x_n) \in \mathbf{U}$ and then to choose any $h \in \mathbf{H}(\mathbf{G})$ such that $h \supseteq h_n$ for each $n < \omega$. As before we have $x_n \to e$ in G, while $h(x_n) \in \mathbf{U}$ and therefore $h(x_n) \nrightarrow 1$. □

1.2. Remark. (a) For generalizations (based on other techniques) of Lemma 1.1 to arbitrary LCA groups, the reader may consult [DPS](3.4.1) or [V](p. 484, Corollary 1). The approach in [Re] is similar to ours; see also [Fl].

(b) While no point of bG is the limit of a sequence drawn from G (for G a discrete Abelian group), the compact groups bG do contain convergent sequences. Indeed it is known that bG, like every compact group, is dyadic [V], [K], and every infinite dyadic space contains a non-trivial convergent sequence [E]; further, it is a theorem of Šapirovskiĭ [Š] (see also Juhász [Ju] for an expository account) that every compact group of infinite weight α contains up to homeomorphism a copy of the space $\{0, 1\}^{\alpha}$.

1.3. Theorem. If $\mathbf{G} \in \mathbf{AG}$ then every compact subspace of $\mathbf{G}^{\#}$ is finite.

Proof. Let $\mathbf{K}$ be a compact subspace of $\mathbf{G}^{\#}$ such that $|\mathbf{K}| \geq \omega$, let $\mathbf{A} \subseteq \mathbf{K}$ satisfy $|\mathbf{A}| = \omega$, and set $\mathbf{N} = \langle \mathbf{A} \rangle$. Then $\mathbf{N}$ is closed in $\mathbf{G}^{\#}$ by 0.2(b), so $\mathbf{N} \cap \mathbf{K}$ is a countably infinite compact subspace of $\mathbf{G}^{\#}$. Like every countably infinite compact Hausdorff space, the space $\mathbf{N} \cap \mathbf{K}$ contains a non-trivial convergent sequence. This contradicts Lemma 1.1. □

1.4. Theorem. If $\mathbf{G} \in \mathbf{AG}$ then every pseudocompact subspace of $\mathbf{G}^{\#}$ is finite.

Proof. Let $\mathbf{A}$ be a pseudocompact subspace of $\mathbf{G}^{\#}$ satisfying $|\mathbf{A}| \geq \omega$. Since a pseudocompact, Lindelöf space is compact [GJ](5H2), we have $|\mathbf{A}| > \omega$. Let $\mathbf{H}_0 = \langle a_0 \rangle$ with $a_0 \in \mathbf{A}$ and recursively, if $n < \omega$ and a_k has been chosen for each $k < n$, let $\mathbf{H}_{n-1} = \langle \{a_k : k \leq n-1\} \rangle$ and, using $|\mathbf{A}| > \omega$, choose $a_n \in \mathbf{A} \setminus \mathbf{H}_{n-1}$. Since $\mathbf{H}_{n-1} \in \mathbf{S}(\langle \mathbf{A} \rangle)$ and $a_n \in \langle \mathbf{A} \rangle \setminus \mathbf{H}_{n-1}$, there is $h_n \in \mathbf{H}(\langle \mathbf{A} \rangle)$ such that $h_n \equiv 1$ on $\mathbf{H}_{n-1}$ and $h_n(a_n) \neq 1$. We define $h: \cup_{k<\omega} \mathbf{H}_k \to \mathbf{T}^{\omega}$ by $h(x)_n = h_n(x)$ and we note that $|\, h[\cup_{k<\omega} \mathbf{H}_k] \,| \leq |\cup_{k<\omega} \mathbf{H}_k| = \omega$ and also that $h(a_n) \neq h(a_k)$ when $0 \leq k < n < \omega$ (since $a_k \in \mathbf{H}_{n-1}$ so $h(a_k)_n = h_n(a_k) = 1$, while $h(a_n)_n = h_n(a_n) \neq 1$). Thus $|\, h[\cup_{k<\omega} \mathbf{H}_k] \,| = \omega$. Let $\mathbf{D}$ be a countable divisible subgroup of $\mathbf{T}^{\omega}$ such that $h[\cup_{k<\omega} \mathbf{H}_k] \subseteq \mathbf{D}$ [HR](Appendix A) and choose $\overline{h} \in \mathrm{Hom}(\langle \mathbf{A} \rangle, \mathbf{D})$ such that $\overline{h} \supseteq h$. Then $\overline{h}$ is continuous from $\langle \mathbf{A} \rangle^{\#}$ into $\mathbf{D}^{\#} \subseteq b\mathbf{D}$ by 0.2(d), so $\mathbf{D}^{\#}$ has a countably infinite pseudocompact (hence compact) subspace (namely $\overline{h}[\mathbf{A}]$), contradicting Theorem 1.3. □

1.5. Corollary. If $\mathbf{G} \in \mathbf{AG}$ then every countably compact subspace of $\mathbf{G}^{\#}$ is finite.

§2. $\mathbf{G}^{\#}$ is zero-dimensional.

As usual, a space is said to be *zero-dimensional* if it has a basis of open-and-closed sets.

In Lemma 2.1 and Theorem 2.2 we denote by u the usual topology on $\mathbf{T}$ or on any of its subsets.

2.1. Lemma. There is a totally bounded, zero-dimensional, topological group topology v on $\mathbf{T}$ such that $v \supseteq u$.

Proof. Write $\mathbf{T} = \mathbf{A}\times\mathbf{B}$ algebraically [HR](A.14) where $\mathbf{A}$ and $\mathbf{B}$ are proper subgroups of $\mathbf{T}$ and note that $\langle\mathbf{A},u\rangle$ and $\langle\mathbf{B},u\rangle$ are totally bounded, zero-dimensional topological groups. Let w be the product topology for $\mathbf{T}$, so that $\langle\mathbf{T},w\rangle = \langle\mathbf{A},u\rangle \times \langle\mathbf{B},u\rangle$, and note that the homomorphism $\phi : \mathbf{T} = \mathbf{A}\times\mathbf{B} \to \mathbf{T}$ given by $\phi(a,b) = ab$ is continuous from $\langle\mathbf{T},w\rangle$ to $\langle\mathbf{T},u\rangle$; indeed ϕ is the restriction to $\langle\mathbf{A},u\rangle \times \langle\mathbf{B},u\rangle$ of the usual multiplication function from $\langle\mathbf{T},u\rangle \times \langle\mathbf{T},u\rangle$ to $\langle\mathbf{T},u\rangle$.

Since ϕ is an isomorphism of $\mathbf{T}$ onto $\mathbf{T}$ there is a (compact, connected) topological group topology t for $\mathbf{T}$ for which $\phi : \langle\mathbf{T},t\rangle \to \langle\mathbf{T},u\rangle$ is a homeomorphism. (In detail: $S \in t$ if and only if there is $U \in u$ such that $S = \phi^{-1}(U)$.) We claim that $t \subseteq w$. Indeed let S be a t-neighborhood of 1, let U be the u-neighborhood of 1 such that $S = \phi^{-1}(U)$, let V be a u-neighborhood of 1 such that $V \cdot V \subseteq U$, and define $W = (V \cap \mathbf{A}) \times (V \cap \mathbf{B})$; since $W \in w$, the claim will be proved when $W \subseteq S$ is shown. For $(a,b) \in W$ we have $ab \in V \cdot V \subseteq U$ and hence $(a,b) \in \phi^{-1}(U) = S$. The proof that $t \subseteq w$ is complete.

Finally let $v = \{\phi[W] : W \in w\}$. Since ϕ is an isomorphism the topology v, like w, is a totally bounded, zero-dimensional topological group topology for $\mathbf{T}$; and $v \supseteq u$ since $w \supseteq t$. □

2.2. Theorem. If $\mathbf{G} \in \mathbf{AG}$, then $\mathbf{G}^{\#}$ is zero-dimensional.

Proof. Let v be a topology for $\mathbf{T}$ as given by Lemma 2.1. The topology t of $\mathbf{G}^{\#} = \langle\mathbf{G},t\rangle$ is the smallest topology for $\mathbf{G}$ such that every $h \in \mathbf{H}(\mathbf{G})$ is continuous from $\mathbf{G}$ to $\langle\mathbf{T},u\rangle$. Since $u \subseteq v$ and every $h \in \mathbf{H}(\mathbf{G})$ is continuous from $\langle\mathbf{G},t\rangle$ to $\langle\mathbf{T},v\rangle$ (by 0.3(a)), the topology t of $\mathbf{G}^{\#} = \langle\mathbf{G},t\rangle$ is the smallest topology for $\mathbf{G}$ such that every $h \in \mathbf{H}(\mathbf{G})$ is continuous from $\mathbf{G}$ to $\langle\mathbf{T},v\rangle$. It then follows that

$\{h^{-1}(F) : h \in \mathbf{H}(\mathbf{G}), F$ open-and-closed in $\langle\mathbf{T},v\rangle\}$

is a sub-base for $\mathbf{G}^{\#} = \langle \mathbf{G}, t \rangle$ consisting of t-open-and-closed subsets of G. □

2.3. Remark. We first proved 2.2 some years ago in response to a query from van Douwen [vD1]. Subsequently and independently, van Douwen himself proved 2.2 [vD2]. Van Douwen's more difficult question [vD2], whether every topological Abelian group of the form $G^{\#}$ is strongly zero-dimensional (in the sense that $\dim(G^{\#}) = 0$) has recently been answered affirmatively by Shakhmatov [S1], [S2].

List of References

[C] W. W. Comfort. Topological groups. In: Handbook of General Topology, edited by Kenneth Kunen and Jerry Vaughan, pp. 1143-1263. North-Holland Publ. Co. Amsterdam. 1984.

[CS] W. W. Comfort. and V. Saks. Countably compact groups and finest totally bounded topologies. Pacific J. Math. 49 (1973). 33-44.

[CT] W. W. Comfort and F. J. Trigos. The maximal totally bounded group topology. Abstracts Amer. Math. Soc. 9 (1988), 420-421 (= Abstract #88T-22-195). [Abstract]

[DPS] D. N. Dikranjan, I. R. Prodanov and L. N. Stoyanov. Topological Groups. Pure and applied Mathematics series, Vol. 130. Marcel Dekker. New York. 1989.

[vD1] Eric van Douwen. Letters to W. W. Comfort: June 30, 1986, and May 9, 1987.

[vD2] E. van Douwen. The maximal totally bounded group topology on G and the biggest minimal G-space, for Abelian groups G. Topology and its Applications 34 (1990), 69-91.

[E] R. Engelking. Cartesian products and dyadic spaces. Fundamenta Math. 57 (1965), 287-304.

[Fl] Peter Flor. Zur Bohr-Konvergenz von Folgen. Math. Scandinavica 23 (1968), 169-170.

[Fu] Lazló Fuchs. Infinite Abelian Groups. Vol. I, Pure and Applied Mathematics, No. 36, Academic Press, New York and London. 1970.

[GJ] L. Gillman and M. Jerison. Rings of Continuous Functions. Graduate Texts in Mathematics, Vol. 43. Springer-Verlag. New York-Heidelberg-Berlin. 1976.

[G] I. Glicksberg. Uniform boundedness for groups. Canadian J. Math. 14 (1962), 269-276.

[HR] Edwin Hewitt and Kenneth A. Ross. Abstract Harmonic Analysis. Volume I. Grundlehren der math. Wissenschaften, Vol. 115. Springer Verlag, Berlin-Göttingen-Heidelberg. 1963.

[Je] Thomas Jech. Set Theory. Academic Press, New York-London. 1978.

[Ju] I. Juhász. Cardinal Functions in Topology--Ten Years Later. Mathematical Centre Tracts 123. Mathematisch Centrum. Amsterdam. 1980.

[K] V. Kuzminov. On a hypothesis of P. S. Alexandroff in the theory of topological groups. Doklady Akad. Nauk SSSR N.S. 125 (1959), 727-729. [In Russian].

[Ra] F. P. Ramsey. On a problem of formal logic. Proc. London Math. Soc. (2) 30, 264-286.

[Re] G. A. Reid. On sequential convergence in groups. Math. Zeitschrift 102 (1967), 227-235.

[Š] B. È. Šapirovskiĭ. On embedding extremally disconnected spaces in compact Hausdorff spaces. b-points and weight of pointwise normal spaces. Doklady Akad. Nauk SSSR 223 (1975), 1083-1086. [In Russian. English translation: Soviet Math. Doklady 16 (1975), 1056-1061.]

[S1]. Dmitrii Shakhmatov. A survey of current research and open problems in the dimension theory of topological groups. Q&A in General Topology, Vol. 8, (1990), 101-128.

[S2] D. B. Shakhmatov. Imbeddings into topological groups preserving dimensions. Topology and Its Applications. To appear.

[T1] F. Javier Trigos-Arrieta. Convergence Modulo 1 and an application to the Bohr compactification of **Z**. Manuscript submitted for publication.

[T2] F. Javier Trigos-Arrieta. Pseudocompactness on groups. This volume.

[T3] F. Javier Trigos-Arrieta. Continuity, boundedness, connectedness and the Lindelöf property for topological groups. Journal of Pure and Applied Algebra 69 (1991) (=Proceedings of the 1989 Curaçao Conference on Locales and Topological Groups). To appear.

[V] N. Th. Varopoulos. Studies in harmonic analysis. Proc. Camb. Phil. Soc. (1964), 60, 465-516.

[Vi] N. Ya. Vilenkin. On the dyadicity of the group space of bicompact commutative groups. Uspehi Mat. Nauk N.S. 13(6) (84) (1958), 79-80. [In Russian.]

Semiregularization Spaces and the Semiclosure Operator, and s-Closed Spaces

Charles Dorsett Louisiana Tech University, Ruston, Louisiana

I. INTRODUCTION

In 1937, regular open sets were introduced. Let (X,T) be a space and let A, B $\subset$ X. Then A is regular open, denoted by A $\in$ RO(X,T), iff A = Int((cl(A)) (Stone, 1937). In the 1937 investigation it was shown that for the space (X,T), RO(X,T) is a base for a topology T_s on X coarser than T and (X,T_s) is called the semi regularization space of (X,T). In 1963, the closure operator was used to define semi open sets. The set A is semi open, denoted by A $\in$ SO(X,T), iff there exists O $\in$ T such that O $\subset$ A $\subset$ Cl(O) (Levine, 1963). In 1970, semi open sets were used to define semi closed sets, which were used to define the semi closure of a set. The subset A is semi closed iff X − A is semi open and the semi closure of B, denoted by sclB, is the intersection of all semi closed sets containing B (Biswas, 1970). Since their introduction, each of regular open sets, semi open sets, and the semi closure operator have been used to define and investigate many new properties of topological spaces. In 1978 (Cameron,

1978) the subset A was defined to be regular semi open, denoted by A ∈ RSO(X,T), iff there exists a regular open set U such that U ⊂ A ⊂ Cl(U). In 1984 (Di Maio, 1984), the semi interior and semi closure operators were used to define semi regular open sets. The semiinterior of A, denoted by sint A, is the union of all semi open sets contained in A and B is semi regular open, denoted by B ∈ SRO(X,T), iff B = sint(sclB). Then in 1987 (Di Maio and Noiri, 1987), semi regular sets were defined and used to define s-closedness. The subset A is semi regular, denoted by A ∈ SR(X,T), iff A is both semi open and semi closed and (X,T) is s–closed iff every cover of X by semi open sets has a finite subcollection whose semi closures cover X. In the 1987 investigation (Di Maio and Noiri, 1987), it was shown that RSO(X,T) = SRO(X,T) = SR(X,T). In this paper the role of semi closure operator in semi regularization spaces is further examined and s–closedness is further investigated.

II. SEMIREGULARIZATION SPACES AND THE SEMI CLOSURE OPERATOR

In 1978 (Maheshwari and Tapi, 1978), the semi closure operator was used to define feebly open sets. Let (X,T) be a space and let A ⊂ X. Then A is feebly open, denoted by A ∈ FO(X,T), iff there exists O ∈ T such that O ⊂ A ⊂ sclO. Further investigations of feebly open sets led to the discovery that FO(X,T) is a topology on X (Dorsett, 1985a) and that RO(X,T) = {sclO | O ∈ FO(X,T)} (Dorsett, 1985b) = {sclO | O ∈ T} (Dorsett, 1988). These results raised questions about the behavior of the semi closure operator on finite intersections of open sets, which led to the following discoveries.

Theorem 2.1. Let (X,T) be a space and let $\{O_i \mid i = 1,\cdots,n\} \subset T$, where n is a natural number. Then $\bigcap_{i=1}^{n} (\text{Int}(\text{Cl}(O_i)) = \text{Int}(\text{Cl}(\bigcap_{i=1}^{n} O_i))$.

Proof: The proof is by mathematical induction. Clearly, the statement is true for $n = 1$. Thus, consider the case $n = 2$. Let $\{O_1,O_2\} \subset T$. Since $\text{Cl}(O_1 \cap O_2) \subset \text{Cl}(O_i)$, $i = 1,2$, then $\text{Int}(\text{Cl}(O_1 \cap O_2)) \subset \text{Int}(\text{Cl}(O_1) \cap \text{Int}(\text{Cl})O_2))$. Also, $\text{Int}(\text{Cl}(O_1)) \cap \text{Int}(\text{Cl}(O_2)) \subset \text{Cl}(O_1 \cap O_2)$, for suppose not. Let $U = \cup \text{Int}(\text{Cl}(O_1)) \cap \text{Int}(\text{Cl}(O_2))] - \text{Cl}(O_1 \cap O_2)$, let $U_1 = O_1 - \text{Cl}(O_1 \cap O_2)$, let $U_2 = O_2 - \text{Cl}(O_1 \cap O_2)$, and let $x \in U$. Let $V \in T$ such that $x \in V$. since $x \in \text{Int}(\text{Cl}(O_1)) \subset \text{Cl}(O_1)$ and $U \cap V \in T$ such that $x \in U \cap V$, then $U \cap V \cap O_1 = U \cap V \cap U_1 \neq \emptyset$. Thus $x \in \text{Cl}(U_1)$. Hence $U \subset \text{Cl}(U_1)$ and $U \subset \text{Int}(\text{Cl}(U_1))$. By a similar argument $U \subset \text{Int}(\text{Cl}(U_2))$, but since U_1 and U_2 are disjoint open sets, then $\text{Int}(\text{Cl}(U_1)) \cap \text{Int}(\text{Cl}(U_2)) = \emptyset$, which is a contradiction. Thus $\text{Int}(\text{Cl}(O_1)) \cap \text{Int}(\text{Cl}(O_2)) \subset \text{Cl}(O_1 \cap O_2)$, which implies $\text{Int}(\text{Cl}(O_1)) \cap \text{Int}(\text{Cl}(O_2)) \subset \text{Int}(\text{Cl}(O_1 \cap O_2))$.

Assume the statement is true for $n = k$. Let $\{O_i \mid i = 1,\cdots,k+1\} \subset T$. Then $\text{Int}(\text{Cl}(\bigcap_{i=1}^{k+1} O_i)) = \text{Int}(\text{Cl}(\bigcap_{i=1}^{k} O_i)) \cap \text{Int}(\text{Cl}(O_{k+1})) = \bigcap_{i=1}^{k+1} \text{Int}(\text{Cl}(O_i))$. Thus the statement is true for $n = k + 1$, which implies the statement is true for each $n \in N$.

Combining Theorem 2.1 with the fact that for each space (X,T) and each $O \in T$, $\text{sclO} = \text{Int}(\text{Cl}(O))$ (Dorsett, 1987a) gives the next result.

Corollary 2.1. Let (X,T) be a space and let $\{O_i \mid i = 1,\cdots,n\} \subset T$, where n is a natural number. Then $\mathrm{scl} \bigcap_{i=1}^{n} O_i = \bigcap_{i=1}^{n} \mathrm{scl} O_i$.

Combining the results above with the fact that for a space (X,T) and $A \subset X$, $\mathrm{scl}_T A = \mathrm{scl}_{FO(X,T)} A$ (Dorsett, 1985a) gives the following result.

Corollary 2.2. Let (X,T) be a space and let $\{O_i \mid i = 1,\cdots,n\} \subset FO(X,T)$, where n is a natural number. Then $\mathrm{scl} \bigcap_{i=1}^{n} O_i = \bigcap_{i=1}^{n} \mathrm{scl}\ O_i$.

Examples can be constructed showing that Theorem 2.1, Corollary 2.1, and Corollary 2.2 cannot be extended to a denumerable set of open sets.

The results above raised questions about what would happen if open in the results above were replaced by semi open, which led to the investigation below.

Theorem 2.2. Let (X,T) be a space and let $\{O_\alpha \mid \alpha \in A\}$ be a collection of subsets of X. Then $\mathrm{Int}(\mathrm{Cl}(\bigcap_{\alpha \in A} O_\alpha)) \subset \bigcap_{\alpha \in A} \mathrm{Int}(\mathrm{Cl}(O_\alpha))$.

The straightforward proof is omitted.

Theorem 2.3. Let (X,T) be a space and let $\{O_i \mid i = 1,\cdots,n\} \subset SO(X,T)$, where n is a natural number. Then $\mathrm{Int}(\mathrm{Cl}(\bigcap_{i=1}^{n} O_i)) = \bigcap_{i=1}^{n} \mathrm{Int}(\mathrm{Cl}(O_i))$.

Proof: Since $O_i \in SO(X,T)$ for each $i \in \{1,\cdots,n\}$, then $\mathrm{Cl}(O_i) =$

$Cl(Int(O_i))$, $i = 1,\cdots,n$. Thus $\bigcap_{i=1}^{n} Int(Cl(O_i)) = \bigcap_{i=1}^{n} Int(Cl(Int(O_i))) =$

$Int(Cl\bigcap_{i=1}^{n} Int(O_i))) \subset Int(Cl\bigcap_{i=1}^{n} O_i)) \subset \bigcap_{i=1}^{n} Int(Cl(O_i))$.

The following example shows that Corollary 2.1 cannot be extended to semi open sets.

Example 2.1. Let $X = \{a,b,c\}$, let $T = \{\emptyset,X,\{a\},\{b\},\{a,b\}\}$, let $U_1 = \{a,b\}$, and let $U_2 = \{a,c\}$. Then $U_i \in SO(X,T)$, $i = 1,2$, and $sclU_1 \cap sclU_2 \neq scl(U_1 \cap U_2)$.

Combining the facts that for a space (X,T) and $U \in SO(X,T)$, $SO(X,T) = SO(X,FO(X,T))$ (Dorsett, 1987b) and $sclU \in SR(X,T)$ (Di Maio and Noiri, 1987) with the definition of semi regular sets and results above give the next result.

Corollary 2.3. Let (X,T) be a space. Then $SR(X,T) = \{sclO \mid O \in SO(X,T)\} = SR(X,FO(X,T))$.

The fact that for a space (X,T), $RO(X,T_S) = RO(X,T)$ (Dorsett, 1985b) and $RO(X,T)$ is a base for a topology on X raised similar questions about $SR(X,T_S)$ and $SR(X,T)$. The fact that for a space (X,T), $U \in SO(X,T)$, and $A \subset X$; $scl_T U = scl_{T_S} U \in SO(X,T_S)$ (Dorsett, 1989) and $scl(sclA) = sclA$ can be combined with Corollary 2.3 to eaisly resolve one of the questions.

Corollary 2.4. Let (X,T) be a space. Then $SR(X,T) = \{sclU \mid U \in SO(X,T_s)\} = SR(X,T_s)$.

The following example shows that for a space (X,T), $SR(X,T)$ need not be a base for a topology on X.

Example 2.2. Let $\mathbb{N}$ denote the natural numbers, let $X = \mathbb{N} \cup \{0\}$, and let T be the topology on X with base $\{X\} \cup \{\{n\} | n \in \mathbb{N}\}$. Then $A = \{0,1\}$ and $B = \{0,2\}$ are semi regular sets and $A \cap B = \{\emptyset\} \notin SR(X,T)$.

Examples can be easily given showing that for a space (X,T) and closed, semi open subset A of X, $SR(A,T_A)$ need not equal $SR(X,T)_A$. Thus questions were raised about conditions that could be imposed on the subset A of the space (X,T) to obtain $SR(A,T_A) = SR(X,T)_A$, which led to the work below.

In 1981 (Mashhour, El-Monsef, and El-Deeb, 1981), pre-open sets were introduced. A subset A of a space (X,T) is pre-open, denoted by $A \in PO(X,T)$, iff $A \subset Int(Cl(A))$. Below semi regular sets are further investigated using pre-open sets.

Theorem 2.4. Let (X,T) be a space and let $A \in PO(X,T)$. Then $SR(A,T_A) = SR(X,T)_A$.

Proof: Let $O \in SR(X,T)_A$. Let $W \in SR(X,T)$ such that $O = W \cap A$. Since $SO(A,T_A) = SO(X,T)_A$ (Dorsett, 1987c), then $O \in SO(A,T_A)$ and since $scl_{T_A} O = A \cap scl_T O$ (Dorsett, 1987c) $\subset A \cap scl_T W = O$, then $O \in SR(A,T_A)$. Thus $SR(S,T)_A \subset SR(A,T_A)$. Let $V \in SR(A,T_A)$. Then V, $A - V \in SO(A,T_A)$ and there exist Y, Z $\in$ SO(X,T) such that V = A $\cap$ Y and A − V = A $\cap$ Z. Let B = Int(Y) $\cap$ Int(Z). Then B $\cap$ Int(Cl(A)) = $\emptyset$, which implies Cl(B) $\cap$ Int(Cl(A)) = $\emptyset$. since A $\subset$ Int(Cl(A)), then A $\cap$ Cl(B) = $\emptyset$. Then scl(Y − Cl(B)) $\in$ SR(X,T) and V = A $\cap$ scl(Y − Cl(B)), which implies V $\in$ $SR(X,T)_A$. Thus $SR(A,T_A) \subset SR(X,T)_A$ and $SR(A,T_A) = SR(X,T)_A$.

In a 1988 investigation of pre-open sets (Noiri, 1988), it was shown that for a space (X,T) and A $\in$ PO(X,T), sclA = Int(Cl(A)). Combining this result with results above give the last result in this section.

Corollary 2.5. Let (X,T) be a space. Then RO(X,T) = {sclA | A $\in$ PO(X,T)}.

III. S-CLOSED SPACES

In the 1987 investigation (Di Maio and Noiri, 1987), it was shown that a space (X,T) is s-closed iff every cover of X by elements of SR(X,T) has a finite subcover. This result can be combined with results above to obtain the following result.

Corollary 3.1. Let (X,T) be a space. Then the following are equivalent:

(a) (X,T) is s-closed,

(b) (X,T_S) is s-closed, and

(c) (X,FO(X,T)) is s-closed.

Since s-closedness is preserved by continuous, open, onto functions (Di Maio and Noiri, 1987), then s-closedness is a topological property. In 1972 (Crossley and Hildebrand, 1972), semi homeomorphisms were defined by replacing open in the definition of homeomorphisms by semi open and properties preserved by semi homeomorphisms were called semi topological properties. In 1978 (Cameron, 1978), topological properties simultaneously shared by both a space (X,T) and its semiregularization space (X,T_s) were called semi regular properties. Then in 1985 (Dorsett, 1985c), topological properties simultaneously shared by both a space (X,T) and (X,FO(X,T)) were called feeble properties and were shown to be equivalent to the semi topological perperties. Thus the results above show that s-closedness is a semi regular, semi topological property.

Theorem 3.1. Let (X,T) be an s-closed space. Then (X,T_s) is compact and for each $O \in PO(X,T)$, $Fr(O) - scl\ O$ is finite.

Proof: Let $\theta = \{O_\alpha \mid \alpha \in A\}$ be a cover of X by regular open sets. since $RO(X,T) \subset SO(X,T)$, then there exists a finite subcollection $\{O_{\alpha_i} \mid i = 1,\cdots,n\}$ whose semi closures cover X and since $O_{\alpha_i} = Int(Cl(O_{\alpha_i})) = sclO_\alpha$, for each $i \in \{1,\cdots,n\}$, then $\{O_{\alpha_i} \mid i = 1,\cdots,n\}$ is a finite subcollection of θ that covers X. Since RO(X,T) is a subbase for T_s, then (X,T_s) is compact by Alexander's subbase theorem. Let $O \in PO(X,T)$. Then $F = Fr(O) - sclO$ is finite, for suppose not. For each $x \in F$, let $O_x = \{x\} \cup sclO \in SO(X,T)$. Then $\{O_x \mid x \in F\} \cup \{Ext(sclO)\}$ is a cover of (X,T) by semi regular sets with no finite subcover, which contradicts (X,T) is s-closed.

In 1969 (Singal and Mathur, 1969), compact spaces were generalized to nearly compact spaces. A space (X,T) is nearly compact iff every open cover of X has a finite subcollection the interiors of the closures of which cover X. Further investigation of nearly compactness in 1972 (Carnahan, 1972) showed that a space is nearly compact iff its semi regularization space is compact. Thus Theorem 3.1 shows that s-closedness implies nearly compactness and Example 2.2 can be used to show that the converse statement is not true.

In the introductory 1987 paper (Di Maio and Noiri, 1987), subsets s-closed relative to a space were introduced and investigated. A subset S of the space (X,T) is s-closed relative to (X,T) if for every cover $\{V_a \mid a \in A\}$ of S by T-semi open sets, there exists a finite subcollection whose T-semi closures cover S. The last result in this paper characterizes s-closed spaces using s-closed relative subsets.

Theorem 3.2. Let (X,T) be a space. Then the following are equivalent:

(a) (X,T) is s-closed,

(b) sclA is s-closed relative to (X,T) for each $A \in PO(X,T)$,

(c) $(sclA, T_{sc\ell A})$ is s-closed for each $A \in PO(X,T)$,

(d) (O, T_O) is s-closed for each $O \in RO(X,T)$, and

(e) O is s-closed relative to (X,T) for each $O \in RO(X,T)$.

The proof is straightforward using the results above and is omitted.

REFERENCES

1. Biswas, N. (1970). "On Characterizations of Semi-continuous Functions", **Atti Accad. Naz. Lincei Rend.** cl. Sci. **Fis. Mat. Natur. 8, 48(4)** : 399.

2. Crossley, S. and Hildebrand, S. (1972). "Semi-topological

Properties", **Fund. Math., 74** : 233.

3. Cameron, D. (1978). "Properties of S–closed Spaces", **Proc. Amer. Math. Soc., 72 no. 3** : 581.

4. Carnahan, D. (1972). "Locally Nearly–Compact Spaces", **Boll. U.M.I. (4), 6** : 146.

5. Di Maio, G. (1984). "On semi topological Operators and Semiseparation Axioms", **Preprint no. 3**, Dept. Math. Appl., Naples Univ.

6. Di Maio, G and Noiri, T. (1987). "On s–closed Spaces", **Indian J. Pure Appl. Math., 18** : 226.

7. Dorsett, C. (1985a). "Feeble Separation Axioms, the Feebly Induced Topology, and Separation Axioms and the Feebly Induced Topology", **Karadeniz Un. Math. J., 8** : 43.

8. Dorsett, C. (1985b). "New Characterizations of Topological Properties Using Regular Open Sets and r–Topological Properties", **Bull. Fac. Sci., Assiut Univ., A. Phys. Math., 14 no.** 1 : 75.

9. Dorsett, C. (1985c). "Feebly Open α–Set, and semi closure Induced Topologies and Feeble Properties", **Pure Math. Manuscript, 4** : 107.

10. Dorsett, C. (1987a). "Feebly Continuous Images, Feebly Compact R_1 Spaces, and semi topological Properties", **Pure Math. Manuscript, 6** : 1.

11. Dorsett, C. (1987b). "New Characterizations and the Feebly Induced Topology", **Rend. Mat. Ser. VII, 7 no.1** : 121.

12. Dorsett, C. (1987c). "Pre–open Sets and Feeble Separation Axioms", **An. Univ. Timisoara Ser. Stiinte Mat., 25** : 39.

13. Dorsett, C. (1988). "Properties of Topological Spaces and the Semiregularization Topology", **Proc. Nat. Acad. Sci., India, Sec. A, 58(II)** : 251.

14. Dorsett, C. (1989). "Higher semi separation Axioms", **Granit, 9**.

15. Levine, N. (1963). "semi open Sets and Semi-continuity in Topological Spaces", **Amer. Math. Monthly, 70** : 36.

16. Maheshwari, S. and Tapi, U. (1978). "Note on Some Applications of Feebly Open Sets", **Madhya Bharati, J. Un. Saugar**.

17. Mashhour, A., El-Monsef, M. Abd, and El-Deeb, S. (1981), "On Precontinuous and Weak Precontinuous Mappings", **Proc. Math. and Phys. Soc. Egypt, 51** : 47.

18. Noiri, T. (1988). "Almost α-Continuous Functions", **Kyungpook Math. J., 28 no.** 1 : 71.

19. Singal, M. and Mathur, A. (1969). "On Nearly comapct Spaces", **Boll. U.M.I. (4), 2** : 702.

20. Stone, M. (1937). "Applications of the Theory of Boolean Rings to Topology", **Trans. A.M.S., 41** : 374.

A Wild Fréchet Space

Alan Dow York University, North York, Ontario, Canada

Abstract

Nogura has shown that every countable Fréchet space which has no isolated point and has character less than $\mathfrak{b}$ contains a homeomorphic copy of the rationals. Furthermore, if the Continuum Hypothesis is assumed then there is a countable Fréchet space with no isolated points which does not contain a copy of the rationals. We produce a countable Fréchet space with no isolated points which does not even contain a homeomorphic copy of the ordinal space $\omega^2 + 1$.

A space X is *Fréchet* if it is the case that a point is in the closure of a subset of X iff there is a sequence from that set converging to the point. Van Douwen calls a space *crowded* if it contains no isolated points. Nogura, [Nog], has asked about the existence of a crowded Fréchet space which contains no copies of the rationals. We produce one. Our main idea is borrowed from [Bs80] in that we use a dense subtree of $\mathcal{P}(\omega)$, the power set of the integers, to guide an induction through $\mathfrak{c}$ steps.

The set of functions from ω to ω is denoted by ${}^\omega\omega$ and $[\omega]^\omega$ denotes the set of infinite subsets of ω. Recall that $\mathfrak{b}$ is the minimum cardinality of a subset of ${}^\omega\omega$ which is unbounded with respect to $<^*$ where, as usual, $<^*$ denotes the usual mod finite ordering on ${}^\omega\omega$. For each $n \in \omega$, ${}^n\omega$ denotes the set of functions into ω with domain equal to the ordinal n and ${}^{<\omega}\omega$ is used to denote the set $\bigcup_{n\in\omega} {}^n\omega$. When ${}^{<\omega}\omega$ is ordered by inclusion, $\subset$, it forms a tree. That is, for each $s \in {}^{<\omega}\omega$, the set $\{t \in {}^{<\omega}\omega : t \subseteq s\}$ is well-ordered (in this case finite). Notice that a consequence of this fact is that if $s, s' \in {}^{<\omega}\omega$ are incomparable, then $s^\uparrow \cap s'^\uparrow = \emptyset$, where $s^\uparrow = \{t \in {}^{<\omega}\omega : s \subseteq t\}$. To see this, suppose that $t \in s^\uparrow \cap s'^\uparrow$. Then $s, s' \in \{x \in {}^{<\omega}\omega : x \subseteq t\}$. Since this set is linearly ordered, s and s' must be comparable.

Research supported by NSERC of Canada

Definition 0.1 *It follows immediately from the results in [BPS80] that there is a set $\mathfrak{T} \subset \mathcal{P}(\omega)$ which is a dense subtree of $\mathcal{P}(\omega)$ when ordered by ${}^*\supset$, reverse inclusion mod finite. That is, $\mathfrak{T}$ satisfies:*

1. *$T, T' \in \mathfrak{T}$ are either disjoint or comparable (wrt ${}^*\supset$),*

2. *for each $T \in \mathfrak{T}$, $\{T' \in \mathfrak{T} : T' \,{}^*\supset T\}$ is well-ordered by ${}^*\supset$ and has order-type less than $\mathfrak{b}$.*

3. *for all $Y \in [\omega]^\omega$, there is a $T \in \mathfrak{T}$ such that $T \subset^* Y$.*

The minimum height of a tree $\mathfrak{T}$ as above is usually denoted $\mathfrak{h}$.

We will frequently make use of the following result.

Proposition 0.2 ([Bs80]) *If $\mathfrak{S}$ is any subset of $\mathfrak{T}$ such that $\mathfrak{S}$ has cardinality less than $\mathfrak{c}$, and if Y is any infinite subset of ω, then there is a $T \in \mathfrak{T}$ such that $T \subset^* Y$ and T does not contain (mod finite) any member of $\mathfrak{S}$.*

Proof. Given Y as above choose any $T' \in \mathfrak{T}$ with $T' \subset^* Y$ as in definition 0.1. Let $\{Y_\alpha : \alpha < \mathfrak{c}\}$ be a family of pairwise almost disjoint subsets of T'. For each of these, choose some $T_\alpha \subset^* Y_\alpha$ with $T_\alpha \in \mathfrak{T}$. Clearly, there is an $\alpha < \mathfrak{c}$ so that T_α does not contain (mod finite) any member of $\mathfrak{S}$.

We will construct a new topology on the tree ${}^{<\omega}\omega$ which refines the very natural topology which one might call the **rational topology**. The rational topology is obtained by taking as a subbase (for the clopen sets) the family $\{s^\uparrow : s \in {}^{<\omega}\omega\}$, together with their complements (let us recall that $s^\uparrow = \{t \in {}^{<\omega}\omega : s \subseteq t\}$). Let us also define, for $A \subset {}^{<\omega}\omega$, $A^\uparrow = \bigcup\{t^\uparrow : t \in A\}$. For $x \in {}^{<\omega}\omega$ we shall use the notation $x\frown m$ to denote the function whose domain is $\mathrm{dom}(x) + 1$ and whose value at $\mathrm{dom}(x)$ is m. Similarly, if $t, s \in {}^{<\omega}\omega$, then $t\frown s$ denotes the function h such that $h(n) = t(n)$ for $n \in \mathrm{dom}(t)$ and, for $\mathrm{dom}(t) \leq n < \mathrm{dom}(t) + \mathrm{dom}(s)$, $h(n) = s(n - \mathrm{dom}(t))$. It is easy to see that, for a point $x \in {}^{<\omega}\omega$, the family $\{x^\uparrow - \bigcup\{(x\frown m)^\uparrow : m < n\} : n \in \omega\}$ forms a local base at x.

Proposition 0.3 *If $A \subset {}^{<\omega}\omega$ is closed with respect to the rational topology, then $A^\uparrow$ is clopen in the rational topology.*

Proof. Of course it suffices to prove that $A^\uparrow$ is closed. Let $x \in \overline{A^\uparrow}$. We are to show that $x \in A^\uparrow$. Since A is closed and $A \subset A^\uparrow$, we may assume that $x \notin A$. Choose $n \in \omega$, such that $U = x^\uparrow - \bigcup\{(x\frown m)^\uparrow : m < n\}$ is disjoint from A. Since $U \cap A^\uparrow \neq \emptyset$, choose $t \in A$ such that $t^\uparrow \cap U \neq \emptyset$. Since $U \subseteq x^\uparrow$, $x^\uparrow \cap t^\uparrow \neq \emptyset$. Since ${}^{<\omega}\omega$ is a tree, it follows from our earlier remarks that x and t are comparable. It suffices to show that $t \subseteq x$, since it then follows that

$x \in t^{\uparrow} \subseteq A^{\uparrow}$. Suppose therefore that x is properly contained in t and let j be such that $x^\frown j \subseteq t$. If $j < n$, then $t^{\uparrow} \cap U = \emptyset$. Therefore $n \leq j$ and $t^{\uparrow} \subset U$. However this contradicts that $U \cap A = \emptyset$ since $t \in t^{\uparrow} \cap A$.

1 The Outline of the Construction

We will first concentrate on the point $\emptyset \in {}^{<\omega}\omega$. We let $\{X_\alpha : \alpha < \mathfrak{c}\}$ list all subsets, X, of ${}^{<\omega}\omega$ such that, $\{n : |\{(0,n)\}^{\uparrow} \cap X| = \omega\}$ is infinite. This list contains all potential copies of ω^2+1 which converge to $\emptyset$. Also let $\{A_\alpha : \alpha < \mathfrak{c}\}$ list $\mathcal{P}({}^{<\omega}\omega)$. Our plan is to inductively "kill" copies of ω^2+1 by choosing an infinite $Y_\alpha \subset X_\alpha$ and ensure that $\emptyset$ is not a limit point of Y_α and to choose sets $S_\alpha \subset A_\alpha$ so as to witness Fréchet. That is, it will either be the case that $\emptyset$ will not be a limit point of A_α, in which case S_α will be empty, or S_α will be an infinite subset of A_α and we will ensure that S_α converges to $\emptyset$.

In the end we will have $\{S_\alpha : \alpha < \mathfrak{c}\}$ determining a Fréchet topology at $\emptyset$. We'll show that in this topology $\emptyset$ has a base $\mathfrak{U}$ which refines the usual topology and for each $U \in \mathfrak{U}$ and $n \in \omega$, $U \cap \{(0,n)\}^{\uparrow}$ is a clopen set in the rational topology. That is, we obtain a Fréchet topology in which the neighbourhood bases for points in ${}^{<\omega}\omega - \{\emptyset\}$ are unchanged and the base at $\emptyset$ is $\mathfrak{U}$. Furthermore, there is no copy of ω^2+1 which converges to $\emptyset$.

The construction is completed by simply taking the topology on ${}^{<\omega}\omega$ which is generated by taking $\mathfrak{U}_t = \{U_t : U \in \mathfrak{U}\}$ as a base at $t \in {}^{<\omega}\omega$, where $U_t = \{t^\frown s : s \in U\}$. This topology will be crowded since we will certainly ensure that at least one of the S_α's is infinite.

2 The construction

Fix $\mathfrak{T}$ as in Definition 0.1. For any $A \subset {}^{<\omega}\omega$, define

$$\Pi(A) = \{n \in \omega : \{(0,n)\}^{\uparrow} \cap A \neq \emptyset\}$$

and

$$\Pi^+(A) = \{n \in \omega : |\ \{(0,n)\}^{\uparrow} \cap A| = \omega\}\ .$$

Let us call a set $A \subset {}^{<\omega}\omega$, a *transversal* if A is infinite and, for each n, $|\{(0,n)\}^{\uparrow} \cap A| \leq 1$.

Induction Hypotheses 2.1 *We construct the sets Y_α and S_α, by induction on $\alpha < \mathfrak{c}$, so that:*

1. *$Y_\alpha \subset X_\alpha$ is a transversal,*

2. $\Pi(Y_\alpha) \in \mathfrak{T}$,

3. *for* $\beta < \alpha$, *neither* $\Pi(Y_\beta)$ *nor* $\Pi(S_\beta)$ *is contained (mod finite) in* $\Pi(Y_\alpha)$,

4. $Y_\alpha^\uparrow \cap S_\beta$ *is finite for each* $\beta < \alpha$,

5. $S_\alpha \subset A_\alpha$,

6. *if* $S_\alpha = \emptyset$ *then for every transversal* $S \subset A_\alpha$, *there is a* $\beta \leq \alpha$ *such that* $Y_\beta^\uparrow \cap S$ *is infinite.*

7. *if* $S_\alpha \neq \emptyset$, *then*

   (a) S_α *is a transversal,*

   (b) $\Pi(S_\alpha) \in \mathfrak{T}$ *and does not contain (mod finite) any set from* $\{\Pi(S_\beta) : \beta < \alpha\} \cup \{\Pi(Y_\beta) : \beta \leq \alpha\}$,

   (c) $S_\alpha \cap Y_\beta^\uparrow$ *is finite for all* $\beta \leq \alpha$.

Let $\alpha < \mathfrak{c}$ and suppose that $\{Y_\beta : \beta < \alpha\}$ and $\{S_\beta : \beta < \alpha\}$ have been chosen satisfying conditions 1 through 7 of 2.1. We first define Y_α. Choose any $T \subset \Pi^+(X_\alpha)$ so that $T \in \mathfrak{T}$ and T does not contain (mod finite) any set from $\{\Pi(Y_\beta) : \beta < \alpha\} \cup \{S_\beta : \beta < \alpha\}$. We will choose Y_α so that $\Pi(Y_\alpha) = T$. Let $\{x(n,m) : m \in \omega\}$ list $X_\alpha \cap \{(0,n)\}^\uparrow$ for each $n \in T$.

Let us note that for each $\beta < \alpha$, there is an $f_\beta \in {}^T\omega$ such that, for each $n \in T$ and $m > f_\beta(n)$, $S_\beta \cap (x(n,m))^\uparrow = \emptyset$. Indeed, let $n \in T$ and suppose that $S_\beta \cap \{(0,n)\}^\uparrow \neq \emptyset$ and let $\{s\} = S_\beta \cap \{(0,n)\}^\uparrow$. Since, $\{t \in \{(0,n)\}^\uparrow : t \subseteq s\}$ is finite, so is $\{t \in \{(0,n)\}^\uparrow : s \in t^\uparrow\}$.

Let $\Gamma = \{\beta < \alpha : \Pi(S_\beta) \cap T \neq^* \emptyset\}$. Since, by our construction, Γ also equals the set $\{\beta < \alpha : T \subset^* \Pi(S_\beta)\}$, we have that $|\Gamma| < \mathfrak{b}$. Choose $f \in {}^T\omega$ such that $f_\beta <^* f$ for all $\beta \in \Gamma$ and let $Y_\alpha = \{x(n, f(n)) : n \in T\}$. Since $f_\beta <^* f$, $Y_\alpha^\uparrow \cap S_\beta$ is finite for each for each $\beta \in \Gamma$. For $\beta \in \alpha - \Gamma$, note that $\Pi(S_\beta) \cap \Pi(Y_\alpha)$ is finite. Since each S_β is a transversal and $\Pi(S_\beta \cap Y_\alpha^\uparrow) \subset \Pi(S_\beta) \cap \Pi(Y_\alpha)$, it follows that $(Y_\alpha)^\uparrow \cap S_\beta$ is finite for all $\beta < \alpha$.

This shows that the inductive conditions 2.1 1-4 are satisfied.

Now let us pick S_α.

If it is possible, choose a transversal $S \subset A_\alpha$, such that, for all $\beta \leq \alpha$, $Y_\beta^\uparrow \cap S$ is finite. Let S_α be any infinite subset of S so that $\Pi(S_\alpha)$ has the properties that $\Pi(S_\alpha) \in \mathfrak{T}$ and does not contain any member of $\{\Pi(Y_\beta) : \beta \leq \alpha\} \cup \{\Pi(S_\beta) : \beta < \alpha\}$. Clearly conditions 2.1 5,7 are satisfied. On the other hand, if we cannot pick such an S, then we let $S_\alpha = \emptyset$ and condition 2.1 6 is satisfied. In either case conditions 5-7 of 2.1 are satisfied.

This completes the induction; (but let us remark that the hard part is in the next section). It is useful to make the following observation.

Fact 2.2 $S_\alpha \cap (Y_\beta)^\uparrow$ *is finite, for all* α, $\beta \in \mathfrak{c}$.

3 The Clopen Base

Since we are constructing a Fréchet topology on ${}^{<\omega}\omega$, we intend that for any set $A \subset {}^{<\omega}\omega$, $\emptyset \in \overline{A}$ if and only if there is some S_α which intersects A on an infinite set. Therefore the topology is actually determined but we want to now show that it has the property we mentioned at the end of section 1.

Definition 3.1 *Define* $\mathfrak{U}$ *by* $U \in \mathfrak{U}$ *if and only if*

1. $\emptyset \in U$,

2. *for all* $n \in \omega$, $U \cap \{(0,n)\}^\uparrow$ *is a rational clopen set, and*

3. *for all* $\alpha < \mathfrak{c}$, $S_\alpha - U$ *is finite.*

In this section when we discuss a topology on ${}^{<\omega}\omega$ it will either be the usual rational topology or the topology we get by letting the neighbourhood base at $\emptyset$ be given by $\mathfrak{U}$ and taking the usual neighbourhoods for all other points. The first condition on $\mathfrak{U}$ guarantees that this is a topology.

Claim 1. The new topology is finer than the usual topology.
Indeed, for each $n \in \omega$, $U = {}^{<\omega}\omega - \{(0,n)\}^\uparrow \in \mathfrak{U}$.

Claim 2. There are no copies of $\omega^2 + 1$ which converge to $\emptyset$.
Indeed, let $X = \{x_\xi : \xi < \omega^2\}$ be a topological copy of ω^2 in ${}^{<\omega}\omega$ with respect to the new topology. Since the new topology is finer than the rational topology, we may assume that we have taken a subset of X and have arranged that, for each $\xi < \omega^2$, $\Pi(\{x_\beta : \beta < \xi\})$ is finite. Therefore, if $X \cup \{\emptyset\}$ is to be a copy of $\omega^2 + 1$ every transversal from X must converge to $\emptyset$. Furthermore it follows that $\Pi^+(X)$ is infinite. Now choose $\alpha < \mathfrak{c}$ such that $X_\alpha = X$. We claim that $U = {}^{<\omega}\omega - Y_\alpha^\uparrow$ is in $\mathfrak{U}$. If it is, then we are done since clearly Y_α is a transversal contained in X which does not converge to $\emptyset$.
Now to check that U is in $\mathfrak{U}$. Clearly, since Y_α is a transversal, $Y_\alpha^\uparrow$ satisfies condition 3.1.2 and so does its complement. Induction hypotheses 2.1 4,6,7 guarantee that U satisfies condition 3.1.1 .

Claim 3. The new topology is Fréchet (at $\emptyset$). In fact we show that if $A \subset {}^{<\omega}\omega$ and $\emptyset \in \overline{A}$ then there is an $\alpha < \mathfrak{c}$, such that $S_\alpha \subset A$ is infinite and converges to $\emptyset$.
Let $A \subset {}^{<\omega}\omega - \{\emptyset\}$ and assume that $A \cap S_\alpha =^* \emptyset$ for each $\alpha < \mathfrak{c}$. We wish to show that there is a $U \in \mathfrak{U}$ such that $U \cap A = \emptyset$. Fix $\alpha < \mathfrak{c}$ such that $A_\alpha = A$. Of course, we have that $S_\alpha = \emptyset$ by 7 of 2.1. Therefore, by 6 of 2.1, for every transversal $S \subset A_\alpha$ there is a $\beta < \alpha$ so that $|Y_\beta^\uparrow \cap S|$ is infinite. Let B denote

the set $\overline{A} - \{\emptyset\}$ where the closure is taken with respect to the usual rational topology.

Claim 4. If there is a transversal $S \subset B^\dagger$ such that $S \cap Y_\beta^\dagger =^* \emptyset$ for each $\beta \leq \alpha$, then there is such a transversal $S' \subset B$.

Proof. Suppose that S is such a transversal as in the Claim. Let $J = \Pi(S)$, and for each $n \in J$, choose $t_n \in B$ such that $t_n^\dagger \cap S \neq \emptyset$. We claim that $S' = \{t_n : n \in J\}$ is as required. Indeed, suppose that $\beta \leq \alpha$ and $Y_\beta^\dagger \cap S'$ is infinite. But now, $t_n \in Y_\beta^\dagger$ implies that $t_n^\dagger \subset Y_\beta^\dagger$, hence $Y_\beta^\dagger \cap S$ is infinite.

Claim 5. If there is a transversal $S \subset B$ such that $Y_\beta^\dagger \cap S$ is finite for each $\beta \leq \alpha$, then there is such a transversal $S' \subset A_\alpha$

Proof. Fix S as above and pick, by Proposition 0.2, $J \subset \Pi(S)$ such that $J \in \mathfrak{I}$ and J does not contain $\Pi(Y_\beta)$ for any $\beta \leq \alpha$. For each $n \in J$, let $s_n \in \{(0,n)\}^\dagger \cap S$. Since s_n is a limit point of $A_\alpha \cap \{(0,n)\}^\dagger$, it follows that for infinitely many $k \in \omega$, $(s_n{}^\frown k)^\dagger \cap A_\alpha$ is not empty. For each such k we may pick one element of A_α and let $\{x(n,m) : m \in \omega\} \subset A_\alpha$ enumerate this set so that each is listed only once. For each $\beta \leq \alpha$ and each $n \in J$, there is at most one m such that $x(n,m) \in Y_\beta^\dagger$ since, by condition 1 of 2.1, there is at most one element of Y_β in $\{(0,n)\}^\dagger$ and this element is not below s_n. Therefore there is a function $f_\beta \in {}^J\omega$ such that, for each $n \in J$ and each $m > f_\beta(n)$, $x(n,m) \notin Y_\beta^\dagger$. Again, let $\Gamma = \{\beta \leq \alpha : \mid \Pi(Y_\beta) \cap J \mid = \omega\}$ and, again, we have that $|\Gamma| < \mathfrak{b}$. Choose $f \in {}^J\omega$ so that $f_\beta <^* f$ for all $\beta \in \Gamma$. The set we seek is $S' = \{x(n, f(n)) : n \in J\}$. Indeed, if $\beta \in \Gamma$, then $x(n, f(n)) \notin Y_\beta^\dagger$ for all $n \in J$ such that $f(n) > f_\beta(n)$. While if $\beta \in (\alpha + 1) - J$, then $\Pi(Y_\beta) \cap J$ is finite. This completes the proof of the claim.

Now we complete the proof of Claim 3. Since it was not possible to find an S' as in Claim 5 we deduce that it was not possible to find an S as in Claim 4. That is, we conclude that every transversal, S, which is contained in $B^\dagger$, meets some $(Y_\beta)^\dagger$. Therefore, by 2.2, $B^\dagger$ is almost disjoint from S_β for all $\beta < \mathfrak{c}$. Also $B \cap \{(0,n)\}^\dagger$ is the closure (in the rational topology) of $A_\alpha \cap \{(0,n)\}^\dagger$ for each $n \in \omega$, hence $(B \cap \{(0,n)\}^\dagger)^\dagger = B^\dagger \cap \{(0,n)\}^\dagger$ is clopen by Proposition 0.3. Therefore $U = {}^{<\omega}\omega - B^\dagger$ is a member of $\mathfrak{U}$. It remains only to note that $U \cap A_\alpha$ is empty.

References

[BPS80] B. Balcar, J. Pelant, and P. Simon. The space of ultrafilters on N covered by nowhere dense sets. *Fundamenta Mathematica*, 11–24, 1980.

[Bs80] B. Balcar and P. Vojtáš. Almost disjoint refinements of families of subsets of N. *Proceedings AMS*, 465–470, 1980.

[Nog] T. Nogura. Embedded copies of the rationals. 23rd Annual Spring Topology Conference (Knoxville).

Orthogonal Bases in Topological Algebras: Highlights of Recent Results

Sherif El-Helaly The Catholic University of America, Washington, D.C.
Taqdir Husain McMaster University, Hamilton, Ontario, Canada

0. Preliminaries. Let E be a Hausdorff topological vector space over the field **C** of complex numbers. It is known that the topology of E is generated by a 0-neighborhood base κ consisting of closed circled absorbing sets. If there is a 0-neighborhood base with the additional property that each of its members is convex, E is said to be locally convex. Every convex circled absorbing set V gives rise to a seminorm p_V on E (called the gauge of V) given by $p_V(x)=\inf\{a>0: x \text{ in } aV\}$ and conversely, every seminorm p on E gives rise to a convex circled absorbing set V_p given by $V_p=\{x \text{ in } E: p(x)\leq 1\}$. It follows from this that the topology of a locally convex space can be generated by a family of seminorms.

E is metrizable if and only if a countable 0-neighborhood base exists. A complete metrizable topological vector space is called an F-space and a locally convex F-space is called a Fréchet space.

A topological algebra is a topological vector space with a jointly continuous multiplication. The joint continuity of multiplication in a topological algebra A is equivalent to the following condition: If κ is a 0-neighborhood base in A then for each U in κ there exists V in κ such that VV is contained in U. In a locally convex algebra A with a generating family P of seminorms, the joint continuity of multiplication is also equivalent to the condition that for every p in P there exists q in P such that $p(xy) \leq q(x)q(y)$ for all x,y in A. A locally convex algebra A is said to be locally m-convex if it has a generating family P of seminorms such that $p(xy) \leq p(x)p(y)$ for all p in P and x,y in A (such seminorms are said to be submultiplicative). Equivalently, a locally m-convex algebra is a locally convex algebra that possesses a 0-neighborhood base κ such that each U in κ is idempotent (i.e., UU is contained in U), in addition to being closed, circled, absorbing and convex. A complete metrizable topological algebra is called an F-algebra, a locally convex F-algebra is called a B_0-algebra, and a locally m-convex F-algebra is called a Fréchet algebra. Clearly, Banach spaces and Banach algebras are special cases of Fréchet spaces and Fréchet algebras, respectively.

A topological basis (or simply, a basis) in a topological vector space E is a sequence $(e_n)_n$ in E such that for each x in E there is a unique sequence $(a_n(x))_n$ of scalars such that

$$x = \lim_N \Sigma_{1 \leq n \leq N} a_n(x)e_n \equiv \Sigma_n a_n(x)e_n.$$

Clearly, the a_n's are linear functionals (called the coordinate linear functionals). If each a_n is continuous, the basis $(e_n)_n$ is said to be a Schauder basis. If for each x in E, the convergence of the above series does not depend on the order of the terms, the basis $(e_n)_n$ is said to be unconditional.

In the 1970's, T. Husain and his coworkers started a study of bases in topological algebras. They introduced the following definition to describe the algebraic relations between the basis elements in a topological algebra A: A basis $(e_n)_n$ in A is said to be orthogonal if $e_n e_m = \delta_{mn} e_n$, where δ_{mn} is the Kronecker delta. The following are some of the immediate consequences of this definition.

(1) $xy = \Sigma_n a_n(x)a_n(y)e_n$.

(2) A is commutative.

(3) $a_n(xy) = a_n(x)a_n(y)$, i.e., the the linear functionals a_n are also multiplicative

(4) $a_n(x)e_n = xe_n$.

For a detailed study of orthogonal bases in topological algebras we refer the reader to [2],[3] and [4]. In the present paper, we give a summary of the results (in [2] and [3]) pertinent to the Schauderness and unconditionality of orthogonal bases in topological algebras. The study of unconditionality of such bases leads to the concept of "s-algebras" (in Section 2). Proofs are in general omitted , since they are available in the references.

1. Schauderness of bases. Let E be a locally convex space with a generating family P of seminorms and a basis $(e_n)_n$. For each x in E, p in P and n in the set $\mathbf{N}$ of positive integers we set

$$p'(x) = \sup_n p(S_n(x))$$

where $S_n(x)=\Sigma_{1\leq k\leq n}a_k(x)e_k$, the nth partial sum of the expansion of x in terms of the basis. It follows from the convergence of $S_n(x)$ that $p'(x)<\infty$. It is easy to verify that each p' is a seminorm on E and the family P' of all such seminorms generates a topology on E finer than the topology generated by P. If E is complete and metrizable for P it will also be complete and metrizable for P' and an application of the open mapping theorem shows that P and P' generate the same topology on E. Now we have

$$|a_n(x)|p(e_n) = p(a_n(x)e_n) = p(S_n(x)-S_{n-1}(x)) \leq 2p'(x)$$

which establishes the continuity of the linear functional a_n (take p in P with a nonzero $p(e_n)$). We have just outlined the proof of the following theorem proven by W.F. Newns in 1953 (a detailed proof can be found in Köthe [5], p.249): A basis in a Fréchet space is a Schauder basis. This theorem was extended to F-spaces by V.N. Nikolskii in 1954; and in 1961 Mityagin generalized it further by observing that a basis in a topological vector space for which the open mapping theorem or the closed graph theorem holds is a Schauder basis. For references see Singer [7], p.202.

Without the help of the open mapping (or the closed graph) theorem, T. Husain and S. Watson observed in 1980 that an orthogonal basis in a locally m-convex algebra is a Schauder basis [4]. In [3], the following theorem (which generalizes this result to all topological algebras) was proved. The continuity of the coordinate linear functionals a_n follows from their involvement in the relation $a_n(x)e_n=xe_n$, which is an immediate consequence of the continuity of multiplication.

1.1 Theorem. An orthogonal basis in a topological algebra is a Schauder basis.

Next, consider the algebra $\mathbf{C}^{\mathbf{N}}$ of all $t:\mathbf{N}\to\mathbf{C}$ with the pointwise operations and the product topology. The topology of $\mathbf{C}^{\mathbf{N}}$ is generated by the seminorms

$$p_K(t) = \sup_{k\in K} |t(k)|$$

where K runs through the family of all finite subsets of $\mathbf{N}$. Clearly, such seminorms are submultiplicative, which shows that $\mathbf{C}^{\mathbf{N}}$ is locally m-convex (indeed, it is a Fréchet algebra). $\mathbf{C}^{\mathbf{N}}$ has the canonical orthogonal basis $(h_n)_n$ where $h_n(k)=0$ for $k\neq n$ and $h_n(n)=1$. Now let A be a topological algebra with an orthogonal basis $(e_n)_n$. It follows from the uniqueness of the representation of each x in A in terms of the basis and the continuity of the multiplicative linear functionals a_n that $x\to x^{\#}$ (where $x^{\#}(n)=a_n(x)$) maps A in a continuous one to one fashion onto a dense topological subalgebra $A^{\#}$ of $\mathbf{C}^{\mathbf{N}}$. This shows that the product topology is the coarsest topology on a topological algebra with an orthogonal basis. The following theorem (given in [3]) shows that this coarsest topology is attained when A is locally m-convex and has an identity.

1.2 Theorem. Let A be a locally m-convex algebra with an orthogonal basis $(e_n)_n$ and a generating family P of submultiplicative seminorms. Then each of the following statements implies the succeeding one. Moreover, if A is complete then all the statements are equivalent.

(i) A is topologically isomorphic with $\mathbf{C}^{\mathbf{N}}$.

(ii) A has an identity.

(iii) For each p in P, $p(e_n)$ is nonzero for at most finitely many e_n's.

(iv) A is topologically isomorphic with a dense subalgebra of $\mathbf{C}^{\mathbf{N}}$.

Proof. (iii) follows from (ii) as follows: If A has an identity e, then

$e=\Sigma_n e_n$. The convergence of this series implies that for each p in P, $p(e_n)<\frac{1}{2}$ for all sufficiently large n. Thus for all sufficiently large n and any positive integer k we have

$$p(e_n) = p(e_n^k) \leq (p(e_n))^k < (\tfrac{1}{2})^k.$$

Since the positive integer k is arbitrary, we have $p(e_n)=0$ for all sufficiently large n. All other implications are clear.

2. Unconditionality of bases. Theorem 1.1 above shows that the orthogonality of a basis in any topological algebra is so strong as to force the Schauderness of the basis, In contrast, orthogonality of a basis is not strong enough to force the unconditionality of the basis, even in the very special case of Banach algebras (see Examples 3.2 below). In [2], the concepts of squarely submultiplicative seminorms and s-algebras are introduced and it is shown that among all B_0-algebras, those which are also s-algebras are the only ones in which an orthogonal basis (if it exists) is unconditional. We highlight these results in the present section.

First, we have the following two propositions on the unconditionality of a basis in a Fréchet space. Proofs in the special case of a Banach space are available in [7] (Theorem 16.1, page 461), and their generalizations to the Fréchet space case are given in [2] (Proposition 1.1).

2.1 Proposition. Let E be a Fréchet space whose topology is generated by a family P of seminorms, and let $(e_n)_n$ be a basis in E. The following statements are equivalent:

(i) The basis $(e_n)_n$ is unconditional.

(ii) For every x in E and every finite subset J of **N** set $S_J(x)=\Sigma_{j\in J}\, a_j(x)e_j$. Then the net $(S_J(x))_J$ (where J runs through the set of all finite subsets of **N** ordered by inclusion) converges to x.

(iii) For every f in E' (the topological dual of E) and x in E, $\Sigma_n|a_n(x)f(e_n)|$ converges.

(iv) Let $T_p=\{n \text{ in } \mathbf{N}: p(e_n)\neq 0\}$ for each p in P. Then for every x in E and every sequence $(b_n)_n$ of scalars such that each of the sets $\{b_n: n \text{ in } T_p\}$ is bounded, the series $\Sigma_n b_n a_n(x)e_n$ converges.

(v) The series in (iv) converges for each x in E and any bounded sequence $(b_n)_n$ of scalars.

(vi) The series in (iv) converges for each x in E and any sequence $(b_n)_n$ with $b_n=\pm 1$.

(vii) The series in (iv) converges for each x in E and each sequence $(b_n)_n$ of 0's and 1's. In other words, every subseries of $\Sigma_n a_n(x)e_n$ is convergent.

2.2 Proposition. Let E be a Fréchet space whose topology is generated by a family P of seminorms and let $(e_n)_n$ be an unconditional basis in E. For each p in P set

$$H(p) = \{f \text{ in } E' : \sup_{p(x)\leq 1} |f(x)| \leq 1\}$$

and for each x in E set

$$p^*(x) = \sup_{f\in H(p)} \Sigma_n |f(e_n)a_n(x)|,$$

then each p* is a seminorm on E and the family P* of all such seminorms generates the same topology on E as the original topology generated by P.

2.3 Definitions. Let A be an algebra (no topology is assumed).

(i) A subset S of A is said to be squarely idempotent if xy is in S whenever x^2 and y^2 are in S.

(ii) A seminorm p on A is said to be squarely submultiplicative if $(p(xy))^2 \leq p(x^2)p(y^2)$ for all x and y in A. (Note the resemblance to the Cauchy-Schwarz inequality. Such resemblance will be more obvious in the course of the proof of Theorem 2.6, and in some of the examples in section 3 below).

The relation between squarely idempotent sets and squarely submultiplicative seminorms is similar to the relation between idempotent sets and submultiplicative seminorms. Indeed we have [2]

2.4 Proposition. Let S be a circled, convex and absorbing subset of an algebra A and let p be the gauge of S. Then S is squarely idempotent if and only if p is squarely submultiplicative.

Next we have

2.5 Definition. A topological algebra is said to be an s-algebra if it has a 0-neighborhood base whose members are squarely idempotent, in addition to being closed, circled and absorbing.

It follows that the topology of a locally convex s-algebra is generated by a family of squarely submultiplicative seminorms. As will be shown by examples, the two classes of locally m-convex algebras and locally convex s-algebras overlap, but neither contains the other.

2.6 Theorem. Let A be a B_0-algebra and let $(e_n)_n$ be an orthogonal basis in A. Then $(e_n)_n$ is unconditional if and only if A is an s-algebra.

Proof. First we prove the "if" part. Assume that A is an s-algebra with P as a generating family of squarely submultiplicative seminorms. For every finite set J of positive integers and every sequence $t=(t_n)_n$ of scalars we set

$$S(t,J) = \Sigma_{n\in J}\, t_n e_n.$$

It follows from the orthogonality of $(e_n)_n$ that for any subset I of J,

$$S(t,I) = S(t^{\frac{1}{2}},I)\, S(t^{\frac{1}{2}},J),$$

where $t^{\frac{1}{2}}$ denotes a sequence $(s_n)_n$ of scalars with $s_n{}^2=t_n$ for all n. Since P is a family of squarely submultiplicative seminorms we then have

$$[p(S(t,I))]^2 \leq p(S(t,I))\, p(S(t,J))$$

and consequently

$$p(S(t,I)) \leq p(S(t,J))$$

for each p in P. It follows from the last inequality that for each x in A and each sequence $(b_n)_n$ of 0's and 1's, the finite partial sums of $\Sigma_n b_n a_n(x)e_n$

form a Cauchy sequence in A, which must converge as A is complete. The unconditionality of $(e_n)_n$ now follows from Proposition 2.1.

Now we prove the "only if" part. Assume that $(e_n)_n$ is unconditional, then the family P* of seminorms defined in Proposition 2.2 generate the topology of A. We show that each p* is squarely submultiplicative. Each fixed p in P and f in H(p) (as defined in Proposition 2.2) define a measure μ_f on the power set of **N** by

$$\mu_f(T) = \Sigma_{n\in T}|f(e_n)|.$$

It follows from the equality $a_n(x)a_n(y) = a_n(xy)$ and the Cauchy-Schwarz inequality for the measure μ_f that for x and y in A,

$$(\Sigma_n|a_n(xy)f(e_n)|)^2 \leq \{\Sigma_n|(a_n(x))^2f(e_n)|\} \{\Sigma_n|(a_n(y))^2f(e_n)|\}$$

$$= \{\Sigma_n|(a_n(x^2)f(e_n)|\} \{\Sigma_n|(a_n(y^2)f(e_n)|\}.$$

Taking the supremum of each of the above quantities over f in H(p) we get

$$(p^*(xy))^2 \leq p^*(x^2)\ p^*(y^2)$$

thus proving that each p* is squarely submultiplicative.

Note: From the equivalence of statements (i) and (iii) in Proposition 2.1 we see that the measure defined above is finite if A has an identity, for if e is an identity in A then $a_n(e)=1$ for all n and we may then put x=e in statement (iii). On the other hand, if A does not have an identity, the measure may be finite or infinite as in the examples of c_0 and ℓ_1 (Examples 3.1(1) below) respectively. This is easy to see by observing that $c_0' \simeq \ell_1$ and $\ell_1' \simeq \ell_\infty$.

The concepts of squarely submultiplicative seminorms and s-algebras were introduced by the present authors [2] in the course of their study of orthogonal bases in topological algebras. Bases, however, are not involved in the definitions of these concepts. For a short study of locally convex s-algebras (no bases involved), we refer the reader to section 3 of [2].

3. Examples. In this section we have various examples of topological algebras with orthogonal bases. We give special attention to the classification of these algebras into locally m-convex algebras and s-algebras.

3.1 Algebras which are both locally m-convex and s-algebras.

(1) For $1\leq p<\infty$, the Banach space ℓ_p of all $x:\mathbf{N}\to\mathbf{C}$ with $\Sigma_k|x(k)|^p<\infty$ with the usual ℓ_p-norm has the unconditional basis $(e_n)_n$, where $e_n(k)=\delta_{nk}$, the Kronecker delta. This basis is known as the canonical basis. Under the pointwise multiplication, ℓ_p is a Banach algebra in which the canonical basis is orthogonal. By Theorem 2.6 above, This is an s-algebra, In fact, the "square submultiplicativity" of the ℓ_p-norms is merely the statement of the usual Cauchy-Schwarz inequality. Another example of a Banach algebra which is also an s-algebra is c_0, the algebra of all $x:\mathbf{N}\to\mathbf{C}$ with $\lim_n x(n)=0$, endowed with the pointwise operations and the sup norm. The canonical basis described above is an unconditional orthogonal basis in c_0.

(2) The algebra $\mathbf{C}^{\mathbf{N}}$ is a locally m-convex algebra. Each member of the family of seminorms of $\mathbf{C}^{\mathbf{N}}$ given in Section 1 is squarely submultiplicative, in addition to being submultiplicative. Thus $\mathbf{C}^{\mathbf{N}}$ is also an s-algebra.

3.2 Locally m-convex algebras which are not s-algebras.

(1) The algebra w_0 (also denoted by bv_0) consists of all $x:\mathbf{N}\to\mathbf{C}$ such that $\lim_n x(n)=0$ and $\Sigma_n|x(n)-x(n+1)|<\infty$. The algebra structure of A is given by the pointwise operations and the topology is given by the norm

$$p(x) = \sup_n|x(n)| + \Sigma_n|x(n)-x(n+1)|.$$

With this structure, w_0 is a Banach algebra in which the canonical basis $(e_n)_n$ (as defined above in the case of ℓ_p) is an orthogonal basis. It is easy to see that $x(n)=1/n$ defines an element x in w_0, whose representation in terms of the basis $(e_n)_n$ is $x=\Sigma_n(1/n)e_n$. However, the series $\Sigma b_n(1/n)e_n$ (where $b_n=0$ if n is even and $b_n=1$ if n is odd) does not converge in w_0. By Proposition 2.1, the orthogonal basis $(e_n)_n$ is not unconditional, and consequently by Theorem

2.6, w_0 is not an s-algebra.

(2) For $1\leq p<\infty$, the convolution algebra $L_p(\mathbf{T})$ over the torus group $\mathbf{T}$ (with the convolution as multiplication, and the L_p-norm) is a Banach algebra. For $1<p<\infty$ we have

$$x = \lim_N \Sigma_{-N\leq n\leq N} \hat{x}(n)e^{in(.)}$$

in the L_p-norm, where $\hat{x}:\mathbf{Z}\to\mathbf{C}$ is the Fourier transform of x in $L_p(\mathbf{T})$. If we rearrange the set $\mathbf{Z}$ of all integers in the order 0,1,-1,2,-2,..., the above series will become one-sided and will still converge to x, thus showing that the characters $e^{in(.)}$ of the torus group $\mathbf{T}$ form a basis in $L_p(\mathbf{T})$. This basis is orthogonal in $L_p(\mathbf{T})$ since the convolution of two such characters $e_m=e^{im(.)}$ and $e_n=e^{in(.)}$ is given by

$$e_m * e_n(t) = (1/2\pi) \int_0^{2\pi} e^{im(t-u)} e^{inu} du = \delta_{mn} e_n(t).$$

For $p\neq2$, it is possible [1] to find an x in $L_p(\mathbf{T})$ and some $b:\mathbf{Z}\to\{-1,1\}$ such that $b\hat{x}:\mathbf{Z}\to\mathbf{C}$ is not the Fourier transform of any element in $L_p(\mathbf{T})$, which by Proposition 2.1, shows that the orthogonal basis under consideration is not unconditional. From Theorem 2.6 we now see that for $1<p<\infty$, $p\neq2$, the convolution algebra $L_p(\mathbf{T})$ is not an s-algebra. Note that for p=2 the above mentioned orthogonal basis is unconditional and consequently $L_2(\mathbf{T})$ is an s-algebra.

3.3 Locally convex s-algebras which are not locally m-convex.

(1) Let B be the set of all bounded $x:\mathbf{N}\to\mathbf{C}$ and let H be the subset of B consisting of all h in B with $\lim_n h(n)=0$. B is a locally convex algebra under the pointwise operations and the topology of the seminorms

$$p_h(x) = \sup_n |h(n)x(n)|$$

where h runs through H. One can easily see that the seminorms p_h are squarely submultiplicative and so B is an s-algebra. However, B is not locally m-convex. Indeed, if B were locally m-convex then due to the existence of an identity in B it would follow from Theorem 1.2 above that B has the topology of $\mathbf{C}^{\mathbf{N}}$ (which is the product topology). This is obviously not the case.

(2) Let H(**D**) be the algebra of all analytic functions on the open unit disc **D** with the pointwise addition and scalar multiplication. The multiplication of elements of H(**D**) is the Hadamard multiplication given by

$$(fg)(z) = (2\pi i)^{-1} \int_{|u|=r} f(u)g(zu^{-1})u^{-1}\, du,$$

$|z|<r<1$. The topology of H(**D**) is the compact-open topology. With this structure H(**D**) is a B_0-algebra. The convergence of the Taylor series

$$f(z) = \Sigma\, [f^{(n)}(0)/n!]z^n$$

of f in the compact-open topology shows that the elements e_n given by $e_n(z)=z^n$, n=0,1,2,... form an unconditional basis in H(**D**). It is not difficult to see that this basis is orthogonal with respect to the multiplication of H(**D**) defined above. It is now possible to use the alternative expression

$$fg = \Sigma\, [f^{(n)}(0)/n!][g^{(n)}(0)/n!]e_n$$

for multiplication, which shows that H(**D**) has an identity e given by $e(z)=(1-z)^{-1}=\Sigma z^n=\Sigma e_n$. The existence of an identity can be used, as in the previous example, to establish the fact that H(**D**) is not locally m-convex.

H(**D**) is an s-algebra by Theorem 2.6, since it is a B_0-algebra with an unconditional orthogonal basis. An alternative way of showing that H(**D**) is an s-algebra is by simply considering the equivalent family of seminorms

$$p_r(f) = \sup_n |r^n f^{(n)}(0)/n!|,$$

$0<r<1$. The seminorms p_r are squarely submultiplicative, as it is easy to verify, using the alternative expression for multiplication given above.
Note: A type of locally convex algebras that generalizes locally m-convex algebras is the "A-convex" algebras. An A-convex algebra A is topologized

by a family of "absorbing seminorms". A seminorm p on A is called absorbing if for every a in A there exists positive constants K(p,a) and L(p,a) such that $p(ax) \leq K(p,a)p(x)$ and $p(xa) \leq L(p,a)p(x)$ for all x in A. It is known [6] that a barrelled (in particular, B_0) A-convex algebra is locally m-convex. This shows that H(**D**), which has been shown to be an s-algebra) is not even A-convex.

REFERENCES

1. R.E. Edwards, Fourier series; 2nd edition, Vol. 2, Springer-Verlag, New York, 1981.

2. S. El-Helaly and T. Husain, Unconditionality of orthogonal bases in B_0-algebras, Commentationes Mathematicae, XXVII, 237-251 (1988).

3. S. El-Helaly and T. Husain, Orthogonal bases are Schauder bases and a characterization of Φ-algebras, Pacific J. Math., 132 No. 2, 265-275 (1988).

4. T. Husain and S. Watson, Topological algebras with orthogonal Schauder bases, Pacific J. Math. 91 No. 2, 339-347 (1980).

5. G. Köthe, Topological vector spaces II, Springer-Verlag, New York, 1979.

6. E. Michael, Locally multiplicatively-convex topological algebras, AMS Memoirs #11, 1952.

7. I. Singer, Bases in Banach spaces I, Springer-Verlag, New York, 1970.

On a Decomposition of Continuity

M. Ganster and F. Gressl Graz University of Technology, Graz, Austria

Ivan L. Reilly University of Auckland, Auckland, New Zealand

1 INTRODUCTION.

An early generalization of continuity was given by Blumberg (1922) when he introduced the concept of a real valued function on euclidean space being densely approached at a point of its domain. This notion was generalized to topological spaces by Ptak (1958) who used the term "nearly continuous", and by Husain (1966) under the name of "almost continuity". The concepts of nearly continuous functions and nearly open functions are important in functional analysis, especially in the context of open mapping theorems and closed graph theorems. The interested reader should consult the work of Ptak (1958), Pettis (1974), Wilhelm (1979, 1986) and Noll (1986), for example. It should be pointed out that Mashhour, Abd El-Monsef and El-Deeb (1982) have used the term "precontinuity" for nearly continuity.

An obvious problem is to search for a continuity dual of near continuity, that is to provide a property $\mathcal{P}$ of functions between topological spaces which is weaker than continuity and such that a function is continuous if and only if it is nearly continuous and has property $\mathcal{P}$. Such a property $\mathcal{P}$ has been obtained by Tong (1989) with his definition of $\mathcal{B}$-continuity, and by Ganster and Reilly (1990) who introduced the concept of LC-continuity.

In this paper, we consider the notion of weak $\mathcal{B}$-continuity, and show that it is a continuity dual of near continuity, weaker than either of the two afore-mentioned properties. We are able to use this property to obtain interesting variations of results from functional analysis. Let us quote two.

(i) If G is a Baire topological group and H is a separable (or Hausdorff and Lindelöf) topological group, then a homomorphism $f : G \to H$ is continuous if and only if it is weakly $\mathcal{B}$-continuous, Husain (1977), page 222.

(ii) If X is a Baire topological vector space, Y is a topological vector space and $f : X \to Y$ is linear, then f is continuous if and only if it is weakly $\mathcal{B}$-continuous. Husain (1977), page 224.

A more general question has been addressed by Wilhelm (1981), namely: When is a nearly open relation open? Using the notion of weakly $\mathcal{B}$-open relations, we are able to give a solution to this question for general topological spaces, in our Proposition 7.

2 GENERALIZED OPEN SETS.

Throughout this paper we consider general topological spaces, and no (separation) properties are assumed unless explicitly stated. Let S be a subset of a topological space (X, τ). We denote the interior of S by $intS$ and the closure of S by clS.

DEFINITION 1: A subset S of (X, τ) is called

(i) semi-open if $S \subset cl(intS)$,
(ii) preopen if $S \subset int(clS)$,
(iii) locally closed if $S = U \cap F$ where U is open and F is closed.

Notice that preopen sets are called nearly open sets by several authors. Clearly every open set is preopen and semi-open and locally closed. Closed sets are locally closed. A set whose complement is semi-open is called semi-closed. Thus S is semi-closed if and only if $int(clS) \subset S$. In particular, nowhere dense sets are semi-closed and regular open sets are semi-closed. Recall that S is nowhere dense if $int(clS) = \phi$, and S is regular open if $int(clS) = S$.

DEFINITION 2: If S is a subset of (X, τ) the pre-interior of S, denoted by $pintS$, is the largest preopen set contained in S. The semi-closure of S, denoted by $sclS$, is the smallest semi-closed set containing S.

LEMMA 1: [Andrijević (1987), Theorem 1.6]. Let A be a subset of (X, τ). Then

(i) $pintA = A \cap int(clA)$,
(ii) $sclA = A \cup int(clA)$.

Recently, Tong (1989) has defined two classes of distinguished subsets in a topological space.

DEFINITION 3: A subset S in (X, τ) is said to be a t-set if $int(clS) = intS$. If $S = U \cap A$ where U is open and A is a t-set in (X, τ), then S is said to be a $\mathcal{B}$-set.

Tong (1989) distinguished between t-sets and closed sets, open sets and regular open sets. However, the following result shows that the class of t-sets coincides with an existing class of subsets. The proof is straightforward.

PROPOSITION 1: Let S be a subset of (X, τ). Then S is a t-set if and only if S is semi-closed.

Proposition 1 indicates that a $\mathcal{B}$-set is an intersection of an open set and a semi-closed set.

PROPOSITION 2: Let S be a subset of (X,τ). Then S is a $\mathcal{B}$-set if and only if $S = U \cap sclS$ for some open set U.

PROOF: Sufficiency is clear, since $sclS$ is semi-closed.
For the necessity, let $S = U \cap A$ where U is open and A is semi-closed in (X,τ). Then $S \subset A$ implies $sclS \subset A$, and so $S = S \cap sclS = U \cap A \cap sclS = U \cap sclS$.

In the context of generalized open sets, the question under consideration becomes a search for a property $\mathcal{Q}$ of subsets of topological spaces weaker than openness and such that a subset is open if and only if it is preopen (i.e. nearly open) and has property $\mathcal{Q}$. Two solutions to this question have been obtained independently by Tong (1989), Proposition 9, where $\mathcal{Q}$ is the property "$\mathcal{B}$-set", and by Ganster and Reilly (1990), Theorem 2, where $\mathcal{Q}$ is the property "locally closed". We are able to improve these results by introducing the following notion.

DEFINITION 4: A subset S of (X,τ) is called a weak $\mathcal{B}$-set if $pintS = intS$.

PROPOSITION 3: Let S be a subset of (X,τ).
(i) If S is locally closed, then S is a $\mathcal{B}$-set.
(ii) If S is a $\mathcal{B}$-set, then S is a weak $\mathcal{B}$-set.

PROOF:
(i) Let $S = U \cap F$ where U is open and F is closed. Since F is semi-closed, S is a $\mathcal{B}$-set.
(ii) Let $S = U \cap A$ where U is open and A is semi-closed. Since $S \subset A$ we have $int(clS) \subset int(clA) = intA$, because A is a t-set. By Lemma 1, $pintS = S \cap int(clS)$, so that $pintS \subset U \cap A \cap intA = U \cap intA = int(U \cap A) = intS$. Since $intS \subset pintS$ for every subset S of (X,τ), we have $pintS = intS$, so that S is a weak $\mathcal{B}$-set.

We now provide examples to show that the implications of Proposition 3 cannot be reversed.

EXAMPLE 1: Consider the set $\mathbb{R}$ of real numbers with the usual topology, and let C be the Cantor set. Then C is closed and nowhere dense in $\mathbb{R}$, see Steen and Seebach (1970), Example 29. Let (X,τ) be C with the relative topology. C is resolvable, so there are disjoint dense subsets C_1 and C_2 of C with $C = C_1 \cup C_2$. Now C_1 and C_2 are nowhere dense subsets of (X,τ), and thus semi-closed and hence $\mathcal{B}$-sets.

Now suppose that C_1 is locally closed. Then $C_1 = U \cap clC_1 = U \cap C$ for some open set U. But $clC_2 = C$, so that $U \cap C_2 \neq \phi$. This contradicts the disjointness of C_1 and C_2, so that C_1 is not locally closed.

EXAMPLE 2: Let (X,τ) be $\mathbb{R}$ with the usual topology, let $[0,1]$ denote the closed unit interval, and define $S = [0,1] - \{\frac{1}{n} : n\ is\ a\ positive\ integer\}$. Then $clS = [0,1]$,

$intS = (0,1) - \{\frac{1}{n} : n \text{ is a positive integer}\}$, $int(clS) = (0,1)$, $cl(intS) = [0,1]$, and $sclS = [0,1)$. Hence S is semi-open. By Lemma 1, $pintS = S \cap int(clS)$, so that $pintS = intS$ and thus S is a weak $\mathcal{B}$-set.

Now suppose S is a $\mathcal{B}$-set, so that there is, by Proposition 2, an open set V in $\mathbb{R}$ such that $S = V \cap sclS$. Hence $S = V \cap [0,1)$. Now V is an open set containing the origin 0, so there is a positive integer m such that $\frac{1}{m} \in V \cap [0,1)$. But $\frac{1}{m} \notin S$, a contradiction. Thus S is not a $\mathcal{B}$-set.

Now we show that "weak $\mathcal{B}$-set" is an appropriate property $\mathcal{Q}$.

PROPOSITION 4: If S is a subset of (X,τ), then S is open if and only if S is preopen and a weak $\mathcal{B}$-set.

PROOF: Only the sufficiency requires proof. If S is a weak $\mathcal{B}$-set then $intS = pintS$. By Lemma 1, $pintS = S \cap int(clS) = S$ since S is preopen. Thus $intS = S$, and hence S is open.

3 GENERALIZED CONTINUITIES.

We now consider three variations of continuity.

DEFINITION 5: A function $f : X \to Y$ is called weakly $\mathcal{B}$-continuous (LC-continuous, $\mathcal{B}$-continuous respectively) if the inverse image under f of each open set in Y is a weak $\mathcal{B}$-set (locally closed, $\mathcal{B}$-set respectively) in X.

Proposition 3 provides an immediate proof of the following result.

PROPOSITION 5: Let $f : X \to Y$ be a function.

(i) If f is LC-continuous, then f is $\mathcal{B}$-continuous.
(ii) If f is $\mathcal{B}$-continuous, then f is weakly $\mathcal{B}$-continuous.

The following examples show that the converses of Proposition 5 are false.

EXAMPLE 3: Let (X,τ) be the set $\mathbb{R}$ of real numbers with the usual topology, and (Y,σ) be Sierpinski space. So $Y = \{0,1\}$ and $\sigma = \{\phi, \{1\}, Y\}$. Let S be a $\mathcal{B}$-set in (X,τ) which is not locally closed, see Example 1. Define $f : X \to Y$ by

$$f(x) = \begin{cases} 1 & \text{if } x \in S \\ 0 & \text{if } x \notin S. \end{cases}$$

Then f is $\mathcal{B}$-continuous but not LC-continuous.

EXAMPLE 4: Let (X,τ) be the set $\mathbb{R}$ of real numbers with the usual topology, and (Y,σ) be a two-point discrete space $\{0,1\}$. Let S be a semi-open weak $\mathcal{B}$-set in (X,τ)

which is not a $\mathcal{B}$-set, see Example 2. Then $X - S$ is semi-closed, and thus is a $\mathcal{B}$-set. Define $f : X \to Y$ by

$$f(x) = \begin{cases} 1 & \text{if} \quad x \in S \\ 0 & \text{if} \quad x \notin S. \end{cases}$$

Then f is weakly $\mathcal{B}$-continuous but not $\mathcal{B}$-continuous.

Our decomposition of continuity follows immediately from Proposition 4.

PROPOSITION 6: A function $f : X \to Y$ is continuous if and only if f is precontinuous and weakly $\mathcal{B}$-continuous.

From Proposition 6 we see that weak $\mathcal{B}$-continuity is a continuity dual of near continuity, which is weaker than either LC-continuity or $\mathcal{B}$-continuity.

4 RELATIONS.

Naimpally (1976) has raised, and Wilhelm (1981) has studied the more general problem: when is a nearly open relation open? A relation R between topological spaces (X, τ) and (Y, σ) is called nearly open if for each open subset U of X, $R(U)$ is nearly open (ie preopen) in Y, that is $R(U) \subset int(cl(R(U)))$.

We shall call a relation R between (X, τ) and (Y, σ) a weakly $\mathcal{B}$-open relation if for each open subset U of X, $R(U)$ is a weak $\mathcal{B}$-set in Y, that is $pintR(U) = intR(U)$.

Proposition 4 allows us to answer the question for general topological spaces.

PROPOSITION 7: A relation R between topological spaces is open if and only if it is nearly open and weakly $\mathcal{B}$-open.

REFERENCES

Andrijević, D. (1987). Mat. Vesnik, 39 : 367–376.
Blumberg, H. (1922). Trans. Amer. Math. Soc., 24 : 113–128.
Ganster, M and Reilly, I.L. (1990). Acta Math. Hung., (to appear).
Husain, T (1966). Prace Mat., 10 : 1–7.
Husain, T. (1977). Topology and Maps, Plenum Press, New York and London.
Mashhour, A.S., Abd El-Monsef, M.E., and El-Deeb, S.N. (1982).
Proc. Math. Phys. Soc. Egypt., 53 : 47–53.
Naimpally, S. (1976). Proceedings of Topology Conference, Memphis,
Lecture Notes in Pure and Applied Mathematics, 24 : 77-86.
Noll, D. (1986). Proc. Amer. Math. Soc., 96 : 141–151.
Pettis, B.J. (1974). Bull. London Math. Soc., 6 : 37–41.
Ptak, V. (1958). Bull. Soc. Math. France, 86 : 41–74.
Steen, L.A. and Seebach, J.A. (1970). Counterexamples in Topology, Holt,

Rinehart and Winston, New York.
Tong, J. (1989). Acta Math. Hung., 54 : 51–55.
Wilhelm, M. (1979). Colloq. Math., 42 : 387–394.
Wilhelm, M. (1981). Fund. Math., 114 : 219–228.
Wilhelm, M. (1986). Comment. Math., 26 : 187–194.

A Generalization of Compactness and Perfectness

Eraldo Giuli University of L'Aquila, L'Aquila, Italy

ABSTRACT. Given infinite cardinal numbers $\mathfrak{a} \leq \mathfrak{b}$ let $P[\mathfrak{a},\mathfrak{b}]$ be the class of all Hausdorff spaces of character not greater than $\mathfrak{b}$ and such that every $<\mathfrak{a}$-intersection of open sets is open. It is known that a (Hausdorff) topological space X is $[\mathfrak{a},\mathfrak{b}]$-compact iff for each $P[\mathfrak{a},\mathfrak{b}]$-space Y the projection $p_Y: X \times Y \to Y$ is a closed map. Moreover for $\mathfrak{b} = \infty$, the spaces in $P[\mathfrak{a},\mathfrak{b}]$ are in the same way determined by the class of $[\mathfrak{a},\mathfrak{b}]$-compact spaces.

These facts give rise to the notions of compact class F(**A**) and discrete class G(**A**) relative to a fixed class **A** of (Hausdorff) spaces. Also a relative notion of perfectness (**A**-perfectness) is introduced and investigated.

Among others it is shown that:

(a) Every class F(**A**) is closed hereditary, is closed under continuous images, and is closed by products with compact spaces;

(b) Every class G(**A**) is a hereditary coreflective subcategory of the category of Hausdorff spaces;

Partial financial support by Italian Ministry of Public Education is gratefully acknowledged.

1980 *Mathematics Subject Classification* (1985 *revision*). 54D20, 54G10, 54C10, 54D30, 54B30.

Key words and phrases. $[\mathfrak{a},\mathfrak{b}]$-compact space, $\mathfrak{a}$-Lindelöf space, closed map, $P[\mathfrak{a},\mathfrak{b}]$-space, compact class, discrete class, $\mathfrak{a}$-compact space.

(c) Every **A**-perfect map is closed and point-inverses belong to F(**A**);
(d) If **A** = P[$\mathfrak{a}$,∞] = $\mathbf{P}_{\mathfrak{a}}$, then the converse of (c) is true;

(e) Some product-theorems given by Noble and Vaughan remain true in the above general context.

1. [$\mathfrak{a}$,$\mathfrak{b}$]-COMPACT AND P[$\mathfrak{a}$,$\mathfrak{b}$]-SPACES

Throughout the paper, $\mathfrak{a}$ and $\mathfrak{b}$ denote infinite cardinal numbers with $\mathfrak{a} \leq \mathfrak{b}$ and, if X is a set (topological space), |X| denotes the cardinality of X. **Top, Haus** and **Tych** will denote the category of topological spaces, Hausdorff spaces and Tychonov (= completely regular Hausdorff) spaces, respectively.
A topological space X is said to be [$\mathfrak{a}$,$\mathfrak{b}$]-*compact* (cf. [37]) if every open cover $\mathfrak{U}$ with $|\mathfrak{U}| \leq \mathfrak{b}$ admits a subcover $\mathfrak{U}'$ with $|\mathfrak{U}'| < \mathfrak{a}$. Following Frolík [13], X is said to be $\mathfrak{a}$-*Lindelöf* ([$\mathfrak{a}$,∞]-compact or finally $\mathfrak{a}$-compact in [40]) if it is [$\mathfrak{a}$,$\mathfrak{b}$]-compact for each $\mathfrak{b} \geq \mathfrak{a}$.
$\mathbf{L}_{\mathfrak{a}}$ will denote the class (full subcategory) of $\mathfrak{a}$-Lindelöf spaces.
Notice that $\aleph_0$-Lindelöf is the same as compact, $\aleph_1$-Lindelöf is the same as Lindelöf property, and [$\aleph_0$,$\aleph_0$]-compact is the same as countably compact. A good reference for [$\mathfrak{a}$,$\mathfrak{b}$]-compact spaces is Vaughan's survey [40] (see also [27] for the relationship between the above [$\mathfrak{a}$,$\mathfrak{b}$]-compactness and the older notion of [$\mathfrak{a}$,$\mathfrak{b}$]-compact space, in the sense of complete accumulation points, introduced by Alexandroff and Urysohn [1]).
A collection $\mathfrak{F}$ of subsets of a topological space X is called [$\mathfrak{a}$,$\mathfrak{b}$]-*family* if $|\mathfrak{F}| \leq \mathfrak{b}$ and $\mathfrak{F}$ has the <$\mathfrak{a}$-intersection property, that is : for each $\mathfrak{F}' \subset \mathfrak{F}$ with $|\mathfrak{F}'| < \mathfrak{a}$, we have $\cap \mathfrak{F}' \neq \emptyset$). $\mathfrak{F}$ is non fixed if $\cap \mathfrak{F} = \emptyset$. Gaal [14] (see also [41], Lemma 5.(b)) proved that

LEMMA 1.1. *A topological space* X *is* [$\mathfrak{a}$,$\mathfrak{b}$]-*compact iff every (non fixed)* [$\mathfrak{a}$,$\mathfrak{b}$]-*family* $\mathfrak{F}$ *of subsets of* X *has non empty adherence, i.e.,* $\cap \overline{\mathfrak{F}} \neq \emptyset$.

A topological space X is said to be $\mathrm{P}_{\mathfrak{a}}$-*space* if the family of open sets of X is closed under the formation of <$\mathfrak{a}$-intersections and it is said to be P[$\mathfrak{a}$,$\mathfrak{b}$]-*space* if, moreover, every point $x \in X$ admits a neighbourhood base $\mathfrak{B}_X$ with $|\mathfrak{B}_X| \leq \mathfrak{b}$, i.e. X has character $\leq \mathfrak{b}$.

$\mathbf{P}_{\mathfrak{a}}$ and $\mathbf{P}[\mathfrak{a},\mathfrak{b}]$ will denote the class of all $P_{\mathfrak{a}}$-spaces and the class of all $P[\mathfrak{a},\mathfrak{b}]$-spaces, respectively.
Notice that $P[\aleph_0,\aleph_0]$ is the class of first countable spaces, $\mathbf{P}_{\aleph_0} = \mathbf{Top}$ and $\mathbf{P}_{\aleph_1}$ is the class of P-spaces (for a reference on P-spaces see [**34**]).

Let $\mathcal{F}$ be an $[\mathfrak{a},\mathfrak{b}]$-family of subsets of a topological space X and let $\mathfrak{B}$ be the filter base, obtained from $\mathcal{F}$, by adding the subsets of type $\cap\{F_i : i \in I\}$, with $|I| < \mathfrak{a}$, $\{F_i\} \subset \mathcal{F}$. Denote by $X_{\mathcal{F}}$ the space whose underlying set is $X \cup \{s\}$, $s \notin X$, every $x \in X$ is isolated and the basic neighbourhood of s is $\{s\} \cup B$, $B \in \mathfrak{B}$. $X_{\mathcal{F}}$ is 0-dimensional and, if $\mathcal{F}$ is non fixed, it is Hausdorff. If, moreover, $\mathfrak{a}$ and $\mathfrak{b}$ satisfy the condition $\Sigma\{\mathfrak{b}^{\mathfrak{m}} : \mathfrak{m} < \mathfrak{a}\} = \mathfrak{b}$, then $X_{\mathcal{F}}$ is a $P[\mathfrak{a},\mathfrak{b}]$-space.

For topological spaces X and Y denote by $p_Y: X \times Y \to Y$ the second projection. If $Y = X_{\mathcal{F}}$ we write $p_{\mathcal{F}}$ for $p_{X_{\mathcal{F}}}$.
We say that $f: X \to Y$ is a closed map if f(M) is closed in Y for each closed subset M of X.
Good references for closed mappings are Burke's survey [**4**] and [**35**], [**36**].

THEOREM 1.2. (a) (cf. [**18**, Theorem 3] for the case $\mathfrak{a} = \aleph_0$) *If a topological space* X *is* $[\mathfrak{a},\mathfrak{b}]$*-compact then for each* $P[\mathfrak{a},\mathfrak{b}]$*-space* Y *the projection* p_Y *is a closed map.*
(b) *If* X *is a space such that, for each non fixed* $[\mathfrak{a},\mathfrak{b}]$*-family* $\mathcal{F}$ *of subsets of* X, $p_{\mathcal{F}}$ *is closed, then* X *is* $[\mathfrak{a},\mathfrak{b}]$*-compact.* .

PROOF. (a). Let X be $[\mathfrak{a},\mathfrak{b}]$-compact and let $Y \in \mathbf{P}[\mathfrak{a},\mathfrak{b}]$. If $M \subset X \times Y$ and $y \in Y \setminus p_Y(M)$ then $(X \times \{y\}) \cap M = \varnothing$. If in addition M is closed, then, for each $x \in X$, we can choose a rectangular neighbourhood $U_X \times U_y(x)$ of (x,y) such that $(U_X \times U_y(x)) \cap M = \varnothing$. Let $\mathfrak{B}_y$ be a neighbourhood base of y with $|\mathfrak{B}_y| \le \mathfrak{b}$ and, for each $x \in X$ choose $B_y(x) \in \mathfrak{B}_y$ such that $B_y(x) \subset U_y(x)$. Define $x \sim x'$ in X if $B_y(x) = B_y(x')$ and set $V_{[x]} = \cup\{U_{x'} : x' \sim x\}$. Then the family $\{V_{[x]} \times B_y(x) : [x] \in X/\sim\}$ is an open cover of $X \times \{y\}$ whose cardinality is not greater than $\mathfrak{b}$. Since $X \times \{y\}$ is $[\mathfrak{a},\mathfrak{b}]$-compact, then there is a subcover $\{V_{[x_i]} \times B_y(x_i) : i \in I\}$ with $|I| < \mathfrak{a}$. Since $Y \in P[\mathfrak{a},\mathfrak{b}]$, then $\cap\{B_y(x_i) : i \in I\}$ is a neighbourhood of y and it does not meet $p_Y(M)$.
(b). Let X be a topological space such that for each non fixed $[\mathfrak{a},\mathfrak{b}]$-family $\mathcal{F}$ of

subsets of X the projection $p_{\mathfrak{F}}$ is a closed map. To show that X is $[\mathfrak{a},\mathfrak{b}]$-compact we will show, in virtue of Lemma 1.1, that each non fixed $[\mathfrak{a},\mathfrak{b}]$-family $\mathfrak{F}$ on X has non empty adherence. Set $\Delta_X = \{(x,y) \in X \times X_{\mathfrak{F}} : x=y\}$; then $p_{\mathfrak{F}}(\Delta_X) = X$, so $s \in p_{\mathfrak{F}}(\overline{\Delta_X})$ since $p_{\mathfrak{F}}$ is a closed map and $s \in \overline{X}$. Thus there is a point $x_0 \in X$ such that $(x_0,s) \in \overline{\Delta_X}$. This means that for every $B \in \mathfrak{B}$ and for every neighbourhood U of x_0, $(U \times (B \cup \{s\})) \cap \Delta_X \neq \varnothing$ which is equivalent to $U \cap B \neq \varnothing$. Thus x_0 is (adherent to $\mathfrak{B}$, hence it is) adherent to $\mathfrak{F}$.

Noble [36] introduced $\mathfrak{b}$-$\mathfrak{a}$-sequences and $\mathfrak{b}$-$\mathfrak{a}$-sequential spaces and observed that every $P[\mathfrak{a},\mathfrak{b}]$-space is $\mathfrak{b}$-$\mathfrak{a}$-sequential (the converse is not true, e.g. $\aleph_0$-$\aleph_0$-sequential space = sequential space while $P[\aleph_0,\aleph_0]$-space = first countable space). He showed ([36], Theorem 1.4 , see also [42]) that a topological space X is compact with respect to $\mathfrak{b}$-$\mathfrak{a}$-sequences (i.e., every $\mathfrak{b}$-$\mathfrak{a}$-sequence has a cluster point) iff, for each $\mathfrak{b}$-$\mathfrak{a}$-sequential space Y, the projection p_Y is closed. Since for $\mathfrak{a}$ regular and $\mathfrak{a}$, $\mathfrak{b}$ satisfying the condition $\Sigma\{\mathfrak{b}^{\mathfrak{m}} : \mathfrak{m} < \mathfrak{a}\} = \mathfrak{b}$, $[\mathfrak{a},\mathfrak{b}]$-compact is equivalent to compact with respect to $\mathfrak{b}$-$\mathfrak{a}$-sequences (cf. [41]) and since, in the above conditions, the spaces $X_{\mathfrak{F}}$ are $P[\mathfrak{a},\mathfrak{b}]$-spaces, then we obtain the following

COROLLARY 1.3. *Let $\mathfrak{a}$, $\mathfrak{b}$ be infinite cardinals with $\mathfrak{a}$ regular and $\Sigma\{\mathfrak{b}^{\mathfrak{m}} : \mathfrak{m} < \mathfrak{a}\} = \mathfrak{b}$. Then for a topological space X the following conditions are equivalent:*
(i) X *is* $[\mathfrak{a},\mathfrak{b}]$*-compact;*
(ii) *for each* $\mathfrak{b}$-$\mathfrak{a}$*-sequential space* Y *the projection* p_Y *is closed;*
(iii) *for each* $P[\mathfrak{a},\mathfrak{b}]$*-space* Y *the projection* p_Y *is closed;*
(iv) *for each non fixed* $[\mathfrak{a},\mathfrak{b}]$*-family* $\mathfrak{F}$ *of subsets of* X, $p_{\mathfrak{F}}$ *is closed.*

These two particular cases should be noted

COROLLARY 1.4. (a) X *is countably compact iff, for each sequential space (resp. first countable space)* Y, *the projection* p_Y *is closed.*
(b) X *is* $\mathfrak{a}$*-Lindelöf iff, for each* $P_{\mathfrak{a}}$*-space* Y, *the projection* p_Y *is closed.*

Vaughan introduced in [42] the general notion of Ω-compact space and gave a characterization of these spaces in terms of closed projections (Theorem 1). He also observed that, if Ω is the class of $<\mathfrak{a}$-directed sets D such that $|D| \leq \mathfrak{b}$ and if $\mathfrak{a}$ and $\mathfrak{b}$ satisfy $\Sigma\{\mathfrak{b}^{\mathfrak{m}} : \mathfrak{m} < \mathfrak{a}\} = \mathfrak{b}$, then $[\mathfrak{a},\mathfrak{b}]$-compact = Ω-compact, $\mathfrak{b}$-$\mathfrak{a}$-sequential = Ω-

net space and P[$\mathfrak{a},\mathfrak{b}$]-space = Ω-neighbourhood space. Thus (i) <=> (ii) <=> (iii) in Corollary 1.3 is contained in Theorem 1 of [**42**].

It was shown in [**34**], Theorem 2.1, that a topological space Y is a P-space iff for each Lindelöf space X the projection p_Y is closed. More generally it is shown in [**35**], first part of Corollary 2.3, that

PROPOSITION 1.5. *A topological space* Y *is a* $P_{\mathfrak{a}}$*-space iff, for each* $\mathfrak{a}$*-Lindelöf space* X, *the projection* p_Y *is closed.*

2. COMPACT PAIRS

From now on all topological spaces are Hausdorff, so our base category is the category **Haus** of Hausdorff spaces. **HComp** will denote the class of compact Hausdorff spaces. Notice that the results in §1 remain valid under the Hausdorffness restriction, since the spaces $X_{\mathfrak{F}}$ considered in the proof of Theorem 1.2.(b) are (zero-dimensional) Hausdorff. More generally, the above results remain valid if they are restricted to an arbitrary hereditary, productive and isomorphism-closed class of spaces (i.e., epireflective subcategory of **Top**).

The results in Corollary 1.4.(b) and Proposition 1.5 lead to the following

DEFINITION 2.1. Let **K** and **D** be two non-empty classes of Hausdorff spaces. We say that **K** and **D** form a *compact pair* (**K**,**D**) if
(a) A space X is in **K** iff, for each space Y in **D**, the projection p_Y is closed;
(b) A space Y is in **D** iff, for each space X in **K**, the projection p_Y is closed.
Then **K** is called the *compact class* and **D** the *discrete class* of the compact pair (**K**,**D**).

In other words, a pair (**K**,**D**) is compact iff the following conditions are satisfied
(1) For each X $\in$ **K** and Y $\in$ **D** the projection p_Y is closed;
(2) If X is such that, for each Y $\in$ **D**, p_Y is closed, then X $\in$ **K**;
(3) If Y is such that, for each X $\in$ **K**, p_Y is closed, then Y $\in$ **D**.

(**Haus**,**Discr**), (**HComp**,**Haus**) are trivial examples of compact pairs (**Discr** denotes the class of discrete spaces). More generally, it follows from Corollary 1.3 and Proposition 1.5 that, for each cardinal $\mathfrak{a}$, $(\mathbf{L}_{\mathfrak{a}},\mathbf{P}_{\mathfrak{a}})$ is a compact pair

HComp is the smallest and **Haus** the largest compact class. Anagously **Discr** is the smallest and **Haus** the largest discrete class.
Notice that ({countably compact spaces},{sequential spaces}) is not a compact pair (see Remark 2.7 below).

In a compact pair (**K**,**D**), **K** determines **D**, and conversely **D** determines **K** by means of the Galois equivalence explained below.
Let $(\mathbb{A},\leq)$ be the conglomerate of all non-empty subclasses of the class of Hausdorff spaces ordered by inclusion. For each **A**, **B** $\in \mathbb{A}$ set

$$F(\mathbf{A}) = \{X \in \mathbf{Haus} : \text{for each } Y \in \mathbf{A},\ p_Y \text{ is closed}\}$$
$$G(\mathbf{B}) = \{Y \in \mathbf{Haus} : \text{for each } X \in \mathbf{B},\ p_Y \text{ is closed}\}.$$

The assignements $\mathbf{A} \rightarrow F(\mathbf{A})$ and $\mathbf{B} \rightarrow G(\mathbf{B})$ are order reversing maps and $\mathbf{A} \leq G(F(\mathbf{A}))$ and $\mathbf{B} \leq F(G(\mathbf{B}))$ for each **A**, **B** $\in \mathbb{A}$, i.e., F and G deternine a Galois correspondence (cf. [**26**], 27Q). For each **A**, **B** $\in \mathbb{A}$, F(**A**) is a compact class and G(**B**) is a discrete class. In fact it is easy to see that (F(**A**),G(F(**A**))) and (F(G(**B**)),G(**B**)) are compact pairs. Thus the restrictions of F and G to $\mathbb{K}$ and $\mathbb{D}$ are anti-isomorphisms (cf. [**26**], (2) => (4) in 27Q). This shows part (a) of the following

THEOREM 2.2. (a) $((\mathbb{A},\leq), F, G)$ *is a Galois correspondence whieh is a Galois equivalence between the conglomerate* $\mathbb{K} = \{F(\mathbf{A}) : \mathbf{A} \in \mathbb{A}\}$ *of compact classes and the conglomerate* $\mathbb{D} = \{G(\mathbf{B}) : \mathbf{B} \in \mathbb{A}\}$ *of discrete classes.*
(b) *In both conglomerates, arbitrary infima and suprema exist.*

PROOF. (b). If $\{\mathbf{K}_i : i \in I\}$ is an arbitrary family of compact classes, and $\{\mathbf{D}_i : i \in I\}$ is the corresponding family of discrete classes, then we have

$$\bigvee\{\mathbf{K}_i : i \in I\} = F(G(\cup\{\mathbf{K}_i : i \in I\})) = F(\cap\{\mathbf{D}_i : i \in I\}).$$

Basic properties of discrete classes, of compact classes and of compact pairs, respectively, are collected in the following theorems:

THEOREM 2.3. *Every discrete class* **D** *is (the object-class of) a hereditary coreflective subcategory of* **Haus**, *that is:*

(a) **D** *is isomorphism-closed;*

(b) **D** *is closed under the formation of subspaces;*

(c) **D** *is closed under the formation of disjoint unions (= coproducts);*

(d) **D** *is closed under the formation of Hausdorff quotients.*

PROOF. (a). Trivial.

(b). If Z is a subspace of Y then, for each closed subset $F \subset X \times Z$, if we denote by M a closed subset of $X \times Y$ such that $F = M \cap (X \times Z)$, we have $p_Z(F) = p_Y(M) \cap Z$. Thus p_Z is closed whenever p_Y is closed.

(c). It follows from the well known formula $X \times (\Sigma Y_i) \cong \Sigma(X \times Y_i)$ valid in **Top**.

(d). Let $Y \in$ **D**, $\sim$ be an equivalence in Y such that the quotient $Y/\sim = Y'$ is Hausdorff, and let $q: Y \to Y'$ be the quotient map. To show that $Y' \in$ **D** we have to prove that the projection $p_{Y'}: X \times Y' \to Y'$ is a closed map whenever $X \in F(\mathbf{D})$. Now if $X \in F(\mathbf{D})$, then the projection $p_Y: X \times Y \to Y$ is closed so that, for each closed set F in $X \times Y'$, $p_Y((1_X \times q)^{-1}(F))$ is closed in Y. Since $p_{Y'}(1_X \times q) = qp_Y$, then $p_Y((1_X \times q)^{-1}(F)) = q^{-1}(p_{Y'}(F))$. Since $q^{-1}(p_{Y'}(F))$ is saturated with respect to q and since every quotient map carries saturated closed sets to closed sets, then $q(q^{-1}(p_{Y'}(F))) = p_{Y'}(F)$ is closed in Y'.

For a class **A** of Hausdorff spaces, denote by C(**A**) the coreflective hull of **A** in **Haus**, i.e., the class of all Hausdorff quotients of coproducts of spaces in **A**.

COROLLARY 2.4. *For each class* **A** *of Hausdorff spaces we have* F(**A**) = F(C(**A**)).

PROOF. Since $\mathbf{A} \leq C(\mathbf{A})$ then $F(C(\mathbf{A})) \leq F(\mathbf{A})$. For the reverse inclusion note that $\mathbf{A} \leq G(F(\mathbf{A}))$ and $G(F(\mathbf{A}))$ is a discrete class so that, by Theorem 2.3, $C(\mathbf{A}) \leq G(F(\mathbf{A}))$, consequently $F(G(F(\mathbf{A}))) \leq F(C(\mathbf{A}))$. Now $F(G(F(\mathbf{A}))) = F(\mathbf{A})$ by Theorem 2.2.(a), then $F(C(\mathbf{A})) \geq F(\mathbf{A})$.

As an application of both Corollary 1.4.(a) and Corollary 2.4 we obtain the following result due to Hanai [**19**]

COROLLARY 2.5. *A Hausdorff topological space* X *is countably compact iff* $p_{\mathbb{N}^*}: X \times \mathbb{N}^* \to \mathbb{N}^*$ *is closed* ($\mathbb{N}^*$ is the Alexandroff one-point compactification of the discrete space of natural numbers).

PROOF. It is well known that $C(\{\mathbb{N}^*\}) = \{$sequential Hausdorff spaces$\}$.

COROLLARY 2.6. *No discrete class* **D** $\neq$ **Haus** *is closed under the formation of arbitrary products.*

PROOF. If **D** is closed under the formation of arbitrary products then, by Theorem 2.3, it is both coreflective and epireflective in **Haus** so that **D** = **Haus** by a result of Kannan [**32**].

REMARK 2.7. The pair ({countably compact spaces},{sequential spaces}) is not a compact pair even if the countably compact spaces form a compact class since they coincide with F({sequential spaces}) (see Corollary 1.4.(a)). In fact the class of sequential Hausdorff spaces does not satisfy property (b) in Theorem 2.3.

Isiwata [**30**] asked for a characterization of the discrete counterpart **D** of the class of countably compact spaces. It is clear from Theorem 2.3.(b) and Corollary 1.4.(a) that **D** contains all subspaces of Hausdorff sequential spaces (this fact was first observed by Fleischer and Franklin [**10**] and by Isiwata [**30**]; see also [**16**]). On the other hand, in Corollary 1.4.(a) sequential cannot be replaced by compactly generated (cf. [**9**], Example 8.10.16).

THEOREM 2.8. (a) *Every compact class* **K** *is closed-hereditary ;*
(b) *Every compact class is preserved by continuous maps, i.e., if* $f: X \to Z$ *is a continuous onto map and* $X \in$ **K**, *then* $Z \in$ **K**.

PROOF. (a). Let M be a closed subset of a space X. Since every closed subset $F \subset M \times Y$ is also closed in $X \times Y$, then $p_Y: M \times Y \to Y$ is closed whenever $p_Y: X \times Y \to Y$ is a closed map. Consequently $M \in$ **K** whenever $X \in$ **K**.
(b). Denote by $p: X \times Y \to Y$ and $p': Z \times Y \to Y$ the projections. Since $p'(f \times 1_Y) = p$ and $f \times 1_Y$ is onto then, for each $M \subset Z \times Y$, $p'(M) = p((f \times 1_Y)^{-1}(M))$. Consequently p' is

closed whenever p is closed.

PROPOSITION 2.9. *Let* (**K**,**D**) *be a compact pair.*
(a) *If* Y ∈ **D** *and* X *is a subspace of* Y *such that* X ∈ **K**, *then* X *is closed in* Y.
(b) *Every continuous map* f:X→Y *with* X ∈ **K** *and* Y ∈ **D** *is a closed map.*
In particular f is a homeomorphism whenever it is bijective.

PROOF. (a). Observe that $\Delta_X = \{(x,x) : x \in X\} = \Delta_Y \cap (X \times Y)$, so that Δ_X is closed in $X \times Y$ since Δ_Y is closed in $Y \times Y$, by the Hausdorffness of Y. If X ∈ **K** and Y ∈ **D** then $p_Y: X \times Y \to Y$ is closed, so that, in particular, $p_Y(\Delta_X) = X$ is closed in Y.
(b). If M ⊂ X is a closed subset then, by X ∈ **K** and Theorem 2.8.(a), M ∈ **K**, consequently, by Theorem 2.8.(b), f(M) ∈ **K**. Thus it follows from (a) and from Y ∈ **D** that f(M) is closed in Y.

COROLLARY 2.10. (Cf. [**34**] and [**5**] for $\mathbf{K} = \mathrm{L}_{\aleph_1}$, $\mathbf{D} = \mathrm{P}_{\aleph_1}$). *If* (**K**,**D**) *is a compact pair then* **K**∩**D** *is both the class of maximal-***K** *spaces and the class of minimal-***D** *spaces.*

PROPOSITION 2.11. *Let* (**K**,**D**) *be a compact pair.*
(a) *If* **D** *is finitely productive, then* X×Y ∈ **K** *whenever* X ∈ **K** *and* Y ∈ **K**∩**D**;
(b) *If* X ∈ **HComp** *and* Z ∈ **K** *then* X×Z ∈ **K**.

PROOF. If we denote by $p:(X \times Z) \times Y \to Y$, $p':X \times (Z \times Y) \to Z \times Y$ and $p'':Z \times Y \to Y$ the projections then $p = p''p'h$, where $h:(X \times Z) \times Y \to X \times (Z \times Y)$ is the canonical homeomorphism.
(a). If X, Z ∈ **K** and Z, Y ∈ **D** then p" is closed and, by the productivity of **D**, p' is also closed so that p, as composition of closed maps, is closed.
(b). If X ∈ **HComp** then p' is closed; if Z ∈ **K** and Y ∈ **D**, then p" is closed.

COROLLARY 2.12. (a) (cf. [**34**] Proposition 4.2.(4) and [**9**] Corollary 3.8.10 for Lindelöf spaces). *The product of an* $\mathfrak{a}$*-Lindelöf space with either an* $\mathfrak{a}$*-Lindelöf* $P_{\mathfrak{a}}$*-space or a compact Hausdorff space has the* $\mathfrak{a}$*-Lindelöf property.*
(b). (cf. [**9**] Corollary 3.10.15 and 3.10.14). *The product of a countably compact space with either a countably compact subspace of a sequential space or a compact Hausdorff space is countably compact.*

We will make use of the following result due to Noble ([**36**], Theorem 1.8)

LEMMA 2.13. *Let* $\mathfrak{c}$ *be an infinite cardinal and let* $X = \prod\{X_{\mathfrak{d}} : \mathfrak{d} < \mathfrak{c}\}$. *If for each* $\mathfrak{d}_0 < \mathfrak{c}$ *the projection from* $\prod\{X_{\mathfrak{d}} : \mathfrak{d} \leqslant \mathfrak{d}_0\}$ onto $\prod\{X_{\mathfrak{d}} : \mathfrak{d} < \mathfrak{d}_0\}$ *is closed, then for each* $\mathfrak{d}_0$, *the projection from* X *onto* $\prod\{X_{\mathfrak{d}} : \mathfrak{d} < \mathfrak{d}_0\}$ *is closed.*

The following theorem is a simple extension to compact pairs of a result given by Vaughan for the case of Ω-compact spaces.(cf. [**42**], Theorem 2 and Corollary 1):

THEOREM 2.14. *Let* (**K**,**D**) *be a compact pair,* $\mathfrak{c}$ *an infinite cardinal, and assume that every* $< \mathfrak{c}$*-fold product of* **D**-*spaces belongs to* **D**. *Then, for each family* $\{X_{\mathfrak{d}} : \mathfrak{d} < \mathfrak{c}\}$ *of spaces in* $\mathbf{K} \cap \mathbf{D}$, *the product* $\prod\{X_{\mathfrak{d}} : \mathfrak{d} < \mathfrak{c}\}$ *belongs to* **K**.

PROOF. Let $\{X_{\mathfrak{d}} : \mathfrak{d} < \mathfrak{c}\}$ be a family of spaces in $\mathbf{K} \cap \mathbf{D}$. To show that $X = \prod\{X_{\mathfrak{d}} : \mathfrak{d} < \mathfrak{c}\}$ belongs to **K** we will prove that, for each $Y \in \mathbf{D}$, the projection $p_Y : X \times Y \to Y$ is closed. Set

$$Z_{\mathfrak{d}} = \begin{cases} Y & \text{if } \mathfrak{d} = 0 \\ X_{\mathfrak{d}-1} & \text{if } 0 < \mathfrak{d} < \aleph_0 \\ X_{\mathfrak{d}} & \text{if } \aleph_0 \leqslant \mathfrak{d} < \mathfrak{c}. \end{cases}$$

For each $0 < \mathfrak{d}_0 < \mathfrak{c}$ the projection from $\prod\{Z_{\mathfrak{d}} : \mathfrak{d} \leqslant \mathfrak{d}_0\}$ onto $\prod\{X_{\mathfrak{d}} : \mathfrak{d} < \mathfrak{d}_0\}$ is closed since $Z_{\mathfrak{d}_0} \in \mathbf{K}$ and $\prod\{Z_{\mathfrak{d}} : \mathfrak{d} < \mathfrak{d}_0\}$, by assumption, belongs to **D**. Thus, by Lemma 2.13, (for each $\mathfrak{d}_0 < \mathfrak{c}$, hence in particular) for $\mathfrak{d}_0 = 1$, the projection from $\prod\{X_{\mathfrak{d}} : \mathfrak{d} < \mathfrak{c}\} = X \times Y$ onto $\prod\{X_{\mathfrak{d}} : \mathfrak{d} < 1\} = Y$ is closed. Consequently $X \in \mathbf{K}$.

Notice that the above result remains valid if **D** is replaced by a class **A** such that F(**A**) = **K** (these classes where called **K**-test classes in [**42**]).
Since, for each cardinal $\mathfrak{a}$, the class $\mathbf{P}_{\mathfrak{a}}$ of $P_{\mathfrak{a}}$-spaces is finitely productive and since the class of subsequential spaces (= subspaces of sequential spaces) is countably productive ([**36**], Theorem 4.5) then we obtain

COROLLARY 2.15. (a) ([**36**], Theorem 4.1). *Every countably product of* $\mathfrak{a}$-*Lindelöf,* $P_{\mathfrak{a}}$*-spaces is* $\mathfrak{a}$*-Lindelöf.*

(b) ([36], Theorem 4.7). *Every $\aleph_1$-fold product of countably compact subsequential spaces is countably compact.*

Notice that there exist compact classes different from **Haus** and from **HComp** which are closed under the formation of arbitrary products. For example Stephenson and Vaughan [38] proved that for each singular cardinal $\mathfrak{b}$ satisfying the condition $2^{\mathfrak{a}} \leqslant \mathfrak{b}$, for each cardinal $\mathfrak{a} < \mathfrak{b}$, the class of $[\aleph_0, \mathfrak{b}]$-compact (= initially $\mathfrak{b}$-compact in [38]) spaces is a productive class.

3. A-PERFECT MAPS

It is well known that for a continuous map $f:X \to Z$ between Tychonov spaces the following conditions are equivalent and that each of them defines the class of perfect maps (cf. [3] for (i) <=> (ii) and [20] for (ii) <=> (iii)):

(i) for each (Tychonov) space Y, $f \times 1_Y$ is closed;
(ii) f is closed and, for each $y \in Y$, $f^{-1}(y)$ is compact;
(iii) $\beta(f)(\beta(X)\backslash X) \subset \beta(Y)\backslash Y$, where $\beta(X)$ is the Stone-Čech compactification of X.

Tsai [39] used (iii) as a definition of **E**-perfect map in the setting of **E**-compact spaces. Anagously Hager [17] defined **S**-perfect maps for an arbitrary epireflective subcategory **S** of Tychonov spaces. Herrlich [23] explained the strong relation between perfect maps and factorization structures. More information on perfectness and its generalizations can be found in [6]. For further and more recent results see [24] and [8].

Our compact classes are, in general, not epireflective subcategories, so we cannot use Hager's approach. The following definition, which uses (i), seems to be more appropriate for compact classes:

DEFINITION 3.1. (a) Let **A** be a class of spaces containing an one-point space T. A continuous map $f:X \to Z$ is said to be **A**-*perfect* if $f \times p_Y : X \times Y \to Z \times Y$ is closed for each $Y \in$ **A**.

The class of **A**-perfect maps will be denoted by P(**A**). If **M** = P(**A**) for some **A** ⊂

Haus, then **M** is called *perfect class of (closed) maps*

(b) Let **M** be a class of continuous closed maps. A space Y is said to be **M**-*perfect* if $f \times 1_Y$ is closed for each $f \in$ **M**.

Q(**M**) will denote the class of **M**-perfect spaces. If **A** = Q(**M**) for some class **M** of continuous closed maps, then **A** is called *perfect class of spaces*

Let (**A**,≤) be the conglomerate of all subclasses of the class of Hausdorff spaces, ordered by inclusion and let (**M**,≤) be the conglomerate of all subclasses of the class of closed maps between Hausdorff spaces, ordered by inclusion. Then it is easy to see that

PROPOSITION 3.2 ((**A**,≤),(**M**,≤),P,Q) *is a Galois correspondence which is a Galois equivalence between the conglomerate of all perfect classes of closed maps and the conglomerate of all perfect classes of spaces.*

In both conglomerates arbitrary infima and suprema exist.

The following proposition collects basic properties of **A**-perfect maps.

PROPOSITION 3.3. (a) *If* $f:X \to Z$ *is* **A**-*perfect, then f is closed and, for each* $z \in Z$, $f^{-1}(z) \in F(\mathbf{A})$

(b) *The composition of* **A**-*perfect maps is* **A**-*perfect.*

(c) $X \in F(\mathbf{A})$ *iff the constant map* $t:X \to T$ *is* **A**-*perfect.*

(d) *If* $f:X \to Y$ is **A**-*perfect and* F *is a closed subset of* X *then* $f|_F:F \to Y$ *is* **A**-*perfect.*

PROOF. (a). Since $T \in$ **A**, then $f \cong f \times 1_T : X \times T \to Z \times T$ is closed. Since $f \times 1_Y$ is closed for each $Y \in$ **A**, then for each $z \in Z$, the restriction $f \times 1_Y : f^{-1}(z) \times Y \to \{z\} \times Y \cong Y$ is closed as well. Consequently $f^{-1}(z) \in F(\mathbf{A})$.

(b). $(gf) \times 1_Y = (g \times 1_Y)(f \times 1_Y)$ and the composition of closed maps is closed.

(c). The fact that t is **A**-perfect is equivalent to $t \times 1_Y : X \times Y \to T \times Y \cong Y$ is closed, for each $Y \in$ **A**. On the other hand $t \times 1_Y \cong p_Y$.

(d). It follows from (b) and the fact that the inclusion of a closed subset is (perfect, hence) **A**-perfect, for each $\mathbf{A} \subset$ **Haus**.

It is well known that the cartesian product of an arbitrary family of perfect maps is perfect as well (cf. [**12**]). For **A**-perfect maps the above statement is false even for two factors. In fact the finite productivity of **A**-perfect maps trivially implies the finite productivity of the class F(**A**) and, in general, F(**A**) is not finitely productive.

We noted that, for **A** = **Haus** the converse of Proposition 3.3.(a) is true. More generally we have

THEOREM 3.4. *A continuous map* $f:X \to Z$ *is* $\mathbf{P}_{\mathfrak{a}}$*-perfect iff it is closed and, for each* $z \in Z$, $f^{-1}(z)$ *is* $\mathfrak{a}$*-Lindelöf.*

PROOF. Let $f:X \to Z$ be a closed map such that, for each $z \in Z$, $f^{-1}(z)$ is $\mathfrak{a}$-Lindelöf and let $Y \in \mathbf{P}_{\mathfrak{a}}$. To prove that $f \times 1_Y$ is a closed map, we will show that for each $(z,y) \in Z \times Y$ and open set $U \subset X \times Y$ containing $f^{-1}(z) \times \{y\}$, there is an open set $V \subset Z \times Y$, containing (z,y) and such that $(f \times 1_Y)^{-1}(V) \subset U$ (cf. [**9**], Theorem 1.4.12). If $f^{-1}(z) \times \{y\} \subset U$, then for each $x \in f^{-1}(z)$, there exist an open neighbourhood W_x of x and an open neighbourhood $W_y(x)$ of y, such that $W_x \times W_y(x) \subset U$. The family $\{W_x \times W_y(x) : x \in f^{-1}(z)\}$ is an open cover of $f^{-1}(z) \times \{y\}$ admitting, since $f^{-1}(z) \times \{y\}$ is $\mathfrak{a}$-Lindelöf, a subcover $\{W_{x_i} \times W_y(x_i) : i \in I\}$ with $|I| < \mathfrak{a}$. Set $W_1 \times V_2 = (\cup \{W_{x_i}: i \in I\} \times (\cap \{W_y(x_i) : i \in I\})$. Since $Y \in \mathbf{P}_{\mathfrak{a}}$, then V_2 is an open neighbourhood of y, so $W_1 \times V_2$ is an open neighbourhood of $f^{-1}(z) \times \{y\}$ and, by construction, it is contained in U. Since f is a closed map, then there is an open neighbourhood V_1 of z such that $f^{-1}(V_1) \subset W_1$. $V = V_1 \times V_2$ is the needed open set.

COROLLARY 3.5. (a) *If* X *is an* $\mathfrak{a}$*-Lindelöf space and* Y *is a* $\mathbf{P}_{\mathfrak{a}}$*-space then:*
(1) *The projection* $p_Y:X \times Y \to Y$ *is* $\mathbf{P}_{\mathfrak{a}}$*-perfect;*
(2) *Every continuous map* $f:X \to Y$ *is* $\mathbf{P}_{\mathfrak{a}}$*-perfect.*
(b) *If* $f:X \to Y$ is $\mathbf{P}_{\mathfrak{a}}$*-perfect then, for each subset* M *of* Y, *the restriction* $f: f^{-1}(M) \to M$ is $\mathbf{P}_{\mathfrak{a}}$*-perfect.*

PROOF. (a).(1). Follows directly from Theorem 3.4.
(a).(2). By Proposition 2.9.(b) f is closed and, by the Hausdorffness of Y and Proposition 2.8.(a) every fibre of f is $\mathfrak{a}$-Lindelöf, so that Theorem 3.4 applies.
(b). Since f is closed, then $f:f^{-1}(M) \to M$ is closed (cf. [**9**], Proposition 2.1.4).

The following is an extension to α-Lindelöf spaces of a result known as Wallace theorem

THEOREM 3.6. *If* $X' \subset X$ *and* $Y' \subset Y$ *are* α*-Lindelöf subspaces of* $\mathbf{P}_\alpha$*-spaces* X *and* Y, *and* U *is an open set in* $X \times Y$ *containing* $X' \times Y'$, *then there is a rectangular neighbourhood* $U_1 \times U_2$ *of* $X' \times Y'$ *which is contained in* U.

PROOF. If $X' \times Y' \subset U$ then, for each $y' \in Y'$, we can find, as in the proof of Theorem 3.4, a rectangular open neighbourhood $W_1(y') \times V_2(y')$ of $X' \times \{y'\}$ which is contained in U. Since Y' is α-Lindelöf, then the open cover $\{V_2(y') : y' \in Y'\}$ of Y' admits a subcover $\{V_2(y'_i) : i \in I\}$ with $|I| < \alpha$. Since X is a $\mathbf{P}_\alpha$-space then $U_1 = \cap \{W_1(y'_i) : i \in I\}$ is an open neighbourhood of X'. Set $U_2 = \cup \{V_2(y'_i) : i \in I\}$, then $U_1 \times U_2$ has the needed properties.

REMARKS 3.7. (a) Since $\mathbf{P}_\alpha$ is finitely productive and hereditary, then the result above can be extended to every finite family of $\mathbf{P}_\alpha$-spaces by a standard argument (cf. [**9**], 3.2.10).

(b) Some interesting classes of "weak-compact" Hausdorff spaces are not compact classes in the previous sense. For instance, the class of H-closed spaces is not closed hereditary (if every closed subspace of an H-closed space is H-closed, then the space in question is compact [**2**]). The same is true for the class of w-compact spaces (a topological space is called w-compact if every collection of closed sets, closed under finite intersections and such that each member contains a non empty cozero set, has non-empty intersection). In spite of all that, the above classes are compact classes with respect to suitable closure operators different from the ordinary closure. Precisely: X is H-closed iff for each space Y the projection p_Y is θ-closed, that is: p_Y sends θ-closed sets to θ-closed sets (cf. [**31**]); X is w-compact iff, for each space Y, p_Y as a continuous map from the completely regular reflection of $X \times Y$ onto the completely regular reflection of Y is a closed map (cf.[**29**], [**8**]).

The result in [**31**] mentioned above is generalized in [**15**] as follows: a Hausdorff space X is α-θ-Lindelöf (i.e., every open cover admits a $<\alpha$-subfamily whose closures cover X) iff for each $\mathbf{P}_\alpha$- space Y the projection p_Y is θ-closed.

A compactness notion in categories of convergence spaces, which depends on a closure operator, is introduced and investigated in [**7**] (see also [**24**] and [**33**]). Compactness and perfectness in a transportable construct **X**, depending on a closure operator C and on a class **A** of **X**-objects ((C,**A**)-compactness and (C,**A**)-perfectness), is introduced and investigated in [**8**].

Following [**22**] we say that a class **H** of Hausdorff spaces is (**A**)-*left-fitting* if whenever $f:X \to Z$ is an (**A**)-perfect map and $Z \in \mathbf{H}$ then $X \in \mathbf{H}$. It is shown in [**25**] that a class **H** of Tychonov spaces is left-fitting iff: (1) **H** is closed-hereditary and, (2) $X \times Y$ belongs to **H** whenever $X \in \mathbf{H}$ and Y is compact. Thus, in virtue of Proposition 2.8.(a) and Proposition 2.11.(b), every compact class is left-fitting. Moreover

PROPOSITION 3.8. *Every compact class* **K** *is* G(**K**)-*left-fitting.*

PROOF. If $f:X \to Z$ is G(**K**)-perfect then $f \times 1_Y$ is closed whenever $Y \in G(\mathbf{K})$. If $Z \in \mathbf{K}$ then, for each $Y \in G(\mathbf{K})$, the projection $p:Z \times Y \to Y$ is closed. Thus, for each $Y \in G(\mathbf{K})$, $p(f \times 1_Y) = p_Y : X \times Y \to Y$ is closed, i.e. $X \in \mathbf{K}$.

It follows from the above result and from Corollary 3.5.(b) that, for each $\mathbf{P}_{\mathfrak{a}}$-perfect map $f:X \to Z$ and $\mathfrak{a}$-Lindelöf subspace M of Z , the inverse image $f^{-1}(M)$ is $\mathfrak{a}$-Lindelöf. More generally we have

PROPOSITION 3.9. (cf. [**9**], Theorems 3.8.8 and 3.10.9 for Lindelöf spaces and for countably compact spaces). *Let* $\mathfrak{a}$, $\mathfrak{b}$ *be infinite cardinals with* $\mathfrak{a}$ *regular and* $\Sigma\{\mathfrak{b}^{\mathfrak{m}} : \mathfrak{m} < \mathfrak{a}\} = \mathfrak{b}$.
If $f:X \to Z$ *is* $\mathbf{P}[\mathfrak{a},\mathfrak{b}]$-*perfect, then the inverse image* $f^{-1}(M)$ *of every* $[\mathfrak{a},\mathfrak{b}]$-*compact subspace* M *of* Z *is* $[\mathfrak{a},\mathfrak{b}]$-*compact.*

PROOF. To show that $f^{-1}(M)$ is $[\mathfrak{a},\mathfrak{b}]$-compact we shall prove that, for each $[\mathfrak{a},\mathfrak{b}]$-family $\mathfrak{F}$ of closed subsets of $f^{-1}(M)$, $\cap\mathfrak{F} \neq \varnothing$. In virtue of the assumption on the cardinals $\mathfrak{a}$ and $\mathfrak{b}$ it is not restrictive to assume that the family $\mathfrak{F}$ is closed with respect to $<\mathfrak{a}$-intersections. Since the map f is $\mathbf{P}[\mathfrak{a},\mathfrak{b}]$-perfect then, by Proposition 3.3.(a) it is closed, consequently $f:f^{-1}(M) \to M$ is also closed (cf. [**9**], Proposition 2.1.4). Thus $f(\mathfrak{F}) = \{f(F) : F \in \mathfrak{F}\}$ is an $[\mathfrak{a},\mathfrak{b}]$-family of closed subsets of the $[\mathfrak{a},\mathfrak{b}]$-compact space M. Consequently there is a point $z \in \cap f(\mathfrak{F})$. This yields $f^{-1}(z) \cap F \neq \varnothing$ for each $F \in \mathfrak{F}$ and, since $f^{-1}(z)$ is $[\mathfrak{a},\mathfrak{b}]$-compact by Proposition 3.3.(a) and $\mathfrak{F}$ is closed with respect to $<\mathfrak{a}$-intersections, then there is a point belonging to $\cap\{f^{-1}(z) \cap F : F \in \mathfrak{F}\}$.

Thus $\cap \mathcal{F} \neq \emptyset$.

As particular case of the above corollary we have: if $f:X \to Z$ is $\mathbf{P}[\mathfrak{a},\mathfrak{b}]$-perfect and Z is $[\mathfrak{a},\mathfrak{b}]$-compact then X is $[\mathfrak{a},\mathfrak{b}]$-compact. In the case $\mathfrak{a} = \aleph_0$ this result is part of Theorem 1 of [**19**], by Proposition 3.3.(a).
Proposition 3.8 togheter with the Corollary in [**11**] gives the following

PROPOSITION 3.10. *Let* **K** $\subset$ **Haus** *be a compact class and let* E(**K**) be *the epireflective hull of* **K** *in* **Haus.** *Then the* E(**K**)-*reflection* rX *of a Hausdorff space* X *is the intersection of all subspaces of* $\beta(t(X))$, *containing* t(X) *and belonging to* **K**, *where* t(X) *is the Tychonov reflection of* X.

In Theorem 3.4 we have shown that the analogous of (i) <=> (ii) is true for compact pairs $(\mathbf{L}_{\mathfrak{a}},\mathbf{P}_{\mathfrak{a}})$. In what follows we will make use of a result due to Comfort and Retta [**6**] in order to show that the analogous of (i) <=> (iii) is also true, under some restrictions, for compact pairs $(\mathbf{L}_{\mathfrak{a}},\mathbf{P}_{\mathfrak{a}})$.

Let $\mathfrak{a}$ be an infinite cardinal; following Herrlich [**21**] we say that a Tychonov space X is $\mathfrak{a}$-*compact* if $X = \mathfrak{a}(X)$ where $\mathfrak{a}(X) = \{p \in \beta(X) : p$ has $<\mathfrak{a}$-intersection property$\}$. It is well known that the class of $\mathfrak{a}$-compact spaces is (the object class of) an epireflective subcategory of **Tych**. Moreover the full subcategory $\mathbf{H}_{\mathfrak{a}}$ of **Tych**, consisting of all $\mathfrak{a}$-compact spaces, is the epireflective hull of $\mathbf{L}_{\mathfrak{a}}$. Indeed every $\mathfrak{a}$-Lindelöf space is trivially $\mathfrak{a}$-compact and it is also easy to see that the cogenerator found by Hušek (cf. [27]) for the subcategory $\mathbf{H}_{\mathfrak{a}}$ is $\mathfrak{a}$-Lindelöf.
The $\mathbf{H}_{\mathfrak{a}}$-reflection of a Tychonov space X is precisely $\mathfrak{a}(X)$.
Comfort and Retta [**6**] introduced the following concept of $\mathfrak{a}$-perfect map: $f:X \to Y$ is $\mathfrak{a}$-perfect if, for each $y \in Y$, $f^{-1}(y)$ is $\mathfrak{a}$-Lindelöf and $f: X \to Y'$ is a closed map, where Y' is the space having the same underlying set of Y and the topology determined by the family of $<\mathfrak{a}$-intersections of open sets of Y. It is clear that every $\mathbf{P}_{\mathfrak{a}}$-perfect map is $\mathfrak{a}$-perfect, and that the converse is true whenever Y is an $\mathbf{P}_{\mathfrak{a}}$-space. Thus it follows from Theorem 4.2 and Theorem 4.4 of [**6**] that

THEOREM 3.11. *Let* $f:X \to Y$ *be a continuous map between Tychonov spaces.*
(a) *If* f *is* $\mathbf{P}_{\mathfrak{a}}$-*perfect then* $\mathfrak{a}(f)(\mathfrak{a}(X)\setminus X) \subset \mathfrak{a}(Y)\setminus Y$.
(b) *If* $\mathfrak{a}(X)$ *is* $\mathfrak{a}$-*Lindelöf*, $Y \in \mathbf{P}_{\mathfrak{a}}$ *and* f *satisfies* $\mathfrak{a}(f)(\mathfrak{a}(X)\setminus X) \subset \mathfrak{a}(Y)\setminus Y$, *then* f

is $\mathbf{P}_{\mathfrak{a}}$*-perfect.*

COROLLARY 3.12. *The* $\mathfrak{a}$*-compactness is inversely preserved by* $\mathbf{P}_{\mathfrak{a}}$*-perfect maps.*

4. QUESTIONS

Question A. (Isiwata problem) Prove or disprove that the discrete counterpart of the class of countably compact spaces is the class of subsequential Hausdorff spaces. More generally give an explicit description of the discrete class associated to the class of $[\mathfrak{a},\mathfrak{b}]$-compact spaces for $\mathfrak{b} \neq \infty$.

Recall that the classes $\mathbf{P}_{\mathfrak{a}}$ are finitely productive and that the class of subsequential spaces is countably productive (cf. [36]). Moreover no discrete class except **Haus** is closed under the formation of arbitrary products (see Corollary 2.6).

Question B. Find a discrete class which is not closed under the formation of finite products.

Notice that the isomorphism-closed class consisting of the empty space is a non discrete hereditary coreflective subcategory of **Haus**.

Question C. Find a non discrete class containing a non-empty space which is hereditary coreflective in **Haus** (and it is finitely productive, if the answer to the question B is 'not').

Question D. Prove or disprove that Theorem 3.4 still holds if $\mathbf{P}_{\mathfrak{a}}$ is replaced by $P[\mathfrak{a},\mathfrak{b}]$ and $\mathfrak{a}$-Lindelöf by $[\mathfrak{a},\mathfrak{b}]$-compact.
More generally, prove or disprove that Theorem 3.4 holds for each compact pair.

It follows from Proposition 2.8.(b) and Proposition 3.8 that every compact class is closed under continuous images and it is left fitting. On the other hand the isomorphism-closed class consisting of the empty space is a non compact class having the above properties.

Question E. Find a non compact class containing a non-empty space which has the above properties.

Question E is related to the following

Question F. Give an internal characterization of compact classes.

Question G. Give an internal characterization of perfect classes of closed maps.

Question H. Give an internal characterization of perfect classes of spaces.

ACKNOWLEDGEMENTS. I would like to thank D. Dikranjan, G.E. Strecker and J.E. Vaughan for some valuable discussions on the subject of this paper.

REFERENCES

[1] P.S. Alexandroff and P. Urysohn, Mémoire sur les espaces topologiques compacts, *Verh. Koninkl. Akad. Wentesch. Amsterdam* 14 (1929) 1-96.

[2] M.P. Berri, J.E. Porter and R.M. Stephenson, A survey of minimal topological spaces, in: Gen. Topology and its Relations to Modern Analysis and Algebra (Proc. Kanpur Top. Conf., Academic Press, 1970) 93-114.

[3] N. Bourbaki, Topologie générale ch. I et II (third ed.), Paris 1961.

[4] D.K.Burke, Closed mappings, Surveys in General Topology, Edited by G.M. Reed, Academic Press 1980, 1-34.

[5] D.E. Cameron, Maximal and minimal topologies, *Trans. Amer. Math. Soc.* 160 (1971) 229-248.

[6] W.W. Comfort and T. Retta, Generalized Perfect Maps and a Theorem of Juhasz, in: Rings of continuous functions, Lecture Notes in Pure and Applied Mathematics vol. 95 (Marcel Dekker, New York, 1985) 79-102.

[7] D. Dikranjan and E. Giuli, Compactness, minimality and closedness with respect to a closed operator, Proc. CAT-TOP Prague 1989, Edited by J.Adamek and S. MacLane, World Sci. Publ., Singapore, 1990.

[8] D. Dikranjan and E. Giuli, C-perfect morphisms and C-compactness. Preprint.

[9] R. Engelking, General Topology, (PWN, Warszawa, 1977).

[10] I. Fleischer and S.P. Franklin, On compactness and projections, Report 67-20, Dept. of Math. Carnegie Inst. of Tech., 1967

[11] J.P. Franklin, On epi-reflective hulls, *Gen. Topology Appl.*1 (1971) 29-31.

[12] Z. Frolik, On the topological products of paracompact spaces, *Bull. Acad. Pol. Sci. Sér. Math.* 8 (1960) 747-750.

[13] Z. Frolik, Genaralizations of compact and Lindelöf spaces, *Czech. Math. Journ.* 9 (1959) 172-217.

[14] I.S. Gaal, On the theory of (m,n)-compact spaces, *Pacific J. Math.* 8 (1958) 721-734.

[15] E. Giuli, On a-θ-Lindelöf spaces. Preprint.

[16] A.W. Hager, Projections of zero sets (and the fine uniformity in a product), *Trans. Amer. Math. Soc.* 140 (1969) 87-94.

[17] A.W. Hager, Perfect maps and epireflective hulls, *Canadian J. Math.* 27 (1975) 11-24.

[18] A.W. Hager and S. Mrowka, Unpublished manuscript, 1965.

[19] S. Hanai, Inverse images of closed mappings I, *Proc. Japan Acad.* 37 (1961) 298-301.

[20] H. Henriksen and J.R. Isbell, Some properties of compactifications, *Duke Math. Journ.* 25 (1958) 83-106.

[21] H. Herrlich, Fortsetzgarkeit stetiger Abbildungen und Kompaktheitsgrad topologischer Räume, *Math. Zeitschr* 96 (1967), 64-72.

[22] H. Herrlich, Categorical topology, *Gen. Topology Appl.* 1 (1971) 1-15.

[23] H. Herrlich, A generalization of perfect maps, Proc. Third Prague Symposium on General Topology 1971 (Academia, Prague 1972) 187-191.

[24] H.Herrlich, G.Salicrup and G.E.Strecker, Factorizations, denseness,separation, and relatively compact objects, *Topology Appl.* 27 (1987) 157-169.

[25] H. Herrlich and J. van der Slot, Properties which are closely related to compactness, *Indag. Math.* 29 (1967) 524-529.

[26] H. Herrlich and G.E. Strecker, Category Theory, (Heldermann Verlag, Berlin, 1979).

[27] R.E. Hodel and J.E. Vaughan, A note on [a,b]-compactness, *Gen. Topology Appl.* 4 (1974) 179-189.

[28] M. Hušek, The class of k-compact spaces is simple, *Math. Zeithschr.* 110 (1969) 123-126.

[29] T. Ishii, On the Tychonoff functor and w-compactness, *Topology and Appl.* 11 (1980) 173-187.

[30] T. Isiwata, Normality and perfect mappings, Proc. Japan Acad. 39 (1963) 95-97.

[31] J. Joseph, On H-closed spaces, *Proc. Amer. Math. Soc.* 55 (1976) 223-226.

[32] V. Kannan, Reflexive cum coreflexive subcategories in topology, *Math. Ann.* 195 (1972) 168-174.

[33] E.G. Manes, Compact Hausdorff objects, *Gen. Topology Appl.* 4 (1974) 341-360.

[34] A.K. Misra, A topological view of P-spaces, *Gen. Topology Appl.* 2 (1972) 349-362.

[35] N. Noble, Products with closed projections, *Trans. Amer. Math. Soc.* 141 (1969] 381-391.

[36] N. Noble, Products with closed projections II, *Trans. Amer. Math. Soc.* 160 (1971) 169-183.

[37] Ju. Smirnov, On topological spaces, compact in a given interval of powers, *Izv. Akad. Nauk. SSSR Ser. Mat.* 14 (1950) 155-178.

[38] R.M. Stephenson, JR. and J.E. Vaughan, Products of initially m-compact spaces, *Trans. Amer. Math. Soc.* 196 (1974) 177-189.

[39] J.H. Tsai, On E-compact spaces and generalizations of perfect mappings, *Pacific J. Math.* 46 (1973) 275-282.

[40] J.E. Vaughan, Some recent results in the theory of [a,b]-compactness, Proc. TOPO '72, Lecture Notes in Mathematics 378, Berlin 1974, 534-550.

[41] J.E. Vaughan, Some properties related to [a,b]-compactness, *Fund. Math.* 87 (1975) 251-260.

[42] J.E. Vaughan, Convergence, closed projections and compactness, *Proc. Amer. Math. Soc.* 51 (2) (1975) 469-476.

The Wallace Problem and Continuity of the Inverse in Pseudocompact Groups

Douglass L. Grant University College of Cape Breton, Sydney, Nova Scotia, Canada

ABSTRACT

After Bourbaki, a paratopological group is a group G with a topology such that the multiplication map $m: G \times G \to G$ is continuous. For an infinite cardinal α, a paratopological group G is α-totally bounded if, for each neighbourhood V of the identity element, there is a subset A of G with $|A| < \alpha$ such that $G = AV$. For a paratopological group (G, τ), τ' is the topology consisting of the inverses of τ-open sets. Then (G, τ) is a topological group iff any of the following hold for all subsets A of G: (i) $Cl_{\tau}A = Cl_{\tau'}A$; (ii) $Cl_{\tau}A \subseteq Cl_{\tau'}A$; (iii) $Cl_{\tau}A \supseteq Cl_{\tau'}A$.

For a completely regular, pseudocompact paratopological group G, the following statements are equivalent: (1) G is a topological group; (2) G x G is pseudocompact; (3) G is ω-totally bounded; (4) G is ω_1-totally bounded. Every cancellative topological semigroup which is countably compact and weakly first countable in the sense of Nyikos is a topological group. An example of a T_1 pseudocompact paratopological group which is neither regular nor totally bounded is provided.

0. INTRODUCTION

The question of which combinations of algebraic and topological conditions on a semigroup are sufficient to guarantee that it is in fact a group with (jointly) continuous multiplication and inversion has long been of interest to mathematicians. In a posthumous paper of Banach [1948], the following theorem appeared: Let E be a complete metric space with a continuous associative binary operation and two-sided identity, E* its set of invertible elements; then inversion is continuous iff E* is a G_δ subset of E.

K. Numakura [1952] proved that a compact topological semigroup with two-sided cancellation is a topological group. In his survey paper, A.D. Wallace [1955] asked whether "countably compact" could replace "compact" in the above statement, while pointing out that several published assertions to that effect (presumably including Lemma 2 of Gelbaum, Kalisch, and Ohmsted [1951]) were of doubtful validity. Since the statement of Numakura's theorem is clearly false if "locally compact" replaces "compact" (e.g., (0, 1] with usual topology under multiplication), Wallace went on to inquire whether every locally compact topological semigroup which is algebraically a group (a paratopological group, in the terminology of Bourbaki [1966a], p. 297) is a topological group.

In two papers [1957a,b], R. Ellis proved that every locally compact semitopological group is a topological group, a considerably stronger result. W. Zelazko [1965] showed that every complete metrizable semitopological group has continuous inversion. Brand [1982] showed that every locally Čech-complete (and hence every complete metrizable) paratopological group is a topological group. Pfister [1985] proved the following:

PROPOSITION 0.1 Every regular, locally countably compact paratopological group is a topological group.

Raghaven and Reilly [1978], using slightly different terminology, list nine sufficient conditions under which a T_1 paratopological group has continuous inversion.

Progress on the other question has been much more limited. Mukherjea and Tserpes [1972] established that every countably compact, first countable, cancellative topological semigroup is a topological group (hence metrizable and compact). Apparently unaware of their result, Fletcher and Lindgren [1975] used techniques involving quasi-uniformities to show that every first countable, countably compact paratopological group is a topological group.

In this paper, we establish in Section 1 some elementary properties of paratopological groups. In Section 2, we establish necessary and sufficient conditions for a pseudocompact, completely regular paratopological group to be a topological group. We also show by a counterexample that a T_1 pseudocompact paratopological group need not be a topological group. Finally, in Section 3, we generalize the result from Mukherjea and Tserpes [1972] by establishing that every countably compact, cancellative topological semigroup which is weakly first countable in the sense of Nyikos [1981] is a topological group.

1. ELEMENTARY PROPERTIES OF PARATOPOLOGICAL GROUPS

Let (G, τ) be a group endowed with a topology τ. As on p. 297 of Bourbaki [1966a], we say that G is a paratopological group if the group operation $m : G \times G \to G$ is continuous. If the inversion map is also continuous, G is then said to be a topological group. Continuity of the former does not imply that of the latter, as the example of the Sorgenfrey Line shows, even in the presence of the Lindelöf and Baire properties.

Let $V(G, \tau)$, or simply $V(G)$ or $V(\tau)$, where no confusion can arise, denote the filter of τ-neighbourhoods of the identity element of G. As noted by Raghavan and Reilly [1978], $\{V_{-1} : V \in V(\tau)\}$ is the unit neighbourhood filter for a topology τ' on G such that (G, τ') is a paratopological group. Obviously, (G, τ) is a topological group iff $\tau = \tau'$. Let $Cl_\tau A$ denote the closure of the subset A in the topology τ.

The first two results are essentially reformulations of Proposition 1(i)-(ii) of Raghavan and Reilly [1978].

PROPOSITION 1.1 Let G be a paratopological group, $A \subseteq G$. Then $ClA = \cap \{AV^{-1} : V \in V(G)\} = \cap \{V^{-1}A : V \in V(G)\}$.

Proof: Let $x \in \mathrm{Cl}A$. Then, for any $V \in \mathbf{V}(G)$, $xV \cap A \neq \phi$, and so, for some $v \in V$, $a \in A$, we have $xv = a$. Then $x = av^{-1}$, and so $x \in AV^{-1}$.

Conversely, let $x \in \cap\{AV^{-1} : V \in \mathbf{V}(G)\}$, and let U be any open set containing x. By the continuity of multiplication, there is $W \in \mathbf{V}(G)$ such that $xW \subseteq U$. Then $x \in AW^{-1}$, so $x = aw^{-1}$ for some $a \in A$, $w \in W$. But then $xw = a$, and so $a \in A \cap xW \subseteq A \cap U$.

The proof for the other set is similar.

PROPOSITION 1.2 If G is a paratopological group and $U \in \mathbf{V}(G)$, there is $V \in \mathbf{V}(G)$ such that $V^{-1} \subseteq \mathrm{Cl}(V^{-1}) \subseteq U^{-1}$.

Proof: By Proposition 1.1, $\mathrm{Cl}(V^{-1}) = \cap\{V^{-1}W^{-1} : W \in \mathbf{V}(G)\} \subseteq (V^{-1})^2$. It then suffices to take $V \in \mathbf{V}(G)$ such that $V^2 \subseteq U$.

PROPOSITION 1.3 If G is a paratopological group, $Y = Y^{-1} = \mathrm{Cl}X \subseteq G$, and $V \in \mathbf{V}(G)$, then $(XV^{-1}) \cap (V^{-1}X) = Y = (X^{-1}V) \cap (VX^{-1})$.

Proof: The first equality follows directly from Proposition 1.1, and the second by taking inverses.

In particular, Proposition 1.3 holds where $Y = G$.

PROPOSITION 1.4 If (G, τ) is a paratopological group and $A \subseteq G$, then $(\mathrm{Cl}_{\tau'}A)^{-1} = \mathrm{Cl}_\tau(A^{-1})$ and $\mathrm{Cl}_{\tau'}(A^{-1}) = (\mathrm{Cl}_\tau A)^{-1}$.

Proof: Intersections are taken over all $W \in \mathbf{V}(\tau)$:

$$\mathrm{Cl}_\tau(A^{-1}) = \cap\, A^{-1}W^{-1} = [\cap WA]^{-1} = (\mathrm{Cl}_{\tau'}A)^{-1}.$$

$$\mathrm{Cl}_{\tau'}(A^{-1}) = \cap\, A^{-1}W = [\cap W^{-1}A]^{-1} = (\mathrm{Cl}_\tau A)^{-1}.$$

We now establish some criteria for a paratopological group to be a topological group. A function $f : X \to Y$ of topological spaces is said to be almost continuous if, for every $x \in X$ and for every neighbourhood V of $f(x)$, $Cl_X f^{-1}(V)$ is a neighbourhood of x; almost open if $Cl_Y f(V)$ is a neighbourhood of $f(x)$ whenever V is a neighbourhood of x.

THEOREM 1.5 Let (G, τ) be a paratopological group. Then G is a topological group iff the inversion map is almost open (and hence almost continuous).

Proof: Let $U, W \in V(G)$ with $W^2 \subseteq U$. By Proposition 1.2, we have $Cl_\tau(W^{-1}) \subseteq (W^{-1})^2 \subseteq U^{-1}$. Since inversion is almost open, there is $V \in V(G)$ such that $V \subseteq Cl_\tau(W^{-1})$. Then $V \subseteq U^{-1}$, and so $U^{-1} \in V(\tau)$. Then inversion is continuous, and (G, τ) is a topological group.

The following is an extension of Lemma 2 of Brand [1982] which materially reinterprets that result.

THEOREM 1.6 Let (G, τ) be a paratopological group. Then the following statements are equivalent:

(i) (G, τ) is a topological group;

(ii) $Cl_\tau A = Cl_{\tau'} A$ for all $A \subseteq G$;

(iii) $Cl_\tau A \subseteq Cl_{\tau'} A$ for all $A \subseteq G$;

(iv) $Cl_\tau A \supseteq Cl_{\tau'}(A)$ for all $A \subseteq G$.

Proof: By Proposition 1.1, $Cl_{\tau'} A = \cap\{AU : U \in V(\tau)\}$. The equivalence of (i)-(iii) then follows from the result of Brand [1982] mentioned above. Since (ii) obviously implies (iv), it suffices to show only that (iv) implies (iii).

Assume (iv), let $A \subseteq G$ and observe that $Cl_{\tau}(A^{-1}) \supseteq Cl_{\tau'}(A^{-1})$. By Proposition 1.4, we then have $[Cl_{\tau'}A]^{-1} \subseteq [Cl_{\tau}A]^{-1}$, whence $Cl_{\tau'}A \subseteq Cl_{\tau}A$, which is (iii).

If (G, τ) were a paratopological group and not a topological group, it would then follow that G has subsets on which the failure of (iii) and (iv) can be witnessed, either separately or simultaneously. For example, let s denote the Sorgenfrey topology on the real line; i.e., that having as subbasis all sets of form $[a, b)$, with $a < b$. Let $A = [0, 1)$, $B = (-1, 0]$, and $C = (-1, 1)$. Then $A = Cl_{s}A$, but $Cl_{s'}A = [0, 1]$, so (iv) fails. Conversely, $B = Cl_{s'}B$, while $Cl_{s}B = [-1, 0]$, so (iii) fails. Finally, $Cl_{s}C = [-1, 1)$ and $Cl_{s'}C = (-1, 1]$, so both fail together.

2. PSEUDOCOMPACT PARATOPOLOGICAL GROUPS

In the context both of compactifications and of properties of topological groups, pseudocompactness plays the role of a threshold in the hierarchy of compactness properties. We cite a number of examples, all of which will be of significance in what follows. For topological groups, we have the following three results, all from Comfort and Ross [1966]:

(2.01) A pseudocompact topological group is totally bounded.

(2.02) Any product of pseudocompact topological groups is pseudocompact.

(2.03) If G is a pseudocompact topological group, then βG is a topological group.

Even among completely regular topological spaces, there is no strong analogue of (2.02). Instead, we have from Glicksberg [1959] and Theorem 8.12 of Walker [1974]:

(2.04) If X x Y is pseudocompact and completely regular, then $\beta(X \times Y)$ is equal to $\beta X \times \beta Y$.

Furthermore, from Frolik [1960] and Theorem 3.1 of Noble [1969], we have:

(2.05) A completely regular space Y has the property that X x Y is pseudocompact for every pseudocompact space X iff each infinite disjoint family of non-empty open subsets of Y contains an infinite family $\{U_n\}$ such that, for each filter p consisting of infinite subsets of N, $\cap\{Cl[\cup\{U_n : n \in S\}] : S \in p\} \neq \phi$.

We will also require certain results involving the notion of p-compactness investigated by Bernstein [1970] and by Ginsburg and Saks [1975]. Let p be a free ultrafilter on ω. A point z in a topological space X is said to be a p-limit point of the sequence $\{x_n : n < \omega\}$ if, for every neighbourhood V of Z, $\{n : x_n \in V\} \in p$. When they exist, p-limits are unique, and we will write

$z = p\text{-lim}\{x_n\}$. Alternatively, for any completely regular space X and function $f : \omega \to X$, there exists a unique continuous extension $\beta f : \beta\omega \to \beta X$. The space X is then p-compact iff, for every such f, $(\beta f)(p) \in X$.

The following facts are proved in Ginsburg and Saks [1975]:

(2.06) Every p-compact space is countably compact (and so pseudocompact).

(2.07) Let $f : X \to Y$ be a continuous function, $\{x_n : n < \omega\}$ a sequence in X. If $z = \text{p-lim}\ \{x_n\}$, then $f(z) = \text{p-lim}\ \{f(x_n)\}$.

(2.08) The property of being p-compact is preserved in arbitrary Cartesian products and closed subspaces.

(2.09) Every completely regular space X has a maximal p-compact extension $\beta_p X$ such that $X \subseteq \beta_p X \subseteq \beta X$. For every continuous function $f : X \to Y$ with Y p-compact, there exists a continuous function $\beta_p f : \beta_p X \to Y$ extending f. If X is a topological group, then so is $\beta_p X$.

For details of the construction in (2.09), see Ginsburg and Saks [1975], as well as Section 4 of Vaughan [1984] and Section 5 of Stephenson [1968].

It is a natural question whether Proposition 0.1 extends to pseudocompact paratopological groups. The following example shows that the answer to this question is negative, in general.

EXAMPLE 2.1 As in Exercise III. 1.4b of Bourbaki [1966a], let G be the integers with the topology obtained by letting $V(G) = \{\{0\} \cup \{n : n \geq m\} : m > 0\}$. Then the non-empty open sets are exactly those of form $M \cup \{n : n \geq b\}$, where M is an arbitrary subset of $\{n : n \leq b - 1\}$. Then $\{0\}$ is closed, so G is T_1. However, G is not Hausdorff, since no two non-empty open sets are disjoint. The latter fact further implies that any continuous real-valued function on G is constant, and so G is pseudocompact. However, G is plainly not a topological group.

The above example also shows that T_1 paratopological groups need not be Hausdorff, let alone regular nor completely regular. The question of whether a Hausdorff paratopological group must satisfy the latter two properties remains open, however.

This example also shows that (2.01) fails for T_1 paratopological groups, since G cannot be covered by any finite collection of translates of $\{n : n > 0\}$. Whether (2.01) holds for Hausdorff paratopological groups remains open, as is the status of (2.02). This example satisfies the condition in (2.05) vacuously, but is not completely regular. It is not difficult to see, however, that all powers of G are pseudocompact.

The question of whether or not the square of a pseudocompact paratopological group remains pseudocompact will play a significant role in deciding whether that group is in fact topological.

THEOREM 2.2 Let X x Y be pseudocompact and completely regular. Then $\beta_p(X \times Y)$ is equal to $(\beta_p X) \times (\beta_p Y)$.

Proof: Let $\{(x_n, y_n) : n < \omega\}$ be a sequence in $X \times Y$, p-lim $\{x_n\} = x \in \beta_p X$, p-lim $\{y_n\} = y \in \beta_p Y$. By (2.04), $\beta(X \times Y) = (\beta X) \times (\beta Y)$. Then $(x, y) =$ p-lim $\{(x_n, y_n)\}$, since for any neighbourhood $U \times V$ of (x, y), $\{n : x_n \in U\} \cap \{n : y_n \in V\} = \{n : (x_n, y_n) \in U \times V\} \in p$. Using the notation of Ginsburg and Saks [1975], we then have that $(X \times Y)_1 = X_1 \times Y_1$. Proceeding by transfinite induction as in that paper, we then have that $(X \times Y)_\alpha = (X_\alpha) \times (Y_\alpha)$ for each $\alpha < 2\omega$. Since the spaces $X_\alpha \times Y_\alpha$ form a nested collection, we then have

$$\begin{aligned} \beta_p(X \times Y) &= \{(X \times Y)_\alpha : \alpha < 2\omega\} = \{X_\alpha \times Y_\alpha : \alpha < 2\omega\} \\ &= \{X_\alpha : \alpha < 2\omega\} \times \{Y_\alpha : \alpha < 2\omega\} \end{aligned}$$

$= (\beta_p X) \times (\beta_p Y).$

THEOREM 2.3 If S is a topological semigroup with completely regular topology such that S x S is pseudocompact, then $\beta_p S$ is a p-compact topological semigroup. If X is commutative, then so is $\beta_p S$.

Proof: By Theorem 2.2, $\beta_p(S \times S) = (\beta_p S) \times (\beta_p S)$. By (2.09), there exists a continuous binary operation $\beta_p m : \beta_p S \times \beta_p S \to \beta_p S$ which extends the multiplication of S. We repeat the transfinite induction of Ginsburg and Saks [1975], and obtain a transfinite sequence $\{S_\alpha : \alpha < 2^{\aleph}\}$ of subspaces of βS which are topological semigroups containing S.

Let $S_0 = S$. Assume we have constructed semigroups S_α for $\alpha < \beta$ such that:

(i) $\alpha(1) < \alpha(2) < \beta$ implies $S_{\alpha(1)} \subseteq S_{\alpha(2)} \subseteq \beta S$;

(ii) $\alpha(1) < \alpha(2) < \beta$ implies every sequence in $S_{\alpha(1)}$ has a p-limit in $S_{\alpha(2)}$;

(iii) each S_α, $\alpha < \beta$, is a topological semigroup.

Let $H = \{S_\alpha : \alpha < \beta\}$, and let Σ_β be the set of all sequences in H. For each sequence $\sigma \in \Sigma_\beta$, let x_σ be a p-limit point of σ in βS. Then let $S_\beta = H \cup [\, \cup\{x_\sigma : \sigma \in \Sigma_\beta\}]$. Clearly, (i) and (ii) are thereby satisfied, so we show (iii). Since $\beta_p m$ is continuous, it is sufficient to show that $(\beta_p m)(S_\beta \times S_\beta) \subseteq S_\beta$, and that the resulting operation is associative.

Let x, y $\in S_\beta$. If x, y $\in$ H, then, for some $\alpha < \beta$, x, y $\in S_\alpha$, so $(\beta_p m)(x,y) \in S_\alpha$, by the inductive hypothesis.

Suppose y $\in$ H, $x = x_\sigma$ for some $\sigma \in \Sigma_\beta$. Then x = p-lim $\{x_n$:

$n < \omega \}$. But, for each n, $x_n \in S_{\alpha(n)}$ for some $\alpha(n) < \beta$. Similarly, $y \in S_\alpha$, so x_n, $y \in S_{\delta(n)}$, where $\delta(n) = \max \{\alpha, \alpha(n)\}$. Then $(\beta_p m)(x,y) = (\beta_p m)(p\text{-}\lim \{x_n\},y) = p\text{-}\lim \{(\beta_p m)(x_n, y)\} \in S_\beta$.

Finally, if $x = x_\sigma$, $y = x_\tau$ for some $\sigma, \tau \in \Sigma_\beta$, then $x = p\text{-}\lim \{x_n\}$, $y = p\text{-}\lim \{y_n\}$. For each n, $x_n \in S_{\alpha(n)}$, $y_n \in S_{\epsilon(n)}$, for some $\alpha(n)$, $\epsilon(n) < \beta$. Let $\delta(n) = \max \{\alpha(n), \epsilon(n)\}$. Then $(\beta_p m)(x_n, yn) \in S_{\delta(n)}$, and $\{(\beta_p m)(x_n, y_n) : n < \omega \} \in \Sigma_\beta$. Then

$$\beta_p m(x,y) = (\beta_p m)(p\text{-}\lim \{x_n\}, p\text{-}\lim \{y_n\})$$
$$= p\text{-}\lim \{(\beta_p m)(x_n, y_n)\} \in S_\beta.$$

The associative law follows by a computation similar to the above, and the balance of the proof proceeds as in the basic construction from Ginsburg and Saks [1975].

Now, suppose that for each $\alpha < \beta$ we can also satisfy

(iv) each S_α is commutative.

Let x, $y \in S_\beta$. As above we consider the three cases (a) x, $y \in H$; (b) $x = x_\sigma$, $y \in H$; (c) $x = x_\sigma$, $y = x_\tau$, some $\sigma, \tau \in \Sigma_\beta$. The first follows at once from the inductive hypothesis. The second and third cases follow in a manner similar to the proof of the closure property, noting that x_n, y (resp., x_n, y_n) both lie in a common commutative $S_{\delta(n)}$.

THEOREM 2.4 If G is a completely regular paratopological group such that G x G is pseudocompact, then so is $\beta_p G$.

Proof: By Theorem 2.3, $\beta_p G$ is a topological semigroup. One need only show, therefore, that $\beta_p G$ has an identity element and inverses.

Let e be the identity element of G. If $\{x_n : n < \omega \}$ is a sequence in G, then $(\beta_p m)(p\text{-}\lim \{x_n\}, e) = p\text{-}\lim \{m(x_n, e) =$

p-lim $\{x_n\}$, and similarly with the identity on the other side. It then follows that, if e is an identity for each G_α, $\alpha < \beta$, it must also be for G_β. By induction, e is an identity for $\beta_p G$.

Now consider $\{x_n^{-1} : n < \omega\}$, which is also a sequence in G. [Note that this sequence is simply "une comme les autres"; we do not claim that the inversion function is continuous.] Then $(\beta_p m)(\text{p-lim}\,\{x_n\}, \text{p-lim}\,\{x_n^{-1}\}) = \text{p-lim}\,\{(\beta_p m)(x_n, x_n^{-1})$ $= \text{p-lim}\,\{e\} = e$. Using the same technique, one can then show that, if all elements of G_α, $\alpha < \beta$, have inverses, so does every element of G_β. Then every G_β, $\beta < 2^{\mathfrak{c}}$, is a paratopological group, and so then is $\beta_p G$.

We now turn to our main theorem of the section.

Extending the notation of Grant and Comfort [1986] in a natural way, we say that a paratopological group (G, τ) is α-totally bounded if, for every $U \in V(\tau)$, there is $S \subseteq G$ with $|S| < \alpha$ such that $G = US$. A group which is ω-totally bounded will simply be called totally bounded.

THEOREM 2.5 Let G be a pseudocompact, completely regular paratopological group. Then the following statements are equivalent:

- (i) G is a topological group;
- (ii) G x G is pseudocompact;
- (iii) G is totally bounded;
- (iv) G is ω_1-totally bounded.

Proof: It is immediate from (2.01) that (i) implies (iii), and from (2.02) that (i) implies (ii). That (iii) implies (iv) is trivial. We therefore show that (ii) and (iv) each imply (i).

Assume (ii). It then follows from Theorem 2.4 that $\beta_p G$ is a p-compact paratopological group. By (2.06) and Proposition 0.1, $\beta_p G$ is a topological group. But G is a subgroup of $\beta_p G$, and so itself a topological group.

Assume (iv), and let $U \in V(G)$. By Proposition 1.2, there is $V \in V(G)$ such that $V^{-1} \subseteq Cl_\tau V^{-1} \subseteq U^{-1}$. Since G is ω_1-totally bounded, there is $X \subseteq G$ such that $|X| \leq \omega$ and $G = XV = V^{-1}X^{-1}$. By Exercise IX.5.10 of Bourbaki [1966b], G is a Baire space, and so $Cl_\tau(V^{-1}x^{-1})$ contains an open set for some $x \in X$. Then so does $Cl_\tau(V^{-1})$, and U^{-1}. Then $\tau' \subseteq \tau$. Symmetrically, $\tau \subseteq \tau'$, and so G is a topological group.

We can now establish an extensive list of conditions sufficient for a pseudocompact paratopological group to be a topological group. With apologies, we must now define the less familiar of our terms. Sources cited are not necessarily the first where the terminology appeared.

DEFINITION 2.6 (a) (Ginsburg and Saks [1975]) A space X is p-pseudocompact for a free ultrafilter p on ω if, for every sequence $\{S_n\}$ of non-empty open sets in X, there is a point z such that, for every neighbourhood W of z, $\{n : S_n \cap W \neq \phi\} \in p$.

(b) (Noble [1969]) A space X is a P-space if intersections of countable collections of open sets are open.

(c) (Noble [1969]) A space X is a k_R-space if the real-valued continuous functions on X are exactly those whose restrictions to compact subsets of X are continuous.

(d) (Stephenson [1968]) If K is an infinite cardinal, a space X is K-pseudocompact if, whenever **F** is a filter base on X containing at most K sets and each such set in **F** is a union of cozero sets, then **F** has an adherent point.

(e) (Scarborough and Stone [1966]) A space X is an H(i) space if every open filter base has an adherent point.

COROLLARY 2.7 Let G be a pseudocompact, completely regular paratopological group. Then G is a topological group if any of the following conditions is satisfied:

(i) G is p-pseudocompact;

(ii) G is a P-space;

(iii) G is a $k_{\mathbb{R}}$-space;

(iv) G is Lindelöf;

(v) G is separable;

(vi) G is first countable;

(vii) G is sequential;

(viii) G is sequentially compact;

(ix) G is K-pseudocompact and every point has a fundamental system of neighbourhoods containing at most K sets;

(x) G is an H(i) space;

(xi) Each point of G is a G_δ;

(xii) G is a member of any class **C** of pseudocompact paratopological groups for which any bijective homomorphism from one group in **C** to another with closed graph is continuous, and is therefore a topological isomorphism.

Proof: The sufficiency of (i) follows from Ginsburg and Saks [1975], where it is shown that p-pseudocompactness is preserved in arbitrary products. Similarly, it is shown in Noble [1969] that pseudocompactness is preserved in finite products of pseudocompact spaces satisfying (ii) or (iii). Each of (iv) and (v) implies that G is ω_1-totally bounded, the latter using Proposition 1.3. The sufficiency of each of (vi)-(x), inclusive, to establish that G x G is pseudocompact is discussed in Stephenson [1969], with attribution to several sources. Glicksberg established the sufficiency of (xi) as Theorem 4(b) of his 1959 paper.

As for (xii), one can observe that the graph $\{(x, x^{-1}) : x \in G\}$ of the inversion map is the inverse image of the (closed singleton) identity element under the continuous multiplication map. The inversion function can be considered a homomorphism from G to the group having the same underlying set and topology, and operation $x * y = yx$.

REMARK. Condition (xii) seems to offer a new approach to the problem. Unfortunately, the entire class of all pseudocompact, completely regular paratopological groups does not satisfy the condition given therein. In the context of topological groups, it is shown in Comfort and Robertson [1988] that examples abound of pseudocompact topologies τ, μ on a common group G such that
$\tau \subset \mu$, and (G, τ), (G, μ) are both topological groups. The identity function $(G, \mu) \to (G, \tau)$ is then continuous and so has a closed graph, but is plainly not a topological isomorphism.

3. COUNTABLY COMPACT, WEAKLY FIRST COUNTABLE, CANCELLATIVE TOPOLOGICAL SEMIGROUPS

As noted in the introduction, Mukherjea and Tserpes [1972] showed that every first countable, countably compact, cancellative topological semigroup is a topological group. Such semigroups are then metrizable (Theorem 1.8 of Comfort [1984]) and so compact by Theorem IX.5.3 and Corollary XI.3.4 of Dugundji [1966].

In Mukherjea and Tserpes [1972], the following terminology is introduced: If a topological semigroup S is left simple ($Sx = S$ for all $x \in S$) and contains an idempotent e, then S is called a pre-topological left group. The following results are then Lemma 3(b) and Theorem 4 of that paper:

PROPOSITION 3.1 A pretopological left group S is topological iff any maximal subgroup (in particular eS) is topological.

PROPOSITION 3.2 If S is sequentially compact with right cancellation, then S is a pretopological left group.

Nyikos [1981] defined a space X to be weakly first countable if, for each point $x \in X$, it is possible to assign a sequence $\{B(n,x) : n < \omega\}$ of subsets of X containing x in such a way that $B(n, x) \subseteq B(n + 1, x)$ and so that a set U is open iff for each $y \in U$, there exists $n < \omega$ such that $B(n, y) \subseteq U$. By Lemma 1 of that paper, we then have the following.

PROPOSITION 3.3 Every weakly first countable space is sequential.

The following appears, for example, as Proposition 5.3 of Vaughan [1984].

PROPOSITION 3.4 Every countably compact, sequential space is sequentially compact.

We can now establish the following generalization of the main result from Mukherjea and Tserpes [1972].

THEOREM 3.5 Every weakly first countable, countably compact, cancellative topological semigroup S is a topological group.

Proof: By Propositions 3.3 and 3.4, S is sequentially compact, and so, by Proposition 3.2, S is a pretopological left group with idempotent e. Since multiplication is continuous, the maximal subgroup eS is countably compact, and so a topological group by Proposition 0.1. By Proposition 3.1, S is then a topological group.

The original question of Wallace remains open, but we can cite an example which shows that pseudocompact, cancellative topological semigroups are not necessarily topological groups.

EXAMPLE 3.6 Let T be the circle group, U its torsion subgroup, α any cardinal greater than ω, b any irrational number, $H = \{(t_\eta) \in T^\alpha : |\{\eta : t_\eta \in U\}| \le \omega\}$, $z = (z_\eta)$ the element of T^α such that $z_\eta = \exp(2\pi bi)$ for each $\eta < \alpha$. As observed in Kister [1962], $\beta H = T^\alpha$. Then let $E = H\{z^n : n \ge 0\}$. Then E is clearly a topological semigroup (indeed, a monoid) which is not a group. Moreover, as observed in Kister [1962], E is pseudocompact since it contains the dense countably compact subgroup $\{(t_\eta) : |\{\eta : t_\eta \neq 1\}| \le \omega\}$.

However, E is not countably compact. Suppose $\{r_j\}$ is a sequence of rational numbers converging to an irrational number which is not a rational multiple of b. Now each r_j determines an element $p_j \in H$, where each coordinate of p_j is $\exp(2\pi r_j i)$, and $\{p_j\}$ clearly has no limit point in E.

ACKNOWLEDGEMENTS

Thanks are due to Neil Shell for pointing out the reference Banach [1948], to Jane Day and Bob Stephenson for helpful consultations, and to Wistar Comfort and Neil Hindman for pointing out an error in an earlier version of Theorem 2.2. The financial assistance of the Natural Science and Engineering Research Council of Canada (up to 1986) and, subsequently, of the Committee for the Evaluation of Research Proposals of the University College of Cape Breton is also gratefully acknowledged.

BIBLIOGRAPHY

[1948] S. Banach, Remarques sur les groupes et les corps métriques, Studia Math., v. 10(1948), 178-181.

[1970] A.R. Bernstein, A new kind of compactness for topological spaces, Fund. Math., v. 66(1970), 185-193.

[1966a] N. Bourbaki, General Topology, Volume I, Addison-Wesley, Reading, 1966.

[1966b] N. Bourbaki, General Topology, Volume II, Addison-Wesley, Reading, 1966.

[1982] N. Brand, Another note on the continuity of the inverse, Archiv der Math., v. 39(1982), 241-5.

[1984] W.W. Comfort, Topological groups, in "Handbook of Set-Theoretic Topology", p. 1143-1263.

[1988] W.W. Comfort and L.C. Robertson, Extremal phenomena in certain classes of totally bounded groups, Diss. Math. (Rozprawy Mat.), v. 272, Warsaw, 1988.

[1966] W.W. Comfort and K.A. Ross, Pseudocompactness and uniform continuity in topological groups, Pac. J. Math., v.16 (1966), 483-496.

[1966] J. Dugundji, "Topology", Allyn and Bacon, Boston, 1966.

[1957a] R. Ellis, A note on the continuity of the inverse, Proc. Amer. Math. Soc., v. 8(1957), 372-3.

[1957b] R. Ellis, Locally compact transformation groups, Duke Math. J., v. 24(1957), 119-125.

[1975] P. Fletcher and W.F. Lindgren, Some unsolved problems concerning countably compact spaces, Rocky Mtn. J. Math., v. 5(1975), 95-106.

[1960] Z. Frolik, The topological product of two pseudocompact spaces, Czech. Math. J., v. 10(85) (1960), 339-349.

[1959] I. Glicksberg, Stone-Cech compactification of products, Trans. Amer. Math. Soc., v. 90(1959), 369-382.

[1986] D.L. Grant and W.W. Comfort, Products and cardinal invariants of minimal topological groups, Can. Math. Bull., v. 29(1986), 44-49.

[1951] B. Gelbaum, G.K. Kalisch and J.M.H. Ohmsted, On the embedding of topological semigroups and integral domains, Proc. Amer. Math. Soc., v. 2(1951), 807-821.

[1975] J. Ginsburg and V. Saks, Some applications of ultrafilters in topology, Pac. J. Math., v. 57(1975), 403-418.

[1984] "Handbook of Set-Theoretic Topology", ed. K. Kunen and J. E. Vaughan, North-Holland, 1984.

[1962] J.M. Kister, Uniform continuity and compactness in topological groups, Proc. Amer. Math. Soc., v. 13(1962), 37-40.

[1972] A. Mukherjea and N.A. Tserpes, A note on countably compact semigroups, J. Aust. Math. Soc. v. 13(1972), 180-184.

[1969] N. Noble, Countably compact and pseudocompact products, Czech. Math. J., v. 19(94)(1969), 390-397.

[1952] K. Numakura, On bicompact semigroups, Math J. Okayama Univ., v.1 (1952), 99-108.

[1981] P. Nyikos, Metrizability and the Frechet-Urysohn property in topological groups, Proc. Amer. Math. Soc., v. 83 (1981), 793-801.

[1985] H. Pfister, Continuity of the inverse, Proc. Amer. Math. Soc., v. 95 (1985), 312-4.

[1978] T.G. Raghavan and I.L. Reilly, On the continuity of group operations, Indian J. pure appl. Math., v. 9 (1978), 747-752.

[1966] C.T. Scarborough and A.H. Stone, Products of nearly compact spaces, Trans. Amer. Math. Soc., v. 124(1966), 131-147.

[1968] R.M. Stephenson, Jr., Pseudocompact spaces, Trans. Amer. Math. Soc., v. 134(1968), 437-448.

[1984] J.E. Vaughan, Countably compact and sequentially compact spaces, in "Handbook of Set-Theoretic Topology", 569-602.

[1974] R.C. Walker, "The Stone-Čech Compactification", Ergebnisse der Math. und ihrer Grenz., v. 83, Springer-Verlag, 1974.

[1955] A.D. Wallace, The structure of topological semigroups, Bull. Amer. Math. Soc., v. 61(1955), 95-112.

[1965] W. Zelazko, Metric generalizations of Banach algebras, Rozprawy Math., v. 47, Warsaw, 1965.

Topologies and Cotopologies Generated by Sets of Functions: A Preview

Melvin Henriksen Harvey Mudd College, Claremont, California

1. Introduction

In what follows, the reader will be given a preview of some joint research efforts by R. Kopperman, R.G. Woods and me whose central theme is concerned with describing conditions under which disjoint or separated subspaces of a topological space can be completely separated by a continuous [0,1]-valued or {0,1}-valued function, and deriving consequences of this latter property. We apply this mainly, but not exclusively to the study of subspaces of F-spaces (as defined in Chap. 14 of [GJ]) and to spaces of minimal prime ideals of certain kinds of commutative rings.

In this class of spaces, weakly Lindelöf spaces are C^*-embedded, and with more restrictions, they also have compact closures. They are defined in terms of special bases which in turn are determined by certain kinds of betweeness conditions on families of functions into [0,1] or {0,1}. The additional restrictions involve the use of cotopologies; where the *cotopology* derived from a base *B* for a space is the topology generated by the complements of the closures of the members of *B*.

This study is motivated by two considerations. The first is to try to determine to what extent it is possible to derive properties of topological spaces from the existence of bases derived from special collections of functions with values in [0,1] or {0,1}. This paper may be a pioneering effort along these lines and follows a path begun in [HK]. What is more important to me (but not necessarily to my collaborators) is the application to spaces of minimal prime ideals of a class of commutative rings that include uniformly closed ring of continuous functions. My interest in spaces of minimal prime ideals dates back to the 1960's (see [HJ] and [Ki]) and this paper is part of a continuing effort to characterize such spaces topologically and to answer open questions about their properties. The fact that these techniques apply also to subspaces of F-spaces may be regarded as both a bonus and a way of showing that some of the conditions on bases are not strong enough to characterize spaces of minimal prime ideals of uniformly closed rings of continuous functions. Terminology not defined below may be found in [GJ] or [PW].

Proofs of theorems have been omitted. A number of unsolved problems are posed.

2. Topologies generated by families of functions

Throughout, L will denote either the space [0,1] with its usual order and topology or the space {0,1} with the discrete topology. If X is any set, then L^X denotes the (complete) lattice of functions from X to L under the usual pointwise lattice operations $\vee$ and $\wedge$. We generalize notations used for elements of L^X to subsets; e.g., $A \le B$ means $a \le b$ whenever $a \in A$ and $b \in B$, and $\underline{1} - B = \{\underline{1} - b : b \in B\}$, where, as usual, $\underline{0}$ and $\underline{1}$ denote the appropriate constant functions in L^X. Throughout, H will denote a subfamily of L^X, and if X is a topological space, C(X,L) will denote the sub-lattice of continuous functions on X to L.

2.1 Definitions

(a) A subset of X of the form $\{x \in X : f(x) > 0\}$ for some f in H is called an *H-cozeroset of* X, and is denoted by cz f. As usual, X \ cz f is denoted by Z(f).

(b) H is called *basic* if it satisfies:

(i) $B(H) = \{\text{cz } f : f \in H\}$ is a base for a topology $\tau(H)$ on X,

(ii) if $x \in V \in \tau(H)$, there exist $g, h \in H$ such that

$x \in \text{cz } g \subset Z(\underline{1} - h) \subset V$, and

(c) The topology $\tau^*(H)$ generated by $\{X \setminus \text{Cl}(\text{cz } h) : h \in H\} = \{\text{int } Z(h) : h \in H\}$ is called the *cotopology on X derived from H.*

(d) A basic family H is called *regular* if $\text{cz } (\underline{1} - f) \in \tau(H)$ whenever $f \in H$.

(e) A basic family is called *completely regular* if $H \subset C(X,L)$, where X has the topology $\tau(H)$.

(f) H is *complemented* if $h \in H$ implies $(\underline{1} - h) \in H$.

(g) The pair (Y,Z) is *completely-H-separated* if there is an h in H such that $h[Y] = \{0\}$ and $h[Z] = \{1\}$.

Each of these concepts has a different meaning according as L is $[0,1]$ or $\{0,1\}$.

If $W \subset X$, let χ_W denote the characteristic function of W.

Many topological properties can be described with the aid of basic families of functions. For example:

2.2 It is not too difficult to verify that a topology τ on a set X is (completely) regular if and only if there is a (completely) regular $H \subset [0,1]^X$ for which $\tau = \tau(H)$.

(In the completely regular case, one need only consider $H = C(X,[0,1])$. The regular case needs more argument.)

2.3 If (X,σ) is any topological space and $H = \{\chi_W : W \in \sigma\}$, then $H \subset \{0,1\}^X$ and $\sigma = \tau(H)$.

2.4 <u>Definition</u> Suppose m,n are cardinal numbers, and $H \subset L^X$. If whenever $C,D \subset H$, $|C| < m$, $|D| < n$, and $C < \underline{1} - D$, there is an f in H such that $C < f < \underline{1} - D$, then H is said to be *(m,n)-nested.* If H is (m,n)-nested for every cardinal n, then it is said to be *(m,∞)-nested.*

If m is a cardinal number, then m^+ denotes its immediate successor.

2.5 It is not difficult to see that a basic family H is (m,∞)-nested if and only if it is $(m,|H|^{+})$-nested.

2.6 Comments and Examples

(a) Recall from [CN] that a zero-dimensional Hausdorff space is *m-basically disconnected*, where m is an infinite cardinal, if the closure of the union of fewer than m clopen sets is clopen. (A different definition is given in [Sw] for arbitrary Tychonoff spaces which agrees with this one in strongly zero-dimensional spaces.) If $m = \omega$, this holds automatically, and if $m = \omega^{+}$, then such spaces are called *basically disconnected.* It is easy to see that a space X is m-basically disconnected if and only if $C(X,\{0,1\})$ is (m,∞)-nested.

More generally, if X has an m-basically disconnected compactification αX, and $H = \{f|_X : f \in C(\alpha X,\{0,1\})\}$, then H is a completely regular complemented (m,∞)-nested subset of $\{0,1\}^X$, and any space X with a completely regular complemented (m,∞)-nested subfamily of $\{0,1\}^X$ has an m-basically disconnected compactification.

X is called *extremally disconnected* if the closure of each of its open sets is open, or equivalently, if X is m-basically disconnected for every infinite cardinal m. It is shown in Example 1 of [Ko] that not every (dense) subspace of a basically disconnected space need be basically disconnected.

(b) Recall from [GJ] or [PW] that X is called an F-*space if* each cozeroset of X is C^{*}-embedded in X. It was shown in [Se] that X is an F-space if and only if whenever C and D are countable subsets of $C^{*}(X)$ such that $C < D$, there is an $f \in C^{*}(X)$ such that $C < f < D$. From this it follows easily that X is an F-space if and only if $C(X,[0,1])$ is (ω^{+},ω^{+})-nested.

More generally, if X is a subspace of Y and Y is an F-space, then so is its Stone-Cech compactification βY (by 14.25 of [GJ]) and also the C^{*}-embedded subspace $S = Cl_{\beta Y} X$. Then the family H of restrictions to X of the elements of $C(S,[0,1])$ is a completely regular complemented (ω^{+},ω^{+})-nested subfamily of $[0,1]^X$.

Conversely, if $[0,1]^X$ has a subfamily with these latter properties, then X is a dense subspace of a compact F-space.

In Example 3 of [Ko], an extremally disconnected space (and hence an F-space) is exhibited that has a closed subspace that is not an F-space. (This corrects an unessential error in [Se]).

In Section 3 we will consider spaces with a completely regular family of functions that is (ω^+,∞)-nested which fails to be complemented. These include the space of minimal prime ideals of the ring C(X) of continuous functions on a (Tychonoff) space X.

2.7 <u>Definition</u> Suppose X is a topological space and p is an infinite cardinal. X is called *p-feebly compact* if every open cover of X has a subfamily of cardinality less than p whose union is dense in X. An ω^+-feebly compact space is usually said to be *weakly Lindelöf*.

The proof of the next theorem is long and complicated. Recall that two subsets of a space are said to be *separated* if each of them is disjoint from the closure of the other. If m is any infinite cardinal, a topological space (X,τ) is said to be *m-additive* if the intersection of fewer than m members of τ is in τ. All topologies are ω-additive and an ω^+-additive topological space is usually called a P-space; see also [CN].

2.8 <u>Theorem</u> Suppose H is a completely regular (m,∞)-nested, (m^+,∞)-nested subfamily of L^X, and let $\tau(H)$ be m-additive. Then m^+-feebly compact separated subspaces of X are completely H-separated.

2.9 <u>Corollary</u> If X and H are as above, then m^+-feebly compact subspaces of X are C^*-embedded.

The corollary follows easily from the theorem if one recalls from 1.17 of [GJ] that a subset S of X is C^*-embedded if completely separated subsets of S are completely separated in X. If there is an f $\in C(X,[0,1])$ such that $f[A] = \{0\}$ and $f[B] = \{1\}$, then $\{x \in S : f(x) < 1/3\}$ and $\{x \in S : f(x) > 2/3\}$ are separated m^+-feebly compact subspaces of X, the first of which contains A, and the second contains B. So A and B are completely separated in X by the theorem and hence S is C^*-embedded in X.

2.10 Applications

(a) In the case $m = \omega$, for any completely regular subfamily H of L^X, $\tau(H)$ is automatically ω-additive, and H is (ω,∞)-nested if H is an upper semilattice. So Corollary 2.9 yields that if H is basic and is an upper semilattice contained in $C(X,[0,1])$ that is (ω^+,ω^+)-nested, then weakly Lindelöf subspaces of $(X,\tau(H))$ are C^*-embedded in X. Indeed, by 2.6(b) if $H = C(X,[0,1])$, then H is (ω^+,ω^+)-nested if and only if X is an F-space. See also [CHN].

(b) In order to apply Corollary 2.9 to an m-basically disconnected space if $m > \omega$, one must assume that X is a P_m-space, i.e., that an intersection of less than m open sets is open. Such spaces are a special part of the class of m-basically disconnected spaces, and this strong hypothesis cannot be dropped completely. For, since $\beta\omega$ is extremally disconnected, it is m-basically disconnected for every (infinite) m, and every subspace of $\beta\omega$ is m-feebly compact for sufficiently large m, and one may not conclude that every subspace of $\beta\omega$ is C^*-embedded. (See [SW]).

In the next section, it is assumed that $m = \omega$, and noncomplemented completely regular families of functions play a strong role.

3 Applications to spaces of minimal prime ideals and subspaces of F-spaces

In this part of the paper, the theory developed above is applied to give simpler proofs of known results and to derive new results about the class of spaces just described.

3.1 Definition Suppose $H \subset L^X$ is a completely regular subcollection for some topological space X.

(a) If $L = \{0,1\}$ and H is (ω^+,∞)-nested, then $B(H) = \{\text{coz } f : f \in H\}$ is called a *pretty base for* X.

(b) If $L = [0,1]$ and H is (ω^+,ω^+)-nested, then $B(H) = \{\text{coz } f : f \in H\}$ is called a *Seever base for* X.

As noted in Section 2, if X is an F-space and $H = C(X,[0,1])$, then $B(H)$ is a complemented Seever base.

Clearly, if X has a pretty base $B(H)$, where H is an (ω^+,∞)-nested subset of $\{0,1\}^X$, then the elements of $B(H)$ are cozerosets of two-valued elements of $C(X,[0,1])$, so every pretty base may be regarded as a Seever base.

Next, some background on spaces of minimal prime ideals is given. See [HJ], [Hu], or [HK] for more details.

3.2 <u>Definitions and Remarks</u> A commutative ring R whose only nilpotent element is 0 is called *reduced.* If $S \subset R$ let $A(S) = \{a \in R : aS = \{0\}\}$, and let $A(s) = A(\{s\})$ if $\{s\}$ is a singleton. For R reduced let mR denote the set of minimal prime ideals of R. If $S \subset R$, let $h(S) = \{P \in mR : S \subset P\}$, let $h^c(S) = mR \setminus h(S)$, and make the same sort of abbreviations in case of singletons as above. As noted in [HJ], $h^c(a) = h(A(a))$ for any $a \in R$. The *hull-kernel topology* on mR is the topology whose base is

$$\{h^c(a) : a \in R\}.$$

It is assumed always in the sequel that mR has this topology unless the contrary is stated explicitly. As noted in [HJ], mR is a zero-dimensional Hausdorff space.

By the *countable annihilator condition on* R is meant:

(CAC) $\{A(a) : a \in R\}$ is closed under countable intersection.

It is shown in [HJ] that the (reduced) ring C(X) satisfies CAC. More generally, any uniformly closed subalgebra of C(X) satisfies CAC.

It follows from Lemma 4.2 of [HJ] that if R is reduced and satisfies CAC, then $\{h^c(a) : a \in R\} = \{h(A(a)) : a \in R\}$ is a pretty base for mR. It is shown also in Section 6 of [HK] that if R is reduced, then the ring R[[x]] of all formal power series with coefficients in R satisfies CAC.

Spaces with a pretty base are also discussed in [He], where the definition differs superficially from the one given here. It follows also from 3.4 of [HJ] that for reduced CAC rings, the pretty base $\{h^c(a) : a \in R\}$ is complemented if and only if mR is compact.

If H is a completely regular subfamily of L^X, and Y is a subspace of X, let $H_Y = \{f|_Y : f \in H\}$. Then $B(H_Y)$ is called the *trace of* $B(H)$ *on* Y.

The next two propositions enlarges the class of spaces that have a Seever base or a pretty base and identifies precisely the weakly Lindelöf spaces with these properties.

3.3 Proposition The trace of a Seever base on any subspace is a Seever base and the trace of a pretty base on any dense or open subspace is a pretty base.

3.4 Proposition Suppose X is a weakly Lindelöf space.

(a) X has a Seever base if and only if it is an F-space.

(b) X has a pretty base if and only if it is basically disconnected.

In Example 1 of [Ko], it is shown that a dense subspace of a basically disconnected space need not be basically disconnected. Thus not every space with a pretty base need be basically disconnected. Indeed, in [DHKV] an example of a space of minimal prime ideals of a ring C(X) that fails to be basically disconnected is given. The compact, (and hence weakly Lindelöf) subspace $\beta\omega \setminus \omega$ of the extremally disconnected space $\beta\omega$ is not basically disconnected by 6W of [GJ], and hence cannot have a pretty base by Proposition 3.4(b). So not every (compact) subspace of a space with a pretty base has a pretty base.

A base *B* of a space X is called *proper* if X is not in *B*. A nonempty closed subset K of a space X is called a *P-set* if every G_δ containing K has K in its interior; equivalently, if whenever coz f is a cozeroset disjoint from K, so is its closure.

3.5 Theorem If X has a pretty base , the following are equivalent.

(a) X has a proper pretty base.

(b) X fails to be weakly Lindelöf.

(c) $\beta X \setminus X$ contains a compact P-set of βX.

3.6 Corollary If X has a pretty base, then either X is weakly Lindelöf or $\beta X \setminus X$ contains a compact P-set of βX.

3.7 Remark In Example 1 of [DF] an F-space X is exhibited that is not weakly Lindelöf while $\beta X \setminus X$ contains no P-set of βX. So the conclusion of the corollary does not follow just from the assumption that X has a Seever base.

A space X is called an *almost-P-space* if each of its zerosets has nonempty interior. The next theorem also appears in [He].

3.8 Theorem A weakly Lindelöf subspace S of an almost-P-space with a pretty base is C-embedded. If S is also realcompact, then it is closed.

As noted in the introduction, part of the purpose of this article is to study the space of minimal prime ideals of a reduced commutative ring in which CAC holds. Such spaces are shown to be countably compact in [HJ]; indeed closures of weakly Lindelöf subspaces in them are known to be compact--see [DHKV 2]. Not every space with a pretty base has this latter property as is witnessed by an uncountable discrete space. In the set mR of minimal prime ideals of a commutative reduced ring R, $\{h(a) : a \in R\}$ generates a quasicompact topology as is shown in Section 5 of [HK]. Next, this latter is placed in a more general context. As in Section 2, if $\mathcal{B}$ is a base for a topological space (X,τ), then the topology $\tau^* = \tau^*(\mathcal{B})$ generated by $\{X \setminus Cl\ B : B \in \mathcal{B}\}$ is called the *cotopology derived from* $\mathcal{B}$. If every open cover of $(X,\tau^*(\mathcal{B}))$ has a finite subcover, then $\mathcal{B}$ is called a *cocompact* base.

3.9 Theorem Suppose X has a cocompact Seever base.

(a) If Y is a weakly Lindelöf subspace of X, then $Cl_X\ Y$ and βY are homeomorphic.

(b) If X is locally weakly Lindelöf, then it is locally compact.

In 5.9 of [HJ] it is shown that no point of the space of minimal prime ideals of $C(\beta\omega \setminus \omega)$ has a compact neighborhood. The next result follows immediately from the Theorems 3.8 and 3.9.

3.10 Corollary An almost P-space with a cocompact pretty base is finite.

A Tychonoff space X is called an *F'-space* if disjoint cozerosets of X have disjoint closures. Every F-space is an F'-space, a normal F'-space is an F-space, and if every point of X has a neighborhood that is an F'-space, then X is an F'-space. Not every F'-space is locally an F-space. For more background and details, see [CHN] and [D1].

3.11 <u>Theorem</u> Suppose every point of a space X has a weakly Lindelöf neighborhood.

(a) If X has a Seever base, then X is locally an F-space and hence is an F'-space.

(b) If X has a cocompact pretty base then X is locally compact and basically disconnected.

It is shown in 3.14 below that the converse of 3.11(b) is false.

Recall from [HW] that a locally compact space is called *substonean* if disjoint σ-compact open subspaces have disjoint compact closures and is called a *Rickart space* if σ-compact open subspaces have compact clopen closures. It is shown in [HW] that a locally compact space is a substonean (resp. Rickart) space if and only if it is a compact F- (resp. basically disconnected) space or its one-point compactification $\alpha X = X \cup \{\infty\}$ is an F- (resp. basically disconnected) space and ∞ is a P-point of βX. That paper contains also an example of a substonean space that is not an F-space. On the other hand, every Rickart space is basically disconnected.

3.12 <u>Theorem</u> If X is a substonean space, then X has a Seever base and for any weakly Lindelöf subspace S of X, Cl_X S and βS are homeomorphic F-spaces. Conversely, if X is locally weakly Lindelöf and has a cocompact Seever base, then X is substonean.

The corresponding result relating Rickart spaces and spaces with a pretty base is:

3.13 <u>Theorem</u> If X is a Rickart space, then X has a pretty base and, for any weakly Lindelöf subspace S, Cl_X S and βS are homeomorphic. Conversely, if X is locally weakly Lindelöf and has a cocompact pretty base, then X is a Rickart space.

A Tychonoff space X is called *almost compact* if $|\beta X \setminus X| \le 1$.

3.14 Examples

(a) Example E of [D1] exhibits an almost compact F'-space X that is not a subspace of an F-space and hence is not substonean.

(b) The almost compact extremally disconnected space $\beta\omega \setminus \{p\}$, where p is a nonisolated point of $\beta\omega$, cannot have a cocompact pretty base by Theorem 3.13 since $\beta\omega$ has no P-points. (For, $\beta\omega \setminus \omega$ is a G_δ of $\beta\omega$ with empty interior). It follows that there is no commutative reduced ring R that satisfies CAC whose space mR of minimal prime ideals is homeomorphic to $\beta\omega \setminus \{p\}$.

The next result is related to Theorem 3.12.

3.15 Theorem If X is a locally compact noncompact space and αX is its one-point compactification, the following are equivalent.

(a) αX is an F-space.
(b) Disjoint σ-compact open subspaces of X have disjoint closures, at least one of which is compact.

The last section of this paper is devoted primarily to examples and open problems.

4. An example of Alan Dow and some open problems

Theorems 3.12 and 3.13 provide characterizations of locally weakly Lindelöf (Tychonoff) spaces that have a cocompact Seever (resp. pretty) base. There are many open problems about such spaces that fail to be locally compact.

There seemed to be circumstantial evidence in support of the conjecture that if a space X has a cocompact pretty base then it has a compactification that is an F-space. Alan Dow supplied us with counter-examples dependent in part on some set-theoretic assumptions. They are described next and their proofs are far from easy.

Martin's Axiom is equivalent to the assertion:

(MA) If X is a ccc-space, then an intersection of less than c dense open subspaces is dense in X.

The continuum hypothesis (CH) implies MA, but not conversely.

A nonempty closed subspace K of a space X is called a P_c-*set* if an intersection of less than c open sets containing K has K in its interior. Clearly, every P_c-set is a P-set, and the converse holds under CH. In [Ku], K. Kunen showed that if MA holds then $\beta\omega \setminus \omega$ contains a nowhere dense P_c-set K and a continuous irreducible map of K onto the Cantor set. Since the preimage under a continuous irreducible map of a dense subspace is dense, it follows that K is separable. With the aid of this latter, Alan Dow was able to show that the space of minimal prime ideals of $\beta\omega \setminus \omega$ has no F-space compactification. Indeed, it follows that:

4.1 <u>Theorem</u> (Dow) If MA holds, then the space of minimal prime ideals of $C(\beta D \setminus D)$ has no F-space compactification for any infinite discrete space D.

It was announced recently in [FSZ] that there is a model in ZFC in which $\beta\omega \setminus \omega$ fails to contain a nowhere dense ccc P-set.

In [DHKV2], an example is given in ZFC of a Tychonoff space X such that mC(X) fails to be basically disconnected. Dow has also shown that for this space X, mC(X) has no embedding into a compact F-space. So having a cocompact pretty base is not enough to guarantee the existence of an F-space compactification.

I close with some unsolved problems which reflect the state of ignorance in this area. Some of them appear also in [He].

The first problem is the most important of them. Recall from [HJ] that for any Tychonoff space Y, mC(Y) and $mC(\beta Y)$ are homeomorphic.

P.1 Suppose X has a cocompact pretty base. Is there a (compact) space Y such that X is homeomorphic to the space of minimal prime ideals of C(Y) or to mR for some commutative reduced ring R that satisfies CAC?

If the answer is negative, what additional conditions are needed to characterize mR for such a ring R?

P.2 Suppose X has a cocompact pretty base and is basically disconnected. Does it follow that X is locally compact?

P.3 The following problems are also posed in [He]

Suppose X has a cocompact pretty base. Must X be:

(a) normal?

(b) a k-space?

(c) strongly zero-dimensional (i.e., must βX be zero-dimensional)?

P.4 Suppose X has a cocompact pretty base and is a subspace of an F-space? Must X be basically disconnected?

References

[CHN] W.Comfort, N.Hindman, and S.Negrepontis, *F'-spaces and their product with P-spaces*, Pacific J. Math. 28(1969), 489-502.

[CN] W.Comfort and S.Negrepontis, *The Theory of Ultrafilters*, Springer-Verlag, New York, 1974.

[D1] A.Dow, *F-spaces and F'-spaces*, Pacific J. Math. 108(1983),275-283.

[D2] _____, *The space of minimal prime ideals of $C(\beta N \setminus N)$ is probably not basically disconnected*, General topology and its Applications, Proceedings of the 1988 Northeast Conference, Marcel Dekker Publ. Co., New York, 1990.

[DF] _____ and O.Forster, *Absolute C^*-embedding of F-spaces*, Pacific J. Math. 98(1982), 63-71.

[DHKV1] A.Dow, M.Henriksen, R.Kopperman, and J. Vermeer, *The space of minimal prime ideals need not be basically disconnected*, Proc. Amer. Math. Soc. 104(1988),317-320.

[DHKV2] ______________________________, *The countable annihilator condition and weakly Lindelöf subspaces of minimal prime ideals*, Proc. IMM conference in Baku, USSR, 1987, to appear.

[FSZ] R.Frankewicz, S.Shelah, and B.Zbierski, Embeddings of Boolean algebras in $\mathcal{P}(\omega)$ mod finite, Abstracts Amer. Math. Soc. 10(1989), 399.

[GJ] L. Gillman and M. Jerison, *Rings of Continuous Functions*, D. Van Nostrand Publ. Co., Princeton, NJ 1960.

[He] M. Henriksen, *Spaces with a pretty base*, Proc. Curacao Conference, 1989, to appear.

[Hu] J. Huckaba, *Commutative Rings with Zero divisors*, Marcel Dekker Publ. Co., New York, 1988.

[HJ] M. Henriksen and M. Jerison, *The space of minimal prime ideals of a commutative ring*, Trans. Amer. Math. Soc. 115(1965), 110-130.

[HK] __________ and R. Kopperman, *A general theory of structure spaces with applications to spaces of prime ideals*, Algebra Universalis, to appear.

[HW] _________ and R. G. Woods, *General Topology as a tool in functional analysis*, Annals of the New York Academy of Sciences 552(1989), 60-69.

[Ki] J. Kist, *Minimal prime ideals in commutative semigroups*, Proc. London Math. Soc. (3) 13 (1963), 31-50.

[Ko] C. Kohls, *Hereditary properties of special spaces*, Archiv Math. (Basel) 12(1961), 129-133.

[Ku] K. Kunen, *Some points in βN*, Proc. Camb. Philos. Soc. 80(1976), 385-398.

[PW] J. Porter and R. G. Woods, *Extensions and Absolutes of Topological Spaces*, Springer-Verlag, New York, 1987.

[Se] G. L. Seever, *Measures on F-spaces*, Trans. Amer. Math. Soc. 133(1968), 267-280.

[Sw] M. Swardsen, *A generalization of F-spaces and some characterizations of GCH*, Trans. Amer. Math. Soc. 279(1983), 661-675.

The Groups in βN

Neil Hindman Howard University, Washington, D.C.

I. INTRODUCTION

It has been known for some time that, given any semigroup $(S,\cdot)$ the operation $\cdot$ extends to βS in such a way that (1) $(\beta S,\cdot)$ is a semigroup, (2) for each $p \in \beta S$, the function λ_p defined by $\lambda_p(q) = p \cdot q$, is continuous, and (3) for each $s \in S$, the function ρ_s defined by $\rho_s(q) = q \cdot s$, is continuous. (Conclusions (1) and (2) make $(\beta S,\cdot)$ a "left topological semigroup".) Further $(\beta S,\iota,\cdot)$ is a universal object among all $(X,\varphi,*)$ for which (1) $(X,*)$ is a semigroup and $\varphi: S \to X$ a homomorphism, (2) for each $x \in X$ λ_x is continuous, and (3) for each $s \in S$, $\rho_{\varphi(s)}$ is continuous. (See [2].) As such, the algebraic properties of $(\beta S,\cdot)$ are of considerable interest.

The semigroup $(\mathbb{N},+)$, where $\mathbb{N}$ is the set of positive integers, is arguably the simplest infinite semigroup. As such, it is amazing that $(\beta\mathbb{N},+)$ has a very rich algebraic structure, including in particular some large subgroups (i.e. subsemigroups which are groups). In addition to the interest in $\beta\mathbb{N}$ for its own sake, numerous combinatorial consequences about $\mathbb{N}$ have been obtained using the algebraic structure of $\beta\mathbb{N}$. See [8] and [9] for surveys of some of these results.

The author gratefully acknowledges support received from the National Science Foundation via grant DMS-8901058.

Given an idempotent e in a semigroup S, it is an elementary fact that there is a largest subsemigroup of S which is a group with e as its identity, which we denote by H(e). (To see this one simply lets H(e) = ∪{G: G is a subgroup of S with e as identity}. The only difficulty is to show H(e) is algebraically closed. For this one observes that if G_1 and G_2 are subgroups of S with e an identity then so is the set of all words with letters from $G_1 \cup G_2$.) In this paper we restrict our attention primarily to the groups in $(\beta\mathbb{N},+)$.

In Section 2 we present an elementary derivation of the basic facts about groups in the smallest ideal of a compact left topological semigroup.

In Section 3 we present a survey of the known results about the groups in $\beta\mathbb{N}$. We said above that $(\mathbb{N},+)$ is arguably the simplest infinite semigroup. A case could be made that the direct sum of countably many copies of $\mathbb{Z}_2$ holds that distinction. It is interesting that, if S is that direct sum, most of the known algebraic structure of $\beta\mathbb{N}$ is also part of the algebraic structure of βS.

In Section 4 we address what we feel to be the major open question about the groups in $\beta\mathbb{N}$. Namely, are there any non trivial finite groups in $\beta\mathbb{N}$? The results of this section are new.

II. THE GROUPS IN A COMPACT LEFT TOPOLOGICAL SEMIGROUP.

There is an extensive structure theorem for finite semigroups, which was shown to hold in any compact topological semigroup. It is a result of Ruppert [16] that in fact most of this structure theorem is valid in any compact left topological semigroup. We present here an elementary derivation of those portions of the structure theorem which deal with groups.

Recall that a right ideal R of a semigroup S is a non-empty subset of S such that $R \cdot S \subseteq R$. Similarly a left ideal is a non-empty subset L with $S \cdot L \subseteq L$. Finally an ideal is both a left and a right ideal.

2.1 LEMMA. Let S be any semigroup. If S has a minimal right ideal (resp. a minimal left ideal) then $\mathbf{M} = \cup\{\mathbf{R}$: $\mathbf{R}$ is a minimal right ideal of S$\}$ (respectively $\mathbf{M} = \cup\{L$: L is a minimal left ideal of S$\}$) is a (two-sided) ideal of S and $\mathbf{M}$ is the smallest ideal of S.

Proof. One has immediately that $\mathbf{M}$ is a right ideal. To see that $\mathbf{M}$ is a left ideal let $a \in \mathbf{M}$, $b \in S$ and pick a minimal right ideal $\mathbf{R}$ of S with $a \in \mathbf{R}$. Since $ba \in b\mathbf{R}$, it suffices to show that $b\mathbf{R}$ is a minimal right ideal of S. Now $(b\mathbf{R})S = b(\mathbf{R}S) \subset b\mathbf{R}$ so $b\mathbf{R}$ is a right ideal. Assume T is a right ideal with $T \subseteq b\mathbf{R}$ and let $U = \{x \in \mathbf{R}: bx \in T\}$. Since $T \neq \emptyset$ we have $U \neq \emptyset$. Given $x \in U$ and $y \in S$ we have $bx \in T$ so $bxy \in T$ so $xy \in U$. Thus U is a right ideal contained in $\mathbf{R}$ so $U = \mathbf{R}$ so $T = b\mathbf{R}$.

To see that $\mathbf{M}$ is the smallest ideal of S, let I be any ideal of S. To see that $\mathbf{M} \subseteq I$, we let $\mathbf{R}$ be a minimal right ideal of S and show $\mathbf{R} \subseteq I$. Now $\mathbf{R}$ is a right ideal and I is a left ideal so $\mathbf{R} \cap I \neq \emptyset$ (since if $a \in \mathbf{R}$ and $b \in I$ then $ab \in \mathbf{R} \cap I$). Thus $\mathbf{R} \cap I$ is a right ideal so $\mathbf{R} \cap I = \mathbf{R}$. □

2.2 LEMMA. Let S be any semigroup and let e be an idempotent of S.

(a) For all $x \in Se$, $xe = x$.

(b) For all $x \in eS$, $ex = x$.

(c) If $\mathbf{R}$ is a minimal right ideal of S and $x \in \mathbf{R}$, then $x\mathbf{R} = xS = \mathbf{R}$.

(d) If L is a minimal left ideal of S and $x \in L$, then $Lx = Sx = L$.

Proof. (a). Pick $s \in S$ with $x = se$. Then $xe = (se)e = s(ee) = se = x$.

(c) Immediately $x\mathbf{R} \subseteq xS \subseteq \mathbf{R}S \subseteq \mathbf{R}$. Since $(x\mathbf{R})S = x(\mathbf{R}S) \subseteq x\mathbf{R}$, we have right ideal contained in $\mathbf{R}$. □

2.3 LEMMA. Let S be any semigroup, let $\mathbf{R}$ be a minimal right ideal of S and let e be an idempotent in $\mathbf{R}$. Then Se is a minimal left ideal of S.

Proof. We have immediately that $S(Se) = (SS)e \subseteq Se$ so that Se is a left ideal. Let L be a left ideal with $L \subseteq Se$. To see that $L = Se$, it suffices to show that $e \in L$, for then $Se \subseteq SL \subseteq L$.

Now $L \neq \emptyset$ so pick $b \in L$. Then $eb \in e\mathbf{R} = \mathbf{R}$ so $eb\mathbf{R} = \mathbf{R}$, using Lemma 2.2(c) twice. Pick $y \in \mathbf{R}$ with $eby = e$. Since $y \in \mathbf{R}$, $y\mathbf{R} = \mathbf{R}$ so pick $z \in \mathbf{R}$

with $yz = e$. Now $b \in L$, so $(ye)b \in L \subseteq Se$, so by Lemma 2.2(a)

$$\begin{aligned}(ye)b &= ((ye)b)e \\ &= ((ye)b)(yz) \\ &= y(eby)z \\ &= yez.\end{aligned}$$

Since $z \in R = eS$ we have by Lemma 2.2(b) that $ez = z$. Thus $(ye)b = yez = yz = e$ so $e \in L$ as required. □

The following theorem can be found in [4]. It is apparently originally due to Green [6].

2.4 THEOREM. Let S be any semigroup and assume that S contains a minimal right ideal and each minimal right ideal contains an idempotent. Then S has a smallest (two-sided) ideal M which is the union of all of the minimal right ideals of S and is also the union of all of the minimal left ideals of S. If R is any minimal right ideal and L is any minimal left ideal, then $R \cap L$ is a group and $R \cap L$ is the largest subgroup of S containing the identity of $R \cap L$.

Proof. The first assertions follow immediately from Lemmas 2.1 and 2.3. Let R and L be minimal right and left ideals respectively. We first observe that $R \cap L = RL$. Immediately $RL \subseteq RS \subseteq R$ and $RL \subseteq SL \subseteq L$ so $RL \subset R \cap L$. Let e be an idempotent in R. By Lemma 2.2(c), $eS = R$. To see that $R \cap L \subseteq RL$, let $a \in R \cap L$. Then $a \in R = eS$ so by Lemma 2.2(b), $ea = a$. Thus $a = ea \in RL$. Note that $(RL)(RL) \subseteq RL$.

Pick $a \in R \cap L$. By Lemma 2.2, $aR = R$ and $La = L$. Thus $aRL = RL$ so pick $g \in RL$ with $ag = a$. We claim g is a right identity for RL. To see this, let $y \in RL$ and, since $RLa = RL$, pick $t \in RL$ with $ta = y$. Then $yg = (ta)g = t(ag) = ta = y$. To complete the proof that RL is a group we show that every element of RL has a right inverse with respect to g. Given any $x \in RL = R \cap L$ we have by Lemma 2.2(c) that $xR = R$ so $xRL = RL$ so there is some $z \in RL$ with $xz = g$, as required.

Finally, given g as above we have immediately that $R \cap L \subseteq H(g)$, the largest group with g as identity. On the other hand, given $x \in H(g)$ we have $x = xg \in xL \subseteq L$ and $x = gx \in Rx \subseteq R$ so $x \in R \cap L$. □

In order for Theorem 2.4 to be useful to us we need to know that a compact Hausdorff left topological semigroup has minimal right ideals and each minimal right ideal has idempotents. Part of this result is provided by a classical result of Ellis [5] which was proved under the assumption of joint continuity by Numakura [14] and Wallace [17].

2.5 **THEOREM** (Ellis). If S is a compact Hausdorff left topological semigroup, then S has an idempotent.

Proof. Let $\mathcal{A} = \{A \subseteq S\colon\ A \neq \emptyset,\ AA \subseteq A$, and A is closed$\}$. Then $\mathcal{A} \neq \emptyset$ since $S \in \mathcal{A}$. Any chain in $\mathcal{A}$ is a collection of closed sets with the finite intersection property so one gets easily that its intersection is in $\mathcal{A}$. By Zorn's Lemma choose a minimal member A of $\mathcal{A}$, and pick $a \in A$. (It will turn out since we will show that $aa = a$ that $A = \{a\}$, but we never really bother to establish this fact.)

Let $B = aA$. Now $aa \in B$ so $B \neq \emptyset$ also $BB = aAaA \subseteq aAAA \subseteq aA = B$. Since $B = \lambda_a[A]$, B is the continuous image of a compact space, hence compact. Further as a compact subset of a Hausdorff space, B is closed. Thus $B \in \mathcal{A}$. Since $B = aA \subseteq AA \subseteq A$, we have $B = A$.

Let $C = \{x \in A\colon ax = a\}$. Since $B = A$, we have $C \neq \emptyset$. Since $C = A \cap \lambda_a^{-1}[\{a\}]$, C is closed. To see that $CC \subseteq C$, let $x,y \in C$. Then $xy \in AA \subseteq A$ and $a(xy) = (ax)y = ay = a$. Thus $C \in \mathcal{A}$. Since $C \subseteq A$, $C = A$ so $a \in C$. That is $aa = a$, as required.

We are now in position to prove the portion of the structure theorem which deals with groups.

2.6 **THEOREM**. Let S be a compact Hausdorff left topological semigroup. Then S has a smallest (two-sided) ideal M which is the union of all of the minimal right ideals of S and is also the union of all of the minimal left ideals of S. If R is any minimal right ideal and L is any minimal left ideal then $R \cap L$ is a group and is the largest subgroup of S containing the identity of $R \cap L$. Any two of these maximal groups in M are isomorphic. Any two of these maximal groups contained in the same minimal left ideal are isomorphic and homeomorphic via the same function.

Proof. For the conclusions preceeding the isomorphism conclusions we only need to show that S has a minimal right ideal and each minimal right ideal has an idempotent and then invoke Theorem 2.4. Let $\mathcal{A} = \{R\colon R$ is a closed right ideal of $\mathcal{A}\}$. Since $S \in \mathcal{A}$ and a chain in $\mathcal{A}$ is a family of closed subsets with the finite intersection property (so its intersection is non-empty) we have easily by Zorn's Lemma that $\mathcal{A}$ has a minimal member R. (Then R is a minimal closed right ideal.) To see that R is a minimal right ideal let T be a right ideal with $T \subseteq R$. Pick $a \in T$, then $aS \subseteq TS \subseteq T \subseteq R$ and $aSS \subseteq aS$ and $aS = \lambda_a[S]$ so S is closed. Thus $aS = R$ so $T = R$.

Thus S has minimal right ideals. Further, if R is any minimal right ideal and $a \in R$ we have by Lemma 2.2(c) that $R = aS = \lambda_a[S]$ so R is compact. But then by Theorem 2.5, R has an idempotent.

We now turn our attention to the isomorphism conclusions. Given any idempotent $e \in M$, we have e is in some minimal right ideal R and in some minimal left ideal L. Since e is an idempotent in the group $R \cap L$, e is the identity of $R \cap L = RL$. Also by Lemma 2.2 we have $R = eS$ and $L = Se$ so $R \cap L = RL = eSSe \subseteq eSe$. Since $eSe \subseteq eSS \subseteq eS = R$ and $eSe \subseteq SSe \subseteq Se = L$ we have $eSe = R \cap L$.

Now let e and f be distinct idempotents in M, so that eSe and fSf are maximal groups with identities e and f respectively. Now $efe \in eSe$ so efe has an inverse g in eSe. That is $g \in eSe$ and $efeg = gefe = e$. We define $\varphi\colon eSe \to fSf$ by $\varphi(x) = fgxf$ and show that φ is an isomorphism onto fSf. We show further that if e and f are in the same minimal left ideal then φ and φ^{-1} are continuous.

To see that φ is a homomorphism let $x,y \in eSe$. Then

$$
\begin{aligned}
\varphi(x)\varphi(y) &= fgxffgyf \\
&= fgxfgyf \\
&= fgxefegyf \qquad (xe = x \text{ and } eg = g) \\
&= fgxeyf \qquad (efeg = e) \\
&= fgxyf \\
&= \varphi(xy).
\end{aligned}
$$

To see that φ is one-to-one let x be in the kernel of φ. Then

$$
\begin{aligned}
& fgxf = f \\
& efgxfe = efe \\
& efegxefe = efe \qquad (eg = g \text{ and } xe = x) \\
& exefe = efe \qquad efeg = e \\
& exeefe = eefe
\end{aligned}
$$

Thus by right cancellation in the group eSe we have exe = e so x = e.

To see that φ is onto fSf, let $y \in fSf$ and let h and k be the inverses of fgf and fef respectively in fSf. Then

$$
\begin{aligned}
& ehyke \in eSe \text{ and} \\
& \varphi(ehyke) = fgehykef \\
&\qquad = fghykfef \qquad (ge = g \text{ and } kf = k) \\
&\qquad = fghyf \qquad (kfef = f) \\
&\qquad = fgfhyf \qquad (h = fh) \\
&\qquad = fyf \qquad (fgfh = f) \\
&\qquad = y.
\end{aligned}
$$

Note that we have established that for $y \in fSf$ $\varphi^{-1}(y) = ehyke$.

Now assume there is a minimal left ideal L with $e,f \in L$. Then by Lemma 2.2(d) we have L = Se = Sf so by Lemma 2.2(a) fe = f and ef = e. Thus for $x \in eSe$,

$$
\begin{aligned}
\varphi(x) &= fgxf \\
&= fgxef \qquad (x = xe) \\
&= fgxe \qquad (ef = e) \\
&= fgx \qquad (xe = x)
\end{aligned}
$$

so that φ is the restriction of λ_{fg} to eSe and is hence continuous. Also given $y \in fSf$ we have

$$\begin{aligned}\varphi^{-1}(y) &= ehyke \\ &= ehykfef \qquad (kf = k \text{ and } ef = e) \\ &= ehyf \qquad (kfef = f) \\ &= ehy \qquad (yf = y)\end{aligned}$$

so that φ^{-1} is the restriction of λ_{eh} to fSf and is hence continuous. □

III. THE GROUPS IN $\beta\mathbb{N}$ -- KNOWN FACTS.

In this section we present without proof all major results of which we are aware about the groups in $(\beta\mathbb{N},+)$. Since $\beta\mathbb{N}$ is a compact Hausdorff left topological semigroup, Theorem 2.6 applies to it. We first look then at results about groups in the smallest ideal $\mathbb{M}$ of $\beta\mathbb{N}$. One question which is completely solved is: "How many groups are there in $\mathbb{M}$". (Recall $\beta\mathbb{N}$ has 2^c elements, where c is the cardinality of the continuum.)

3.1 THEOREM. (a) (Chou [3]). $\beta\mathbb{N}$ has 2^c minimal right ideals.

(b) (Baker and Milnes [1]). $\beta\mathbb{N}$ has 2^c minimal left ideals.

(c) Each minimal left ideal of $\beta\mathbb{N}$ contains 2^c distinct maximal groups.

(d) Each minimal right ideal of $\beta\mathbb{N}$ contains 2^c distinct maximal groups.

Observe that (c) and (d) follow immediately from (a) and (b) since any minimal right ideal and any minimal left ideal intersect in a group.

Since we know how many groups are in $\mathbb{M}$ it is natural to ask how big they are and what they look like. The first question is answered and the second addressed by the following result.

3.2 THEOREM. (Hindman and Pym [11]). Let G be a maximal group in $\mathbb{M}$. Then G contains a copy of the free group on 2^c generators. In fact such a group may be found in $G \cap I$ where $I = \bigcap_{n=1}^{\infty} cl_{\beta\mathbb{N}}(\mathbb{N}2^n)$.

Observe that G itself cannot be contained in I. Indeed if e is the identity of G then $e + 1 + e \in G$ and $e + 1 + e \notin cl_{\beta\mathbb{N}}(\mathbb{N}2)$.

With all of these results about groups in the smallest ideal, it is natural to ask about ones which are not in the smallest ideal. Are there any? If so how many? Can they be big? All of these are answered by the following.

3.3 THEOREM. (Lisan [13]). Let $I = \cap_{n=1}^{\infty} cl_{\beta\mathbb{N}}(\mathbb{N}2^n)$. There exist c pairwise disjoint topological and algebraic copies of I which miss $cl\mathbb{M}$. Consequently there are 2^c copies of the free group on 2^c generators which miss $cl\mathbb{M}$.

The question remains as to whether maximal subgroups of $\beta\mathbb{N}$ can be small. Since it is easy to see that all such groups contain a copy of $\mathbb{Z}$, the question becomes: Is it possible to have any maximal groups which are just copies of $\mathbb{Z}$? This question remains open but is partially answered by the following.

3.4 THEOREM. (Hindman [10]). It is relatively consistent with ZFC, (in fact it is a consequence of Martin's Axiom) that there exist maximal groups in $\beta\mathbb{N}$ which are just copies of $\mathbb{Z}$.

We mentioned in the introduction that a case could be made for the direct sum of countably many copies of $\mathbb{Z}_2$ as the simplest infinite semigroup. Since most of what we know about the groups in $\beta\mathbb{N}$ is in fact known about subgroups of I the following result says that such a choice wouldn't have mattered much.

3.5 THEOREM. (Pym [15]). Let S be the direct sum of countably many copies of $\mathbb{Z}_2$ and for $n \in \mathbb{N}$, let H_n be all x in S whose first n coordinates are all zero. Then $\cap_{n=1}^{\infty} cl_{\beta S}(H_n)$ and $\cap_{n=1}^{\infty} cl_{\beta\mathbb{N}}(\mathbb{N}2^n)$ are isomorphic and homeomorphic via the same function.

IV. ARE THERE NONTRIVIAL FINITE GROUPS IN $\beta\mathbb{N}$? The title of this section asks a question which is interesting in its own right. Indeed, given all that is known about the algebraic structure of $\beta\mathbb{N}$ it is amazing that such a simple question remains open. In addition, there are many cases (see for example [12]) where one would like to be able to conclude given $p \in \beta\mathbb{N}$, that $\{p, p + p, p + p + p, \ldots\}$ is infinite based solely on the fact that $p + p \neq p$. In this section we present most of what we know about this question. In particular we show that two natural ways to try to produce non trivial finite groups do not work, not because we aren't smart enough to make them work but because they cannot work.

Until now we have not paid any attention to the particular construction of $\beta\mathbb{N}$ which is used. We now need to be specific. We take the points of $\beta\mathbb{N}$ to be the ultrafilters on $\mathbb{N}$, the principal ultrafilters being identified with the points of $\mathbb{N}$. Given $p \in \beta\mathbb{N}$ a typical basic neighborhood of p is $\bar{A} = \{q \in \beta\mathbb{N}: A \in q\}$. Given p and $q \in \beta\mathbb{N}$ and $A \subseteq \mathbb{N}$, $A \in p + q$ if and only if $\{x \in \mathbb{N}: A - x \in p\} \in q$, where $A - x = \{y \in \mathbb{N}: y + x \in A\}$. See [8] for an elementary construction of $\beta\mathbb{N}$ using this approach.

4.1 DEFINITION. We define $\sigma_p(n)$ for $n \in \mathbb{N}$ and $p \in \beta\mathbb{N}$ inductively by $\sigma_p(n) = p$ and $\sigma_p(n + 1) = \sigma_p(n) + p$.

Thus $\sigma_p(n)$ is $p + p + \ldots + p$, n times. We don't use np for this because np is an object in $(\beta\mathbb{N}, \cdot)$, which is not generally equal to $\sigma_p(n)$.

4.2 DEFINITION. Given $p \in \beta\mathbb{N}$ and $k \in \mathbb{N}$ we say p is of order k in $\beta\mathbb{N}$ provided $\sigma_p(k + 1) = p$ and for $t \in \mathbb{N}$ with $2 \leq t \leq k$, $\sigma_p(t) \neq p$.

Thus if p is of order k in $\beta\mathbb{N}$, then $\{\sigma_p(1), \sigma_p(2), \ldots, \sigma_p(k)\}$ is a group of order k with identity $\sigma_p(k)$. The question we are addressing is whether there is any $p \in \beta\mathbb{N}$ with finite order bigger than 1.

4.3 THEOREM. If p is of finite order in $\beta\mathbb{N}$, then for all $n \in \mathbb{N}$, $\mathbb{N}n \in p$.

Proof. First observe that if $p, q \in \beta\mathbb{N}$ and $\mathbb{N}n + i \in p$ and $\mathbb{N}n + j \in q$ then $\mathbb{N}n + (i + j) \in p + q$. Indeed, $\mathbb{N}n + j \subseteq \{x \in \mathbb{N}: (\mathbb{N}n + (i + j)) - x \in p\}$. Now we note that the theorem is true if p is of order 1, i.e. $p = p + p$. Indeed choose some $i \in \{0, 1, \ldots, n - 1\}$ with $\mathbb{N}n + i \in p$. (We have

$\mathbb{N} = \{1,2,\ldots, n-1\} \cup \bigcup_{i=0}^{n-1} (\mathbb{N}n + i)$ so some $\mathbb{N}n + i \in p$.) Then $\mathbb{N}n + (i + i) \in p + p = p$ so $(\mathbb{N}n + (i + i)) \cap (\mathbb{N}n + i) \neq \emptyset$ so $i + i \equiv i \bmod n$, i.e. $i \equiv 0 \bmod n$. Thus $i = 0$.

Now assume we have $k \geq 2$ and p is of order k. It suffices to show that if $r,\ell \in \mathbb{N}$ and r is a prime then $\mathbb{N}r^\ell \in p$. (Indeed assume we have done this and suppose one has n and $i \in \{1,2,\ldots, n-1\}$ with $\mathbb{N}n + i \in p$. Choose a prime r and an $\ell \in \mathbb{N}$ with $r^\ell | n$ and $r^\ell \nmid i$. Then $\mathbb{N}r^\ell \in p$ so pick $x \in \mathbb{N}r^\ell \cap (\mathbb{N}n + i)$. For some $a,b \in \mathbb{N}$, $x = ar^\ell = bn + i$. But then $r^\ell \mid i$, a contradiction.) Thus we let r be a prime, let $\ell \in \mathbb{N}$ and choose $i \in \{0,1,\ldots, r^\ell - 1\}$ with $\mathbb{N}r^\ell + i \in p$. We show $i = 0$.

Let n be the largest integer (possibly 0) with $r^n \mid k$ and pick $j \in \{0,1,\ldots, r^{\ell+n} - 1\}$ with $\mathbb{N}r^{\ell+n} + j \in p$. Since $(\mathbb{N}r^\ell + i) \cap (\mathbb{N}r^{\ell+n} + j) \in p$ we have $j \equiv i \bmod r^\ell$. As we observed at the start of the proof, $\mathbb{N}r^{\ell+n} + kj \in \sigma_p(k)$. Since $\sigma_p(k)$ is an idempotent we have $\mathbb{N}r^{\ell+n} \in \sigma_p(k)$ so that $kj \equiv 0 \bmod r^{\ell+n}$. By the choice of n we must then have $r^\ell \mid j$ so $i \equiv 0 \bmod r^\ell$ as required. □

This result allows us to show that what is probably the most obvious way to construct elements of finite order bigger than 1 cannot work. Let us say for example that we want to construct an element of order 2. We mimic the Ellis proof of the existence of idempotents (Theorem 2.5) and get a minimal member of $\mathcal{A} = \{A \subseteq \beta\mathbb{N}\colon\ A \neq \emptyset,\ A + A + A \subseteq A$, and A is closed$\}$. However $(\overline{\mathbb{N}2 + 1}) \in \mathcal{A}$ and by Theorem 4.3 no minimal member of $\mathcal{A}$ which is contained in $(\overline{\mathbb{N}2 + 1})$ can contain any element of finite order.

The whole question of the algebraic structure of $\beta\mathbb{N}$ first came to our attention many years ago because of an observation of Galvin's. In today's terminology this observation was: If $p = p + p$, then each member of p contains all finite distinct sums from some infinite sequence. We have a similar, though more elaborate, conclusion about any element of $\beta\mathbb{N}$ of finite order.

4.4 THEOREM. Assume we have $p \in \beta\mathbb{N}$ of finite order $k \geq 2$. For each $t \in \{1,2,\ldots, k\}$ let $f_t\colon \mathbb{N} \to \sigma_p(t)$. There is a sequence $\langle x_n \rangle_{n=0}^{\infty}$ in $\mathbb{N}$ such that, whenever F is a finite non-empty subset of $\mathbb{N}$, $t \in \{1,2,\ldots, k\}$, $|F| \equiv t \bmod k$, and $\ell = \min F$, then $\Sigma_{n \in F}\, x_n \in f_t(x_{\ell-1})$.

Proof. We choose sequences $\langle A_n \rangle_{n=1}^{\infty}$, $\langle x_n \rangle_{n=0}^{\infty}$, and for $t \in \{2,3,\ldots, k\}$, $\langle B_{t,n} \rangle_{n=0}^{\infty}$ satisfying for each n

(1) If $n \geq 1$, $A_n \in p$.
(2) If $n \geq 1$, $A_n \subseteq f_1(x_{n-1})$
(3) If $n \geq 1$, $x_n \in A_n$.
(4) If $t \in \{2,3,\ldots, k\}$, $B_{t,n} \in \sigma_p(t)$.
(5) If $t \in \{2,3,\ldots, k\}$, $B_{t,n} \subseteq f_t(x_n)$.
(6) If $t \in \{2,3,\ldots, k-1\}$ and $n \geq 1$, $B_{t,n} \subseteq (B_{t+1,n-1} - x_n) \cap B_{t,n-1}$.
(7) If $n \geq 1$, $B_{k,n} \subseteq (A_n - x_n) \cap B_{k,n-1}$.
(8) If $n \geq 2$, $A_n \subseteq (B_{2,n-2} - x_{n-1}) \cap A_{n-1}$.
(9) If $n \geq 1$, $B_{2,n-1} - x_n \in p$.

For $n = 0$, let $x_0 = 1$ and for $t \in \{2,3,\ldots, k\}$ let $B_{t,0} = f_t(x_0)$. Then hypotheses (4) and (5) are immediate and the rest are vacuous.

Assume now $n \geq 1$ and the construction has proceeded through $n - 1$. If $n = 1$ let $A_1 = f_1(x_0)$. If $n > 1$, let $A_n = (B_{2,n-2} - x_{n-1}) \cap A_{n-1} \cap f_1(x_{n-1})$. In the latter case we have by (1) and (9) at $n - 1$ that $A_n \in p$. Since $A_n \in p = \sigma_p(k+1)$ we have $\{x \in \mathbb{N}\colon A_n - x \in \sigma_p(k)\} \in p$. Also for $t \in \{2,3,\ldots, k\}$ we have by (4) that $B_{t,n-1} \in \sigma_p(t)$ so that $\{x \in \mathbb{N}\colon B_{t,n-1} - x \in \sigma_p(t-1)\} \in p$. Thus $C_n = \{x \in \mathbb{N}\colon A_n - x \in \sigma_p(k)\} \cap \bigcap_{t=2}^{k} \{x \in \mathbb{N}\colon B_{t,n-1} - x \in \sigma_p(t-1)\} \in p$. Choose $x_n \in A_n \cap C_n$ and note that (9) is satisfied. Let $B_{k,n} = (A_n - x_n) \cap B_{k,n-1} \cap f_k(x_n)$ and for $t \in \{2,3,\ldots, k-1\}$ let $B_{t,n} = (B_{t+1,n-1} - x_n) \cap B_{t,n-1} \cap f_t(x_n)$. All hypotheses can be immediately verified.

The construction being complete we show that for $i \in \{1,2,\ldots, k\}$, $a \in A_n$, and $n > m(1) > m(2) > \ldots > m(i)$,

(i) If $i < k$, then $a + x_{m(1)} + x_{m(2)} + \ldots + x_{m(i)} \in B_{i+1,m(i)-1}$ and

(ii) If $i = k$, then $a + x_{m(1)} + x_{m(2)} + \ldots + x_{m(k)} \in A_{m(k)}$.

We proceed by induction on i. For $i = 1$ we have by (8) $A_n \subseteq A_{m(1)+1}$. Again by (8) (at $m(1) + 1$) we have $A_{m(1)+1} \subseteq B_{2,m(1)-1} - x_{m(1)}$. Since $a \in A_n$ we have $a + x_{m(1)} \in B_{2,m(1)-1}$ as required. Now assume $1 < i < k$ and (i) has been established for $i - 1$. Then using (6), $a + x_{m(1)} + \ldots + x_{m(i-1)} \in B_{i,m(i-1)-1} \subseteq B_{i,m(i)} \subseteq B_{i+1,m(i)-1} - x_{m(i)}$ so $a + x_{m(1)} + \ldots + x_{m(i)} \in B_{i+1,m(i)-1}$. Now let $i = k$ and assume (i) holds for $k - 1$. Then using (6) and (7) we have $a + x_{m(1)} + \ldots + x_{m(k-1)} \in B_{k,m(k-1)-1} \subseteq B_{k,m(k)} \subseteq A_{m(k)} - x_{m(k)}$ so $a + x_{m(1)} + \ldots + x_{m(k)} \in A_{m(k)}$.

To conclude we prove by induction on $|F|$ that if $t \in \{1,2,\ldots, k\}$, $|F| \equiv t \bmod k$, and $\ell = \min F$, then

if $t = 1$, $\Sigma_{n \in F} x_n \in A_\ell$ and

if $t > 1$, $\Sigma_{n \in F} x_n \in B_{t,\ell-1}$.

By (2) and (5) this proves the conclusion of the theorem.

If $|F| = 1$, we have by (3) $x_\ell \in A_\ell$ as required. Next assume $1 < |F| = t \leq k$ and write $F = \{m(t - 1), m(t - 2),\ldots, m(1), n\}$ with the entries in increasing order, so $m(t - 1) = \ell$. Then by (i) we have $x_n + x_{m(1)} + \ldots + x_{m(t-1)} \in B_{t,m(t-1)-1} = B_{t,\ell-1}$.

Now assume $|F| > k$ and $t = 1$. Write $m(k) < m(k - 1) < \ldots < m(1)$ as the k smallest members of F (so $m(k) = \ell$), let $G = F\backslash\{m(1), m(2),\ldots, m(k)\}$ and let $r = \min G$. Then by the induction hypothesis $\Sigma_{n \in G} x_n \in A_r$ so by (ii) $\Sigma_{n \in F} x_n \in A_\ell$.

Finally asume $|F| > k$ and $t > 1$ and write $m(t - 1) < m(t - 2) < \ldots < m(1)$ as the $t - 1$ smallest members of F (so $m(t - 1) = \ell$), let $G = F\backslash\{m(1), m(2),\ldots, m(t - 1)\}$ and let $r = \min G$. Then $|G| \equiv 1 \bmod k$ so by the induction hypothesis $\Sigma_{n \in G} x_n \in A_r$ so by (i) $\Sigma_{n \in F} x_n \in B_{t,m(t-1)-1} = B_{t,\ell-1}$. □

For the remainder of the paper we restrict our attention to elements of order 2 to keep the discussion relatively simple.

4.5 COROLLARY. Assume $p \in \beta\mathbb{N}$ is of order 2. Given any $A \in p$ there exist a sequence $\langle x_n \rangle_{n=1}^{\infty}$ in $\mathbb{N}$ and $B \in p$ with $B \subseteq A$ such that whenever F is a finite non-empty subset of $\mathbb{N}$, we have $\Sigma_{n\in F}\, x_n \in B$ if and only if $|F|$ is odd.

Proof. We have $p \neq p + p$ so pick $C \in p$ with $C \notin p + p$. Given $A \in p$, let $B = C \cap A$ and define $f_1\colon \mathbb{N} \to p$ and $f_2\colon \mathbb{N} \to p + p$ by $f_1(n) = B$ and $f_2(n) = \mathbb{N}\backslash C$ for all n and apply Theorem 4.4. □

The first production of an ultrafilter all of whose members contain the finite sums from some finite sequence was done in [7] with the aid of the continuum hypothesis. It in fact produced an ultrafilter generated by sets of the form $FS(\langle x_n\rangle_{n=1}^{\infty}) = \{\Sigma_{n\in F}\, x_n\colon F$ is a finite non-empty subset of $\mathbb{N}\}$. Motivated by Corollary 4.5 one would be tempted to construct an ultrafilter generated by sets of the form $OFS(\langle x_n\rangle_{n=1}^{\infty}) = \{\Sigma_{n\in F}\, x_n\colon F$ is a finite non-empty subset of $\mathbb{N}$ and $|F|$ is odd$\}$. We shall see in Theorem 4.7 that this would be enough to get elements of order 2.

4.6 LEMMA. Let $p \in \beta\mathbb{N}$ and assume that for every $A \in p$ there exists $\langle x_n\rangle_{n=1}^{\infty}$ with $OFS(\langle x_n\rangle_{n=1}^{\infty}) \subseteq A$ and $OFS(\langle x_n\rangle_{n=1}^{\infty}) \in p$. Then for each $A \in p$ there exists $\langle x_n\rangle_{n=1}^{\infty}$ with $OFS(\langle x_n\rangle_{n=1}^{\infty}) \subseteq A$, and for each $m \in \mathbb{N}$, $OFS(\langle x_n\rangle_{n=m}^{\infty}) \in p$ and such that, whenever F and G are finite non-empty subsets of $\mathbb{N}$, if $\Sigma_{n\in F}\, x_n = \Sigma_{n\in G}\, x_n$ then $F = G$.

Proof. Given any $x \in \mathbb{N}$ choose $n \geq 0$, $k \geq 0$, and $i \in \{1,2,3,4\}$ with $x = 5^n(5k + i)$ and define $f(x) = n$ and $g(x) = i$. Thus if x is written in base 5, $g(x)$ is the right most non-zero digit and it is followed on the right by $f(x)$ zeroes. For the assertions that follow, simply remember to think of the numbers as written in base 5.

For $i \in \{1,2,3,4\}$ let $C_i = \{x \in \mathbb{N}\colon g(x) = i\}$ and pick i with $C_i \in p$. We show first if $\langle y_n\rangle_{n=1}^{\infty}$ is any sequence with $OFS(\langle y_n\rangle_{n=1}^{\infty}) \subseteq C_i$ then for $n \neq m$ one has $f(y_n) \neq f(y_n)$. Suppose not and pick the least t for which there exist $m \neq n$ with $f(y_n) = f(y_m)$. Since for at most t values of j is $f(y_j) < t$, pick $j \notin \{m,n\}$ with $f(y_j) \geq t$. If $f(y_j) = t$ we have $f(y_j + y_n + y_m) = t$ and $g(y_j + y_n + y_m) \equiv 3i \bmod 5$ while if $f(y_j) > t$ we have $f(y_j + y_n + y_m) = t$ and $g(y_j + y_n + y_m) \equiv 2i \bmod 5$. In either case $y_j + y_n + y_m \notin C_i$, a contradiction.

Now let $A \in p$ and choose $\langle x_n\rangle_{n=1}^{\infty}$ with $OFS(\langle x_n\rangle_{n=1}^{\infty}) \subseteq A \cap C_i$ and $OFS(\langle x_n\rangle_{n=1}^{\infty}) \in p$. We may assume that if $n < m$ then $f(x_n) < f(x_m)$. We show now that if F and G are finite non-empty subsets of $\mathbb{N}$ and $\Sigma_{n\in F}\, x_n = \Sigma_{n\in G}\, x_n$, then $F = G$. Suppose not. We may presume $F \cap G = \emptyset$ by subtracting any common terms. Let $m = \min(F \cup G)$ and assume without loss of generality that $m \in F$. Then $f(\Sigma_{n\in F}\, x_n) = f(x_m) < f(\Sigma_{n\in G}\, x_n)$, a contradiction.

Finally let $m \in \mathbb{N}$. We show $OFS(\langle x_n\rangle_{n=m}^{\infty}) \in p$. Indeed $OFS(\langle x_n\rangle_{n=1}^{\infty}) = \cup_{F\subseteq\{1,2,\ldots,m-1\}} \{\Sigma_{n\in F\cup G}\, x_n$: G is a finite subset of $\{m, m+1, \ldots\}$ and $|F \cup G|$ is odd$\}$ so choose $F \subseteq \{1,2,\ldots, ,m-1\}$ such that $B = \{\Sigma_{n\in F\cup G}\, x_n$: G is a finite subset of $\{m, m+1, \ldots\}$ and $|F \cup G|$ is odd$\} \in p$. If $F = \emptyset$ we have $OFS(\langle x_n\rangle_{n=m}^{\infty}) = B \in p$ as required so suppose $F \neq \emptyset$ and let $\ell = \min F$. Then given any $y \in B$ we have $f(y) = f(x_\ell)$. But $B \in p$ so we can choose $\langle y_n\rangle_{n=1}^{\infty}$ with $OFS(\langle y_n\rangle_{n=1}^{\infty}) \subseteq B$. Since $B \subseteq C_i$ we have $f(y_n) \neq f(y_j)$ whenever $n \neq j$, a contradiction. □

In the applications we will use repeatedly without mention the fact that expressions of the form $\Sigma_{n\in F}\, x_n$ are unique.

4.7 THEOREM. Let $p \in \beta\mathbb{N}$ and assume that for any $A \in p$ there exists $\langle x_n\rangle_{n=1}^{\infty}$ with $OFS(\langle x_n\rangle_{n=1}^{\infty}) \subseteq A$ and $OFS(\langle x_n\rangle_{n=1}^{\infty}) \in p$. Then p is of order 2.

Proof. To see that $p = p + p + p$, let $A \in p$ and choose $\langle x_n\rangle_{n=1}^{\infty}$ as guaranteed by Lemma 4.6 for A. We show $OFS(\langle x_n\rangle_{n=1}^{\infty}) \subseteq \{y \in \mathbb{N}: A - y \in p + p\}$. Let $y = \Sigma_{n\in F}\, x_n \in OFS(\langle x_n\rangle_{n=1}^{\infty})$ and let $m = \max F$. Then $OFS(\langle x_n\rangle_{n=m+1}^{\infty}) \in p$. We show $OFS(\langle x_n\rangle_{n=m+1}^{\infty}) \subseteq \{z: (A - y) - z \in p\}$. Let $z = \Sigma_{n\in G}\, x_n \in OFS(\langle x_n\rangle_{n=m+1}^{\infty})$ and let $r = \max G$. Then if $w = \Sigma_{n\in H}\, x_n \in OFS(\langle x_n\rangle_{n=r+1}^{\infty})$ one has $w + z + y = \Sigma_{n\in F\cup G\cup H}\, x_n \in OFS(\langle x_n\rangle_{n=1}^{\infty}) \subseteq A$.

To see that $p \neq p + p$, choose $\langle x_n \rangle_{n=1}^{\infty}$ as guaranteed by Lemma 4.6 for $\mathbb{N} \in p$. Letting $EFS(\langle x_n \rangle_{n=1}^{\infty}) = \{\Sigma_{n \in F} x_n$: F is a finite non-empty subset of $\mathbb{N}$ and $|F|$ is even$\}$, we have $EFS(\langle x_n \rangle_{n=1}^{\infty}) \cap OFS(\langle x_n \rangle_{n=1}^{\infty}) = \emptyset$. We show $EFS(\langle x_n \rangle_{n=1}^{\infty}) \in p + p$ by showing $OFS(\langle x_n \rangle_{n=1}^{\infty}) \subseteq \{y: EFS(\langle x_n \rangle_{n=1}^{\infty}) - y \in p\}$. Indeed if $y = \Sigma_{n \in F} x_n \in OFS(\langle x_n \rangle_{n=1}^{\infty})$, $m = \max F$, and $z = \Sigma_{n \in G} x_n \in OFS(\langle x_n \rangle_{n=m+1}^{\infty})$ then $y + z = \Sigma_{n \in F \cup G} x_n \in EFS(\langle x_n \rangle_{n=1}^{\infty})$. □

Unfortunately, the effect of Theorem 4.7 is blunted by the following alternate proof which shows its hypothesis to be vacuous.

4.8 THEOREM. There is no $p \in \beta\mathbb{N}$ such that for any $A \in p$ there exists $\langle x_n \rangle_{n=1}^{\infty}$ with $OFS(\langle x_n \rangle_{n=1}^{\infty}) \subseteq A$ and $OFS(\langle x_n \rangle_{n=1}^{\infty}) \in p$.

Proof. Suppose we have such and choose $\langle x_n \rangle_{n=1}^{\infty}$ as guaranteed by Lemma 4.6 for $\mathbb{N} \in p$. Let $B_1 = \{\Sigma_{n \in F} x_n\colon |F| \equiv 1 \bmod 4\}$ and $B_3 = \{\Sigma_{n \in F} x_n\colon |F| \equiv 3 \bmod 4\}$. Then $OFS(\langle x_n \rangle_{n=1}^{\infty}) = B_1 \cup B_3$ so choose $i \in \{1,3\}$ with $B_i \in p$. Pick $\langle y_n \rangle_{n=1}^{\infty}$ as guaranteed for B_i by Lemma 4.6. Then $y_1 = \Sigma_{n \in F_1} x_n$ with $|F_1| \equiv i \bmod 4$. Let $m = \max F_1$. Then $OFS(\langle y_n \rangle_{n=2}^{\infty}) \cap OFS(\langle x_n \rangle_{n=m+1}^{\infty}) \in p$ so choose $z = \Sigma_{n \in H_1} y_n = \Sigma_{n \in F_2} x_n$ with $|H_1|$ odd, $\min H_1 \geq 2$, $\min F_2 \geq m + 1$, and, since $OFS(\langle y_n \rangle_{n=2}^{\infty}) \subseteq B_i$, $|F_2| \equiv i \bmod 4$. Let $r = \max F_2$ and $s = \max H_1$. Then $OFS(\langle y_n \rangle_{n=s+1}^{\infty}) \cap OFS(\langle x_n \rangle_{n=r+1}^{\infty}) \in p$ so choose $w = \Sigma_{n \in H_2} y_n = \Sigma_{n \in F_3} x_n$ with $|H_2|$ odd, $\min H_2 \geq s + 1$, $\min F_3 \geq r + 1$, and since $OFS(\langle y_n \rangle_{n=s+1}^{\infty}) \subseteq B_i$, $|F_3| \equiv i \bmod 4$. Now $y_1 + z + w = \Sigma_{n \in \{1\} \cup H_1 \cup H_2} y_n \in OFS(\langle y_n \rangle_{n=1}^{\infty}) \subseteq B_i$ but also $y_1 + z + w = \Sigma_{n \in F_1 \cup F_2 \cup F_3} x_n$ and $|F_1 \cup F_2 \cup F_3| \not\equiv i \bmod 4$, a contradiction. □

REFERENCES

1. Baker, J. and Milnes, P. (1977). The ideal structure of the Stone-Čech compactification of a group, Math. Proc. Cambridge. Phil. Soc., 82: 401-409.

2. Berglund, J., Jungenn,H. and Milnes, P. (1978). Compact right topological semigroups and generalizations of almost periodicity, Lecture Notes in Math. 663, Springer-Verlag, Berlin.

3. Chou, C. (1969). On the size of the set of left invariant means on a semigroup, Proc. Amer. Math. Soc., 23: 199-205.

4. Clifford, A.H., and Preston, G.B. (1961), The algebraic theory of Semigroups, I., The American Mathematical Society, Providence.

5. Ellis, R. (1969). Lectures on topological dynamics, Benjamin, New York.

6. Green, J.A. (1951), On the structure of semigroups, Annals of Math., 54: 163-172.

7. Hindman, N. (1972). The existence of certain ultrafilters on N and a conjecture of Graham and Rothschild, Proc. Amer. Math. Soc., 36: 341-346.

8. Hindman, N. (1979). Ultrafilters and combinatorial number theory, in Number Theory Carbondale (M. Nathanson, ed.) Lecture Notes in Math., 751, Springer-Verlag, Berlin: 119-184.

9. Hindman, N. (1989). Ultrafilters and Ramsey Theory - an update, in Set Theory and its applications, (J. Steprans and S. Watson, eds.), Lecture Notes in Math., 1401, Springer-Verlag, Berlin: 97-118.

10. Hindman, N. (1990). Strongly summable ultrafilters on N and small maximal subgroups of βN, Semigroup Forum, to appear.

11. Hindman, N. and Pym., J. (1984). Free groups and semigroups in $\beta\mathbb{N}$, Semigroup Forum, 30: 177-193.

12. Hindman, N. and Pym, J. (1990). Closures of singly generated subsemigroups of βS, Semigroup Forum, to appear.

13. Lisan, A. (1988). Free groups in βN which miss the minimal ideal, Semigroup Forum, 37: 233-239.

14. Numakura, K. (1952). On bicompact semigroups, Math. J. Okayama Univ., 1: 99-108.

15. Pym, J. (1987). Semigroup structure in Stone-Cech compactifications, J. London Math. Soc., (2) 36: 421-428.

16. Ruppert, W. (1973). Rechstopologische Halbgruppen, J. Reine Angew. Math., 261: 123-133.

17. Wallace, A. (1952). A note on mobs I, An. Acad. Brasil. Ci. 24: 329-334.

Lie Semigroups in Topology and Geometry

Karl H. Hofmann Darmstadt Institute of Technology, Darmstadt, Germany

With these remarks I want to draw attention to the recent development of a Lie theory for semigroups and to indicate what applications motivate such a theory[1]. Let us inspect a typical example. In a Lorentzian manifold with orientation there is a continuous choice of one half of the double cone determined by the Lorentzian metric in each tangent space $T_m(M)$. We call the chosen cone at a point $m \in M$ the *forward light cone* $C(m) \subseteq T_m(M)$. A piecewise differentiable curve $\gamma: I \to M$, where I is any interval in $\mathbb{R}$, is called *a causal trajectory* if the derivative $\gamma'(t)$, whenever it exists, is contained in the forward light cone $C(\gamma(t))$. The manifold M is called *causal* if there are no non-trivial closed causal curves in M. While it is true that simply closed curves are a subject matter of topology, the issue here is not merely a topological one, as one recognizes quickly by considering the manifold $M = \mathbb{R} \times \mathbb{S}^1$, the cylinder. It is then an easy exercise to describe cone fields for which M is causal and other cone fields for which it is not, because curves following a grand circle $\{r\} \times \mathbb{S}^1$ cannot be causal. Once the cone fields are introduced one observes that for questions of global geometry such as causality it is only the cone field which matters. We say that M is a *homogeneous* manifold with a causal structure if there is a transitive orientation preserving Lie group action on M preserving a cone field on M. In that case M is of the form G/H, where G and H are Lie groups, and the cone field on G/H is invariant under the action of G. In particular, the "future light cone" $C(m_0)$ at $m_0 = H$ is invariant under the action of H. We seek criteria for a homogeneous manifold with causal structure to be causal.

If we use the right translations to trivialize the tangent bundle $T(G)$ of G as usual, then the Lie algebra $\mathfrak{g}$ and $T_g(G)$ become identified under the isomorphism $d\rho_g(1): \mathfrak{g} \to T_g(G)$, $\rho_g(x) = xg$, and the vector space automorphism $d\rho_g(1)^{-1} d\lambda_g(1)$ (with $\lambda_g(x) = gx$) is in fact the Lie algebra automorphism $\mathrm{Ad}(g)$. Thus $T(G)$ may be viewed as the semidirect product $\mathfrak{g} \rtimes G$ with

---

[1] In my survey lecture at the conference I selected the topic of chronogeometry and causality of homogeneous manifolds endowed with cone fields such as pseudo-Riemannian manifolds as point of departure for the introduction of relevant semigroups in the defining Lie group (see [6]). A number of recent surveys on the Lie theory of semigroups have appeared [4], [10], [11], [16], Introduction to [12], so that the addition of yet another complete survey at this time is not appropriate.

G acting on the vector group underlying $\mathfrak{g}$ under the adjoint action so that $T_g(G) = \mathfrak{g} \times \{g\}$ and $d\lambda_g(1)(X,1) = (\mathrm{Ad}(g)(X), g)$, $d\rho_g(1)(X,1) = (X,g)$. If $m = gH$ is an arbitrary point of M, then the left translation $\lambda_g: G \to G$ induces a diffeomorphism $\mu_g: M \to M$ given by $\mu_g(g'H) = gg'H$. Any other left translation mapping $m_0 = H$ to m is given by λ_{gh} with some $h \in H$. If $W \subseteq \mathfrak{g}$ is the pull-back of the cone $C(m_0)$ under the map $d\pi(1): T_1(G) \to T_H(G/H)$, $\pi(g) = gH$, then W is a closed convex wedge in the Lie algebra $\mathfrak{g}$ whose edge, i. e., the largest vector subspace contained in it, is exactly $\mathfrak{h} = T_1(H)$, the Lie algebra of H.

Thus we recognize that in discussing the causality question of homogeneous manifolds with causal structure we deal in fact with the following situation:

Definition 1. (a) We say that a triple (G, H, W) is a set of *causal data* if G is a connected Lie group, H a closed subgroup and W is a closed convex cone which need not be pointed in the Lie algebra $\mathfrak{g}$ of G which satisfies the following conditions:

(i) The Lie algebra $\mathfrak{h}$ of H is the largest vector space contained in W.

(ii) The cone W is invariant under H, that is, $\mathrm{Ad}(h)W = W$ for all $h \in H$.

(b) A W-*admissible trajectory* in G (cf. [7], Definition VI.1.7) is a piecewise differentiable curve γ with $\gamma'(t) \in d\lambda_{\gamma(t)}(1)(W)$ for all t for which the derivative exists and with $\gamma(0) = 1$. We shall denote the set of all endpoints of W-admissible trajectories by $S(W)$.

(c) Let $J^+_{m_0}$ denote the set of endpoints of all causal trajectories in M. The crucial concept in our discussion is the pull back

$$S = S_{(G,H,W)} \overset{\mathrm{def}}{=} \pi^{-1}(J^+_{m_0})$$

∎

The following proposition points out how we determine whether M is causal or not:

Proposition 2. (i) *For a set (G, H, W) of causal data, the subsets $S \subseteq G$ and $S(W)$ are subsemigroups, the latter is invariant under all inner automorphisms by elements of H, and $S = HS(W) = S(W)H$.*

(ii) *The manifold M is causal if and only if H is the precise set $S \cap S^{-1}$ of invertible elements of $S \subseteq G$.* ∎

The causality of M is now expressed by an algebraic property of the semigroup S. Our program of characterizing causality of M will succeed in the same measure as we are able to deal with S and its relation to the group H. If H is connected, then $S = S(W)$, and a possible Lie theory of semigroups applies directly to S.

At this point we are satisfied with having indicated one motivation for a Lie theory of semigroups and we do not pursue the question of chronogeometry as such. The reader interested in the application of the Lie theory of semigroup to the area of ordered manifolds and to chronogeometry is referred to [6], [16].

The contours of a *Lie theory of semigroups* have become distinct only recently. It deals, in essence, with the theory of subsemigroups of Lie groups—more specifically, it deals with the question of how semigroups get there in the first place, with their applications, and with their structure theory. Even though [7] is a technical book, its introduction is written for a general audience and can serve as an orientation. The question of embedding semigroups with enough topological or differentiable structure into Lie groups is treated by Lawson in the survey [16]. A survey on applications in analysis has been given by Hilgert in [4]. I have myself overviewd the status of the structure theory of subsemigroups of Lie groups in the survey [10]; the present remarks are largely taken from the introduction to that article. In another paper for a general audience [11], I have attempted to explain how some of the ideas of the theory go back to Sophus Lie himself. (See also [12].)

In a Lie theory of semigroups it would be preposterous to speak about *all* subsemigroups of Lie groups. One does not consider *all* subgroups of Lie groups in Lie group theory either. Rather one concentrates on subgroups which have a Lie algebra and are determined by it. In the Lie group $\mathbb{R}$ this leaves not many subgroups; in $\mathbb{R}^n$ it covers all vector subspaces. In the case of subsemigroups of Lie groups the situation is similar. We shall focus our attention on subsemigroups which are determined by their one-parameter subsemigroups. We must allow a little leeway; the specifics are discussed in [7]. At any rate, in $\mathbb{R}$, the semigroups we find in this way are the two half-lines, and in $\mathbb{R}^n$ we have to take all convex closed cones into consideration. These provide a good preliminary model of our objects. However, certain other types of subsemigroups of Lie groups are quite relevant even though they have no analog in group theory. In a connected Lie group, there are no proper open subgroups, but there are plenty of interesting open subsemigroups, notably those containing the identity in their closure. In $\mathbb{R}^n$ this includes all open half-space semigroups (and their intersections, the open convex cones). A theory of subsemigroups of Lie groups must also deal with open subsemigroups.

The basic feature of Lie group theory is that Lie groups and their subgroups are almost completely described by their infinitesimal structure at the origin, and this infinitesimal structure is encoded into the Lie algebra. Thus a large portion of Lie group theory is Lie algebra theory. In the same spirit, many subsemigroups of Lie groups are determined by their infinitesimal structure, namely, a closed convex cone W in the Lie algebra $\mathfrak{g}$ satisfying an invariance condition such as we have encountered it in Definition 1.2.(a)(ii). The precise condition may be phrased, with the aid of the linear operators $\operatorname{ad} x$ given by $(\operatorname{ad} x)(y) = [x, y]$, in the following form:

$$e^{\operatorname{ad} x} W = W \quad \text{for all} \quad x \in W \cap -W.$$

Such cones are called *Lie wedges*; a vector subspace is a Lie wedge if and only if it is a subalgebra. It is for this reason that a Lie theory of semigroups has to employ the geometry of convex cones in addition to the linear algebra of Lie algebra theory.

Local Lie group theory is a historical episode and a pedagogical device providing preliminary information prior to the more sophisticated global theory. Every local Lie group is the identity neighborhood of a global one. (We stick with finite dimensional Lie groups.) Hence Lie group theory has actually two parts: The infinitesimal theory and the global theory. By contrast, the Lie theory of semigroups has to deal with a local theory for good mathematical reasons; there do exist local semigroups in Lie groups which cannot be identity neighborhoods of a subsemigroup of a Lie group. If we could show that every local semigroup is an identity neighborhood of *some* topological semigroup—be it subsemigroup of a Lie group or not—then this would be a sophisticated result opening up new avenues leading distinctly beyond Lie group theory. Our present knowledge has not carried us that far, but substantial partial results are known [27].

The *history* of topological semigroup has many roots. It is perhaps remarkable that SOPHUS LIE himself addressed transformation *semigroups* under the name of "continuous groups". One can even recognize convex cones in LIE's writings. (See [8], [12], [13].) In functional analysis a large branch of research covers the theory of strongly continuous one-parameter semigroups of operators on Banach spaces going back to E. HILLE in the late forties; this is not our concern here, because the Lie theory of semigroups proposed by us is a multiparameter theory. The study of compact semigroups was initiated by A. D. WALLACE in the fifties. This theory gave much momentum to the investigation of topological semigroups and later entered functional analysis in the context of almost periodic and weakly almost periodic functions. The impact of compact semigroups is fest still today as is, e. g., exemplified by the applications of the compact one-sidedly topological semigroup $\beta\mathbb{N}$ of all ultrafilters on the additive semigroup $\mathbb{N}$ of natural numbers to combinatorial number theory. They provided some novel proofs of classical results such as VAN DER WAERDEN's Theorem on arithmetic progressions (see [15]). A closed subsemigroup of a compact group is a subgroup. Therefore, the flavor of a Lie semigroup theory is quite different from that of the theory of compact semigroups since it is largely, but not exclusively, concerned with subsemigroups of Lie groups. With an eye on the applications one could say that the theory of compact semigroups is applied in studies of the ergodic behavior of semigroups while the Lie theory of semigroups studies the nature of a semigroup around the origin and how its infinitesimal properties influence large scale behavior.

It had not been noticed by workers in the topological and analytical theory of semigroups that certain basic aspects of the Lie theory of semigroups were recognized and developed by LOEWNER whose work originates from the study of functions of one complex variable and merges into the investigation of functions with generalized monotonicity properties and from there into transformation semigroups. Only now, after L. Bers made LOEWNER's work accessible in a volume of collected papers [17], it is possible to follow LOEWNER's access to a Lie theory of semigroups. Remarkably, the idea of what we call a *Lie wedge* and a portion of the basic aspects of an infinitesimal theory of semigroups were clearly formulated by LOEWNER in 1948 in an address to the American Mathematical Society. They were published in 1950 in the Bulletin of the AMS in an article entitled *Some classes of functions defined by difference or differential inequalities* (see [17], pp. 149, 136). Proofs of the results formulated in this paper on basic Lie semigroup theory were not given and their future publication was promised; but they never emerged. From 1955 on, LOEWNER published papers on applications of what we would call the Lie theory of semigroups on topics like the semigroup of totally positive matrices, transformation semigroups, semigroups of holomorphic self-maps of domains, transformation semigroups invariant under the groups of isometries of Euclidean and non-Euclidean geometries, partial orders defined by linear semigroups, semigroups of monotone transformations of higer order on a real interval [17]. He summarized a good deal of his research in an address to the American Mathematical Society in 1962; this overview was published in the Bulletin of the AMS in 1964 under the title *On semigroups in analysis and geometry* (see [17], p. 437–451). It appears that LOEWNER and his work had no immediate followers.

However, in the dissertation of L. J. M. ROTHKRANTZ "Transformatiehalfgroepen van nietcompacte hermitesche symmetrische Ruimten" at Amsterdam in 1980 the issue of semigroups of holomorphic self-maps of domains was resumed and amplified, and new vistas towards the relation of semigroups and the geometry of symmetric spaces were opened. Research on transformation semigroups of manifolds with geometric structure is one of the current applications of Lie semigroup theory. One type of geometric structure is the assignment of a Lorentzian metric; in the introductory remarks we have indicated the applicability of Lie semigroup methods to the investigation of global causality [6].

Considerable momentum in the field of semigroups in Lie groups and of convex cones in Lie algebras is due to investigation of invariant cones in Lie algebras initiated in a seminal paper by E. B. VINBERG on *Invariant cones and orderings in Lie groups* in 1980 [16]. The theory of invariant cones in Lie algebras has become a fairly rich theory as we shall see below. Substantial contributions are due to G. I. OL'SHANSKIĬ, S. PANEITZ, S. KUMARESAN and A. RANJAN, HILGERT and HOFMANN, SPINDLER. For detailed references we refer to [7] and to [22].

In a natural way, the Lie theory of semigroups emerged in systems and geometric control theory in the mid-seventies in the work of GAUTHIER, HIRSCHORN, JURDEVIC, KUPKA, SALLET, SUSSMANN. The reader will find more information on this aspect in KUPKA's survey *Semigroups in systems theory* in [15]. The systematic development of the Lie theory of semigroups [7] shows that its link to systems theory is a two-way street: Significant basic results in the Lie theory of semigroups are established with methods from geometric control theory.

The connection between the Lie theory of semigroups and representation theory of Lie groups, notably the analytic extension of representation was pioneered by G. I. OL'SHANSKIĬ; this relationship continues to be of vital interest as is exemplified by the oscillator semigroup whose significance was discovered recently by HOWE. More details about this link are found in Hilgert's survey [4].

The subsequent references supplement the listing of the sources mentioned in this overview by recent contributions to the Lie theory of semigroups. We draw particular attention to the books [7] and [15] as sources for a general orientation.

References

[1] Dörr, N., *On Ol'shanskiĭ's semigroup*, Math. Ann., 1990, to appear.

[2] Eggert, A., *Zur Klassifikation von Semialgebren*, Thesis, TH Darmstadt 1988.

[3] Gichev, Y. M., *Invariant orderings in solvable Lie groups*, Siberian Math. J. **30** (1989), 57–69.

[4] Hilgert, J., *Applications of Lie semigroups in analysis*, in: The analytical and topological theory of semigroups, K. H. Hofmann, J. D. Lawson, and J. S. Pym, Eds., Walter de Gruyter, Berlin, 1990, vii+395pp., 27–50.

[5] Hilgert, J., and K. H. Hofmann, *On Sophus Lie's Fundamental Theorem*, J. Funct. Analysis **67** (1986), 209–216.

[6] —, *On the causal structure of homogeneous manifolds*, Math. Scand., 1990, to appear.

[7] Hilgert, J., K. H. Hofmann, and J. D. Lawson, "Lie groups, convex cones, and semigroups," Oxford University Press, Oxford 1989, xxxviii+645pp.

[8] Hofmann, K. H., *Semigroups in the* 19th *century?* in: Theory of Semigroups, Conf. Theor. Appl. Semigroups Greifswald, 1984, Proceedings, Math. Gesellschaft d. DDR, Berlin 1985.

[9] —, *A memo on hyperplane subalgebras*, Geom. Dedicata, 1990, to appear.

[10] —, *Lie groups and semigroups*, in: The analytical and topological theory of semigroups, K. H. Hofmann, J. D. Lawson, and J. S. Pym, Eds., Walter de Gruyter, Berlin, 1990, vii+395pp., 3–26.

[11] —, *Einige Ideen Sophus Lies—hundert Jahre danach*, Jahrbuch Überblicke der Mathematik, (BI Mannheim) 1990, to appear.

[12] —, *Zur Geschichte des Halbgruppenbegriffs*, Historia Mathematica, to appear.

[13] Hofmann, K. H., and J. D. Lawson, *On Lie's Fundamental Theorems* I, Indag. Math. **45** (1983), 453-466.

[14] —, *Foundations of Lie semigroups*, in: Lecture Notes in Math. **998** (1983), 128–201.

[15] Hofmann, K. H., J. D. Lawson, and John S. Pym, Eds., "The analytical and topological theory of semigroups", Walter de Gruyter, Berlin, 1990, xii+398 pp.

[16] Lawson, J., *Embedding semigroups into Lie groups*, in: The analytical and topological theory of semigroups, K. H. Hofmann, J. D. Lawson, and J. S. Pym, Eds., Walter de Gruyter, Berlin, 1990, vii+395pp., 51–80.

[17] Loewner, Ch., "Collected Papers", Lipman Bers, Ed., Birkhäuser Boston, Basel, 1988.

[18] Neeb, K.-H., "Globalität von Lie-Keilen", Diplom Thesis, Technische Hochschule Darmstadt, 1988.

[19] —, *Subsemigroups of* SL(2), Semigroup Forum, 1990, to appear.

[20] —, *The duality between subsemigroups of Lie groups and monotone functions*, Trans. Amer. Math. Soc., to appear.

[21] —, *Conal orders on homogeneous spaces with complete Riemannian metrics*, submitted.

[22] —, *Invariant cones in Lie algebras*, Preprint Technische Hochschule Darmstadt, 1990.

[23] Ol'shanskiĭ, G. I., *Invariant orderings on simple Lie groups. The solution to* E. B. VINBERG'S *problem*, Funct. Anal. and Appl. 18 (1984), 28–42.

[24] Spindler, K., "Invariante Kegel in Liealgebren", Mitteilungen Math. Sem. Giessen 188 (1988).

[25] —, *Some remarks on Levi complements and roots in Lie algebras with cone potential*, Preprint, Conf. Anal. Topol. Theory of Semigroups, Oberwolfach 1989.

[26] Vinberg, E. B., *Invariant cones and orderings in Lie groups*, Funct. Anal. and Appl. 14 (1980), 1–13.

[27] Weiss, W., *Local Lie-semigroups and open embeddings into global topological semigroups*, Indag. Math., to appear.

Uniformities and Uniform Continuity on Topological Groups

Gerald L. Itzkowitz Queens College, The City University of New York, Flushing, New York

Our purpose will be to describe an interesting problem involving left and right uniform continuity and the left and right uniformities on a topological group. The general problem is this. On a T_0 topological group are the right and left uniformities equivalent iff every real valued left uniformly continuous function is right uniformly continuous? We note that the implication "$\Longrightarrow$" is elementary and for completeness we prove this fact in Theorem 3. We will describe the method of solution of this problem for the case of a locally compact group. In this case the answer is in the affirmative.

1. History and some model examples.

In 1970 C. Chou asked the author the following question. If a locally compact group G is not discrete or compact is it always possible to construct a real valued function that is continuous on G but not uniformly continuous on G?

This immediately brought to mind the following example from real analysis: Let f_n, n = 2, 3, ... be nonnegative piecewise linear functions with support on $[n + \frac{1}{2} - \frac{1}{n}, n + \frac{1}{2} + \frac{1}{n}]$ and with maximum value 1 at $n + \frac{1}{2}$. Define $f(x) = \sum_{n=2}^{\infty} f_n(x)$. Then f(x) is continuous on $\mathbb{R}$ but not uniformly continuous on $\mathbb{R}$. The crucial points in this example are three conditions.

(1) The cozero-sets of the functions f_n are disjoint open intervals of the form $(n + \frac{1}{2} - \frac{1}{n}, n + \frac{1}{2} + \frac{1}{n})$

(2) $m(n + \frac{1}{2} - \frac{1}{n}, n + \frac{1}{2} + \frac{1}{n}) = \frac{2}{n} \to 0$ $(n \to \infty)$ where m is the Lebesgue measure.

(3) The variation of f_n on its cozero-set is 1.

We note that Chou's question had been answered earlier by Kister [10]. However the above example suggested the following construction that simplifies the proof of Kister's Theorem. The construction imitates (1), (2), and (3) in a locally compact group setting. Since our group is noncompact, if U is any symmetric neighborhood of the identity with compact closure, we inductively choose a sequence $\{x_n\} \subset G$ satisfying $x_n \in \bigcup_{j=1}^{n-1} x_j U$. Then we find a neighborhood V of the identity such that $V^2 \subset U$. The open neighborhoods $x_n V$ are pairwise disjoint. Since G is not discrete, the Haar measure is nonatomic, so there is a sequence of neighborhoods $\{W_n\}$ of the identity such that $\lambda(W_n) \to 0$ where λ is left Haar measure. We may assume that $W_n \subset V$, for all n. We now observe that

(1) The collection of neighborhoods $x_n W_n$ are pairwise disjoint.

(2) $\lambda(x_n W_n) = \lambda(W_n) \to 0$ as $n \to \infty$.

Finally since any T_0 group is completely regular and T_2, for each $n \geq 1$ there is a continuous function f_n such that $f_n(x_n) = 1$, $f_n(x) = 0$ if $x \notin x_n W_n$, and $0 \leq f_n \leq 1$. Thus

(3) The variation of f_n on its cozero-set is 1. Therefore, the function Σf_n is continuous but not left uniformly continuous on G.

We now make an observation and ask a natural question. There are two possible notions of uniform continuity on a topological group namely left and right uniform continuity. Is it possible that a similar construction can be employed to give an example of a function that is left uniformly continuous but not right uniformly continuous? The answer turned out to be yes with some qualifications. First we will state some definitions.

Definitions: Lef $f : G \to \mathbb{R}$.

(1) f is left uniformly continuous if for each $\epsilon > 0$ there is a neighborhood U of e in G such that $|f(x) - f(y)| < \epsilon$ whenever $x \in yU$.

2) f is right uniformly continuous if for each $\epsilon > 0$ there is a neighborhood U of e in G such that $|f(x) - f(y)| < \epsilon$ whenever $x \in Uy$.

(3) $S_\ell(G)$ is the uniformity generated by the entourages L_U, U a neighborhood of e, where $L_U = \{(x,y): y \in xU\}$.

(4) $S_r(G)$ is the uniformity generated by the entourages R_U, U a neighborhood of e, where $R_U = \{(x,y): y \in Ux\}$.

(5) We will say that G has equivalent uniformities if $S_\ell(G) = S_r(G)$.

It is an elementary fact that G has equivalent uniform structures iff for each neighborhood U of e there is a neighborhood V of e such that $Vx \subset xU$ for all $x \in G$.

The first attempts to solve the general problem and the above motivating example appeared in [6] in 1974. In fact there we actually made use of the Haar measure and its associated modular function.

Definition: If λ is the left Haar measure on G then the right modular function is

$$\Delta_r(x) = \frac{\int_G f(tx^{-1})\lambda(dt)}{\int_G f(t)\lambda(dt)}$$

This function is a continuous homomorphism of G into $\mathbb{R}^*$ (multiplicative group of non - negative numbers) satisfying $\lambda(Ax) = \lambda(A)\Delta_r(x)$, $x \in G$, $A \subset G$. In the case where G is compact, discrete, or Abelian $\Delta_r(x) = 1$ for all x. However, there are many cases where $\Delta_r(x) \neq 1$ for some x. When this happens we say G is not unimodular. Otherwise we say G is unimodular.

Theorem 1. Each locally compact, non discrete, non unimodular group admits a function that is left uniformly continuous but not right uniformly continuous.

To prove this theorem let U be a neighborhood of e with compact closure and inductively choose a sequence $\{x_n\} \subset G$ such that

(a) $x_n \notin \bigcup_{i=1}^{n-1} Ux_iU = U_{n-1}$

and

(b) $\Delta_r(x_n) > n$.

We can do this because G is not compact and $\Delta_r(\bar{U}_{n-1})$ is a compact subset of $(0,\infty)$ so there is $x_n \in G$ such that $\Delta_r(x_n) > n$ and $\Delta_r(x_n) \notin \Delta_r(\bar{U}_{n-1})$.

Now let V be a symmetric neighborhood of e such that $V^2 \subset U$, then the following hold.

(1) $x_iV \cap x_jV = \phi$, $i \neq j$.

(2) $Vx_i \cap Vx_j = \phi$, $i \neq j$.

(3) $Vx_i \cap x_jV = \phi$, $i \neq j$.

Since G is completely regular there is a continuous function f_1 satisfying

$$f_1(x_1) = 1,\ f_1(x) = 0 \text{ if } x \notin x_1V, \text{ and } 0 \leq f_1 \leq 1.$$

Define $f_n(x) = f_1(x_1x_n^{-1}x)$. Then

$$f_n(x_n) = f_1(x_1) = 1 \text{ and } f_n(x) = 0 \text{ if } x \notin x_nV.$$

It now follows that $f = \Sigma f_i$ is left uniformly continuous. However we may observe that if $V_n = x_nVx_n^{-1}$ then

$$0 < \lambda(V) = \lambda(x_nV) = \lambda(V_nx_n) = \lambda(V_n)\Delta_r(x_n),$$

and since $\Delta_r(x_n) \to \infty$ it follows that $\lambda(V_n) \to 0$, and so $\cap V_n$ contains no open set. This last observation implies that f cannot be right uniformly continuous: Let $0 < \epsilon < 1$. Suppose a neighborhood W of the identity exists for which $xy^{-1} \in W$ implies that $|f(x) - f(y)| < \epsilon$. Since we may assume that $W \subset V$, if $x \in Wx_n$ it follows that $x \notin Vx_m$ and $x \notin x_mV$, if $m \neq n$. Since $|f(x) - f(x_n)| < \epsilon$, $x \in x_nV = V_nx_n$. Therefore $W \subset V_n$. Since this holds for $n \geq 1$, $W \subset \cap V_n$, a contradiction.

We now note that if G is not unimodular then G has inequivalent right and left uniform structures since the construction used in the proof of Theorem 1 actually shows that $\cap x_nVx_n^{-1}$ is not a neighborhood of the identity. This then implies that $\cap \{xVx^{-1}: x \in G\}$ is not a neighborhood of e. Therefore there is no neighborhood W of the identity satisfying $Wx \subset xV$ for all $x \in G$. This fact was noted much earlier by Braconnier [2] and a proof of it appears in [5], 19.28. However it is well known that there are locally compact groups with inequivalent uniform structures which are unimodular. The simplest such example is $G\ell(2,\mathbb{R})$, the 2 x 2 matrices with non zero determinant (cf. [5], 15.30 (b)). Thus this theorem could not be the most general theorem.

In [5], 8.18, there appear a number of statements equivalent to the statement that a metric group G has inequivalent uniform structures. One of these can be restated as follows: G has inequivalent uniform structures iff there is a neighborhood U of e, a sequence $x_n \to e$, and a sequence $\{y_n\} \subset G$, such that $y_n^{-1}x_ny_n \notin U$, for each n. Since it is well known ([5], 4.9) that if U is a neighborhood of e and if F is compact, there is a neighborhood V of e satisfying $x^{-1}Vx \subset U$, $x \in F$, it follows that $\{y_n\}$ is not contained in any compact set.

After unifying these observations, the same construction achieved through the induction argument of Theorem 1 and definition of the functions f_n yields the following theorem.

Theorem 2. If G is a locally compact metric group with inequivalent uniform structures then G admits a real valued function f that is left uniformly continuous but not right uniformly continuous.

Theorem 3. If G has equivalent uniformities then every real valued left uniformly continuous function is right uniformly continuous.

Proof. Let f be left uniformly continuous and let $\epsilon > 0$. Then there is a neighborhood W of the identity such that $y \in xW$ implies $|f(y) - f(x)| < \epsilon$. Since G has equivalent uniform structures there is a neighborhood V of the identity such that $Vx \subset xW$ for all $x \in G$. We note that if $y \in Vx$ then $y \in xW$ so that $|f(y) - f(x)| < \epsilon$ and so f is right uniformly continuous.

Theorem 2 and the elementary Theorem 3 together yield:

Corollary: If G is locally compact metric group then G has equivalent uniformities iff every real valued left uniformly continuous function is right uniformly continuous.

This settles the question for locally compact metric groups. It was also noted in the paper that the question could be settled without the metrizability assumption if one could find a sequence $\{x_n\} \subset G$ such that $\cap x_n V x_n^{-1}$ contained no open sets. The attempt to find such a sequence in any group with inequivalent uniformities led to the next development in the solution of this problem.

In a joint work of Comfort and Itzkowitz [3] an argument employing a theorem of Kakutani and Kodaira [9] was used. This argument had been suggested to the authors by K. Ross and it was noticed by Itzkowitz that it could be modified to get Theorem 5. We first state the theorem of Kakutani and Kodaira.

Theorem 4. Let G be a σ-compact and locally compact topological group. Then for every countable family $\{U_n : n = 1,2,\cdots\}$ of neighborhoods of e, there is a compact normal subgroup $N \subset G$ such that $N \subset \bigcap_{n=1}^{\infty} U_n$, and G/N has a countable basis for its open sets.

Ross had noticed the following. Each locally compact group contains an open σ-compact subgroup. (Take any symmetric neighborhood U of the identity and form the group $H = \cup U^n$). The Kakutani-Kodaira Theorem implies that H contains a compact normal subgroup N such that H/N is a separable metric group. The homogeneous space G/N is then metrizable. Furthermore if G is a union of α or fewer compact subsets, then G/N is a union of α or fewer compact metric subsets.

Using these facts one can conclude that G/N contains a dense subset $\tilde{J}$ where cardinal $(\tilde{J}) \leq \alpha$. One can then pick a subset of cardinal $\leq \alpha$ in G by selecting one element from each $\phi^{-1}(\tilde{x})$, $\tilde{x} \in \tilde{J}$, where $\phi : G \to G/N$ is the natural map. This leads to the following Theorem.

Theorem 5 [7]: If G is a locally compact α-compact topological group and if V is a neighborhood of e then $\bigcap_{x \in G} xVx^{-1}$ is a neighborhood of e iff for each set A satisfying card(A) $\leq \alpha$ the set $\bigcap_{x \in A} xVx^{-1}$ is a neighborhood of e.

Now we observe that if A is any set we can well order A in the form A $= \{x_\eta : \eta < \alpha\}$ where card (A) $= \alpha$. Therefore if U is a neighborhood of e such that $\bigcap_{x \in A} xUx^{-1}$ is not a neighborhood of e, a transfinite induction allows us to pick an index set $J = \{\eta_i : \eta_i < \alpha\}$ with the properties:

(a) for each j, $x_{\eta_j} \in \bigcup_{i<j} Ux_{\eta_i}$

(b) η_j is the first ordinal greater than or equal to $\sup_{i<j}\{\eta_i\}$ such that (a) holds.

The set $B = \{x_\eta : \eta \in J\}$ then satisfies two properties.

(1) $\bigcap_{x \in B} xUx^{-1}$ is not a neighborhood of e.

(2) B is right uniformly discrete. In fact if W is a symmetric neighborhood of e such that $W^2 \subset U$ then the collection $\{Wx : x \in B\}$ is pairwise disjoint.

This leads to the following definition and Theorem.

Definition: The set B is α-right uniformly discrete if

(a) card (B) $\leq \alpha$

(b) there is a neighborhood V of the identity such that the collection $\{Vx : x \in B\}$ is pairwise disjoint.

Theorem 6 [7]: Let G be a locally compact α-compact group. Then G has equivalent uniformities iff for each α-right uniformly discrete set B and each neighborhood U of the identity, $\bigcap_{x \in B} xUx^{-1}$ is a neighborhood of e.

If G is σ-compact then the transfinite induction may be replaced with an ordinary induction argument and the α-right uniformly discrete set B may be replaced by a right uniformly discrete sequence. Thus we may write the following corollary.

Corollary. Let G be a locally compact σ-compact group. Then G has equivalent uniformities iff for each right uniformly discrete sequence $\{x_n\}$ and for each neighborhood U of the identity $\bigcap_n x_n U x_n^{-1}$ is a neighborhood of the identity.

At this point we can see that at least in the case of σ-compact topological groups, the question of whether or not G has equivalent uniformities can be decided by sequences. It is then an observation [7] that if every locally compact group with inequivalent uniform structures has an open σ-compact subgroup with inequivalent uniform structures then the question of whether or not any locally compact group G has equivalent uniform structures could be decided by sequences. At that time the author did not think that such a situation could be true.

Since then two proofs showing that the answer to this question is in the affirmative have been given. The first was due to Pestov and used a transfinite induction involving some highly technical properties of infinite ordinals. This proof appeared in the Ukrainian Journal of Mathematics in 1988 [12]. The second proof was obtained independently by Itzkowitz, Rothman, Strassberg, and Wu [8]. This makes use of an ordinary induction argument employing Lie groups and yields along the way some interesting Lie group structure theorems and several additional characterizations of equal uniformities in locally compact topological groups.

The solution to this problem came about in an interesting way.

Itzkowitz, Rothman, and Strassberg began working as a group in the Summer of 1988. At that time they were considering the question of how to extend uniformly continuous functions on a subgroup of a locally compact topological group to the entire group. In any event a number of observations were made and by early summer of 1989 the problem of characterizing equal uniformities in σ-compact locally compact groups was solved. It was shown then that G has equal uniformities iff every right uniformly continuous real valued function is left uniformly continuous. Also a number of other important equivalent conditions including the G_δ condition were discovered. In any event at the Northeast Topology Conference in Staten Island in 1989, Itzkowitz and Wu were introduced to each other by Karl Hofmann. Wu immediately took an interest in the problem and with his knowledge of Lie group theory outlined a number of steps that led to a joint solution of the general problem. The solution was completed in September 1989, and is reported in more detail in this survey.

2. The σ-compact case.

The solution depends on the corollary to Theorem 3. The method of solution depends on looking at the open and continuous homomorphic images of the σ-compact group G. First we note that Theorem 6 implies the following result. The details of the proofs will appear in [8].

Theorem 7. Let G and H be locally compact groups. If $\phi : G \to H$ is an open continuous homomorphism onto H, and if H has inequivalent uniform structures, then G has inequivalent uniform structures.

The method of proof used in the Kakutani-Kodaira Theorem (Theorem 4) may be modified slightly to yield.

Theorem 8. Let G be a σ-compact, locally compact group. If G has inequivalent uniformities then there is a compact normal subgroup N such that G/N is metrizable and Lindelöf and has inequivalent uniformities.

Theorems 7 and 8 may be unified to yield the following.

Theorem 9. Let G be a σ-compact locally compact group. Then G has equivalent uniform structures iff for every locally compact metric group H such that there is an open continuous onto homomorphism $\phi : G \to H$, the group H has equivalent uniform structures.

Proof. If there is an open onto continuous homomorphism $\phi : G \to H$ where H has inequivalent uniformities, then by Theorem 7, G has inequivalent uniform structures.

By Theorem 8, if G has inequivalent uniform structures there is a compact normal subgroup N such that G/N is metrizable, Lindelöf, and has inequivalent uniform structures. Evidently the natural map $\phi : G \to G/N$ is an open onto continuous homomorphism and we can take $H = G/N$.

Theorem 9 above has a number of consequences when combined with the corollary of Theorem 3. First we make a definition.

Definition: If there is a sequence $\{y_n\} \subset G$ and a neighborhood V of the identity in G such that

(i) $\{Vy_n : n \geq 1\}$ are pairwise disjoint,

(ii) $\{y_nV : n \geq 1\}$ are pairwise disjoint,

(iii) $y_n V \cap V y_m = \phi$ if $n \neq m$,

(iv) $\bigcap_{n=1}^{\infty} y_n V y_n^{-1}$ is not a neighborhood of e, and

(v) $\bar{V}$ is compact,

then we will say that the sequence is V-uniformly discrete.

If G has inequivalent uniform structures then there is a homomorphism $\phi : G \to H$ where H is a metric group with inequivalent uniform structures. The corollary of Theorem 3 tells us that H has a real valued function f that is left uniformly continuous but not right uniformly continuous. Defining $g = f \circ \phi$ it is not hard to prove the following Theorem.

Theorem 10: Let G be a σ-compact locally compact topological group with inequivalent uniform structures. Then

(a) G contains a V-uniformly discrete sequence.

(b) There is a real valued function g defined on G that is left uniformly continuous but not right uniformly continuous.

If we collect the above results and previously known facts we can summarize them in the following Theorem.

Theorem 11: Let G be a σ-compact locally compact group. Then the following statements are equivalent:

(a) G has equivalent uniform structures.

(b) If $\{x_\alpha\}$ is a net based on the directed set D satisfying $x_\alpha \to e$, and if $\{y_\alpha\}$ in G is any other net based on D, then $y_\alpha x_\alpha y_\alpha^{-1} \to e$.

(c) For each neighborhood U of e, $\bigcap_{x \in G} xUx^{-1}$ is a neighborhood of e.

(d) There is a basis for the neighborhood system of e consisting of sets V satisfying $V = xVx^{-1}$ for all $x \in G$.

(e) For each neighborhood U of e and each right uniformly discrete sequence $\{x_n\}$ in G the set $\bigcap_{n=1}^{\infty} x_n U x_n^{-1}$ is a neighborhood of e.

(f) For each neighborhood U of e and each left uniformly discrete sequence $\{x_n\}$ in G the set $\bigcap_{n=1}^{\infty} x_n^{-1} U x_n$ is a neighborhood of e.

(g) Each locally compact (Lindelöf) metric group H which is an open continuous homomorphic image of G has equivalent uniform structures.

(h) Every real-valued left uniformly continuous function on G is right uniformly continuous.

(i) Every real-valued right uniformly continuous function on G is left uniformly continuous.

(j) Every open subgroup of G has equivalent uniformities.

(k) The identity map $i : G \to G$ is uniformly open for the pairs of structures $(S_\ell(G), S_r(G))$ and $(S_r(G), S_\ell(G))$.

We point out that the equivalence of (a) – (d), (k) in any topological group (not necessarily locally compact or σ–compact) is a classical result though the proofs are not readily available in the literature. Conditions (e) and (f) are results of Itzkowitz [7], while (g)-(i) appear to be new.

3. The G_δ condition.

The investigation of this condition came about because of the desire to prove the conjecture stated by Itzkowitz [7], i.e. if G is locally compact then G has equivalent uniformities iff every open σ–compact subgroup of G has equivalent uniformities.

At this point we backtrack slightly and point out the following. If G is a locally compact metric group then the corollary of Theorem 3 holds and in fact if G has inequivalent uniformities the proof of Theorem 3 shows that G contains a V-uniformly discrete sequence. Since any σ–compact subset of a locally compact group is contained in an open σ–compact subgroup, the observation that a sequence is automatically σ–compact leads to the following.

Theorem 12. Let G be a locally compact metric group. Then G has equivalent uniform structures iff every open σ–compact subgroup of G has equivalent uniform structures.

At this point a technical Theorem is needed.

Theorem 13. Let G be a locally compact α-compact topological group. Suppose $B \subset G$ is (α,U)-right uniformly discrete and that U contains a compact normal G_δ subgroup N. Then G contains an open σ–compact subgroup H with inequivalent uniform structures.

To understand this theorem the following definitions are needed.

Definitions: (1) U is a G_δ set if $U = \bigcap_{k=1}^{\infty} U_k$, where each U_k is open.

(2) G satisfies the G_δ condition iff each neighborhood of the identity contains a compact normal G_δ subgroup.

(3) The set B is (α,U)-right uniformly discrete if U is a

neighborhood of the identity and if

(a) B is α-right uniformly discrete, and if

(b) $\cap \{xUx^{-1} : x \in B\}$ is not a neighborhood of the identity.

As a corollary we obtain the following theorem.

Theorem 14. Let G be a locally compact group which satisfies the G_δ condition. Then the following are equivalent:

(a) G has equivalent uniformities.

(b) Every open σ-compact subgroup of G has equivalent uniformities.

(c) Every real valued left uniformly continuous function defined on G is right uniformly continuous.

It is clear at this point that the G_δ condition is important. It turns out that the combination of this theorem and an induction argument using Lie groups implies the main result in the general case. It might also be mentioned that the G_δ condition and Theorem 14 inspired the renewed attempt to prove the conjecture of Itzkowitz.

4. Projective limits and uniform structures.

An examination of the Kakutani Kodaira Theorem (Theorem 4, this article) and our Theorem 14 shows that projective limits might prove to be a useful tool in settling the problem. We will give the definition of a projective limit of groups, a statement of a classical theorem appearing in Weil [13], and Bourbaki [1], and a reformulation of Theorem 4 using the terminology of projective limits.

Definition. Let A be a set directed by the partial ordering $\leq$. For each $\alpha \in A$, let G_α be a topological group. If $\alpha < \beta$ ($\alpha \leq \beta$ and $\alpha \neq \beta$) there is an open continuous homomorphism $f_{\beta\alpha}$ of G_β into G_α. If $\alpha < \beta < \gamma$, then $f_{\gamma\alpha} = f_{\beta\alpha} \circ f_{\gamma\beta}$. The triple $(A, \{G_\alpha\}, \{f_{\beta\alpha}\})$ is an inverse mapping system. Let $H = \prod_{\alpha\in A} G_\alpha$ and let $G = \varprojlim G_\alpha = \{(x_\alpha) \in H : x_\alpha = f_{\beta\alpha}(x_\beta), \alpha < \beta\}$. Then G is called the projective limit of the inverse mapping system.

Theorem 15 [13]: Let G_α, $\alpha \in A$ be a collection of normal closed subgroups of the topological group G. Suppose the following hold.

(a) Every neighborhood of $e \in G$ contains a G_α.

(b) Given $\alpha, \beta \in A$ there is $\gamma \in A$ such that $G_\gamma \subset G_\alpha \cap G_\beta$.

(c) G is complete or one of the G_α's is compact.

Then $G = \varprojlim G/G_\alpha$, the projective limit of the G_α.

The Kakutani Kodaira Theorem may now be reformulated in the following more general form as a consequence of Theorem 14.

Theorem 16. Let G be a locally compact σ-compact topological group or a locally compact group which satisfies the G_δ condition. Then G is the projective limit of the quotient groups G/N_U, where each U is a neighborhood of $e \in G$, and $N_U \subset U$ is a compact normal G_δ subgroup of G.

The following result appears in [1], III, 3.2.

If $X = \prod_{\alpha\in A} G_\alpha$ is a Cartesian product of the topological groups G_α then X has equivalent uniformities iff every G_α has equivalent uniformities.

It is easily seen that this implies Theorem 17 and Theorem 18 below.

Theorem 17. If $G = \varprojlim G_a$ then G has equivalent uniformities iff every G_a does.

Theorem 18. Let G be a locally compact σ-compact topological group or a locally compact group which satisfies the G_δ condition. Then G has equivalent uniformities iff G/N has equivalent uniformities for each compact normal G_δ subgroup N.

Grosser and Moskowitz [4] showed that if G is a locally compact group with equivalent uniformities then G is a projective limit of Lie groups. We will have more to say about this in the next section. A slightly weaker version of this result may be obtained by making the following observation if G has equivalent uniformities. In this case G has a neighborhood base at the identity consisting of invariant precompact neighborhoods. In this base, $V = xVx^{-1}$ for all $x \in G$.

Starting with any neighborhood V in this base one can choose a sequence of symmetric invariant neighborhoods $\{V_n\}$ from this base satisfying $V_k^2 \subset V_{k-1}$, $k = 2,3,\cdots$, $V_1 = V$. The set $N = \cap V_k$ is a compact normal G_δ group so Theorem 16 implies the following theorem.

Theorem 19. Let G be a locally compact topological group that has equivalent uniform structures. Then G satisfies the G_δ condition. Thus G is a projective limit of locally compact metric groups.

5. Approximation by Lie Groups.

In this section we will describe the general solution to our original problem. We begin with a classical theorem of Montgomery and Zippin [11], Section 4.6.

Theorem 20. [Montgomery and Zippin]. Let G be a locally compact group and let G_0 be the component of the identity in G. Suppose that G/G_0 is compact and let U be an arbitrary neighborhood of e. Then there is a compact normal subgroup $H \subset G$ such that G/H is a Lie group and $H \subset U$.

Since G/H is a Lie group it contains a neighborhood W' of the identity with no subgroups other than the identity. Since H is compact there is a neighborhood V of the identity in G such that $VH \subset U$ (where U, H, G, G_0 are as in the above theorem). If we let $\phi : G \to G/H$ be the natural map then $\phi(VH) = \phi(V)$ is open in G/H. If we let $W = \phi^{-1}(W') \cap VH$ then W has compact closure, it is a neighborhood of $e \in G$, and $H \subset W \subset U$. Furthermore $\phi(W) \subset W'$. Therefore we see that if F is a subgroup of G contained in W then $\phi(F) \subset W'$ and so $\phi(F) = \{e_{G/H}\} = \{H\}$. Therefore $F \subset \phi^{-1}\{e_{G/H}\} = H$. This argument yields a new proof of the following.

Theorem 21. (Grosser and Moskowitz [4]). Let G be a locally compact group such that G/G_0 is compact. Let U be any neighborhood of $e \in G$. Then there is a neighborhood W of $e \in G$, and a compact normal subgroup H of G such that the following hold:

(a) $W \subset U$, and $\bar{W}$ is compact.

(b) $H \subset W$.

(c) G/H is a Lie group.

(d) Every subgroup of G contained in V is contained in H.

The next theorem is a slight generalization of the theorem of Grosser and Moskowitz [4] that says that if G is a SIN group (SIN group = equal uniformities) then G is a projective limit of Lie groups. The basic difference here is that a strong emphasis is put on a maximality condition on H. This condition proved to be critical in the proof of the main Lie group structure theorem used for the characterization of equivalent uniformities on a locally compact group. The proof is a bit technical but does follow from Theorem 21.

<u>Theorem 22</u>. Let G be a locally compact group with equal uniformities and let G_1 be an open subgroup such that $G_0 \subset G_1$ and G_1/G_0 is compact. Let U be a neighborhood of $e \in G$. Then there is a neighborhood V of $e \in G$ and a compact normal subgroup H' of G such that the following hold:

(a) $V \subset U$, and $\overline{V}$ is compact.

(b) $H' \subset V$.

(c) G/H' is a Lie group.

(d) $H' = \bigcap_{g \in G} gHg^{-1}$, where H is the group of Theorem 21 and every normal subgroup of G contained in V is contained in H'.

The maximality condition (d) of the theorem proved to be important in the proof of the Approximation Theorem 23 below. It allowed the construction by induction of two sequences of subgroups of the locally compact group G, where every open σ-compact subgroup of G has equal uniformities. These sequences satisfy

(1) $A_1 \subsetneq A_2 \subsetneq \cdots \subsetneq A_n \subsetneq \cdots$ (A_i is open in G and σ-compact)

(2) $K_1 \supsetneq K_2 \supsetneq \cdots \supsetneq K_n \supsetneq \cdots$ (K_n is maximal normal in A_n)

(3) A_n/K_n is a Lie group.

This construction leads to a contradiction which implies the following.

<u>Theorem 23</u>. (Main Structure Theorem) If G is a locally compact topological group such that every open σ-compact subgroup has equivalent uniformities then the following hold.

(1) G satisfies the G_δ condition.

(2) G can be approximated by Lie groups.

(3) G has equivalent uniformities.

These results now allow the statement of the final theorem which collects all the above results. Again we note that the first four equivalences have been known for a long time to be true for every T_0 group though they do not appear in any readily available reference. Their proofs are given in the joint work of Itzkowitz, Rothman, Strassberg, and Wu [8] for completeness.

<u>Theorem 24</u>. Let G be a locally compact group. Then the following are equivalent:

(1) G has equivalent uniformities.

(2) If $\{x_\alpha\}$ is a net based on the directed set D satisfying $x_\alpha \to e$ and if $\{y_\alpha\}$ in G is any other net based on D then $y_\alpha x_\alpha y_\alpha^{-1} \to e$.

(3) For each neighborhood U of e, the intersection $\cap \{xUx^{-1} : x \in G\}$ is a neighborhood of e.

(4) There is a basis for the neighborhood system of e consisting of

sets W satisfying $W = xWx^{-1}$ for all $x \in G$ (G is a SIN group).

(5) Every open σ-compact subgroup of G has equivalent uniform structures.

(6) For every open neighborhood U of e and each right uniformly discrete sequence $\{x_n\}$ in G, the set $\bigcap_{n=1}^{\infty} x_n U x_n^{-1}$ is a neighborhood of e.

(7) G satisfies the G_δ condition and G is the projective limit of Lie groups with equivalent uniform structures.

(8) G satisfies the G_δ condition and G is the projective limit of metric groups with equivalent uniform structures.

(9) G satisfies the G_δ condition and each open homomorphic image of G onto a locally compact metric group has equivalent uniform structures.

(10) Every real valued left uniformly continuous function on G is right uniformly continuous.

(11) Every open subgroup of G has equivalent uniformities.

Section 6. Some Interesting Questions.

The above work settles the question for all locally compact topological groups. It shows that the uniform structure in these groups is in some sense determined by the classes of uniformly continuous functions on these groups. There are therefore immediately two questions that come to mind.

Question 1. Is this characterization true for all T_0 groups?

Question 2. Let X be a completely regular space that is locally compact but not discrete or compact. Suppose X has two non-equivalent uniformities $\mathcal{U}_1$ and $\mathcal{U}_2$ that give rise to the topology on X. Is the class of real valued functions that are $\mathcal{U}_1$-uniformly continuous different from the class of $\mathcal{U}_2$-uniformly continuous functions?

In the case of Question 1 we conjecture that the answer will be no. In other words we suspect that there are non locally compact topological groups on which the right and left uniformities are different but on which every real valued left uniformly continuous function is right uniformly continuous. The reason for believing this is as follows. It is known that there are P–groups (countable intersections of open sets are open) with inequivalent right and left uniformities. There is an example of such a group in Pestov's paper [12] in which the question of Itzkowitz was answered. Note now the following in such a group G.

If $f : G \to \mathbb{R}$ is left uniformly continuous then for each $n \geq 1$ there is a symmetric neighborhood U_n of the identity for which if $x^{-1}y \in U_n$ then $|f(x) - f(y)| < 1/n$. Consider the intersection of these neighborhoods. This is clearly a zero set and open. Since we could select the U_n's so that $U_n^2 \subset U_{n-1}$ we can conclude that the zero set is an open subgroup J of G. It is clear that since G is a P–space f is right uniformly continuous on any subgroup of G which is a countable union of translates of J.

References

1. Bourbaki, N., General Topology, Addison-Wesley, N.Y. 1966.

2. Braconnier, J., Sur les groupes topologiques localement compacts,

J. Math. Pures appl. N.S. 27, (1948), 1-85.

3. Comfort, W.W., and Itzkowitz, G.L., Density Character in Topological Groups, Math Ann., 226, (1977), 223-227.

4. Grosser, S. and Moskowitz, M., Compactness Conditions in Topological Groups., J. Reine Angew. Math 246 (1971), 1-40.

5. Hewitt, E. and Ross, K.A., Abstract Harmonic Analysis I, 2nd ed., Springer-Verlag, Berlin, (1979).

6. Itzkowitz, G.L., Continuous Measures, Baire Category and Uniform continuity in Topological Groups, Pacific J. Math 54, No. 2, (1974), 115-125.

7. Itzkowitz, G.L., Uniform Structure in Topological Groups, Proc. AMS 57, No. 2, (1976), 363-366.

8. Itzkowitz, G.L., Rothman, S., Strassberg, H., and Wu, T.S., Characterizations of Equivalent Uniformities in Topological Groups, to appear.

9. Kakutani, S. and Kodaira, K., Uber das Haarshe Mass in der lokal bikomppaktan Gruppe, Proc Imp. Acad. Tokyo, 22 (1944), 444-450.

10. Kister, J.M., Uniform Continuity and Compactness in Topological Groups, Proc. Amer. Math. Soc., 13, (1962), 37-40.

11. Montgomery, D. and Zippin, L., Topological Transformation Groups, Interscience, New York (1955).

12. Pestov, V.G., A Test of Balance of a Locally Compact Group. Ukrainian Math. J. 40 (1988), No. 1, 109-11.

13. Weil, Andre, L'Integration dans les Groupes Topologiques et ses Applications, Hermann, Paris, (1940).

A Constructive Theory of Uniform Locales, I: Uniform Covers

Peter T. Johnstone University of Cambridge, Cambridge, England

INTRODUCTION

Some years ago, at the end of a survey article (83) on locales, I expressed an opinion that the theory of uniform locales was ripe for further development. Subsequently, a number of authors (Pultr 84, Kříž 86, Frith 87, Isbell et al. 88, Wraith 87) took up the challenge and made contributions to the study of uniform locales and/or localic groups — the most striking development, perhaps, being the result that a localic subgroup of a localic group is necessarily closed (Isbell et al. 88). However, none of these authors took to heart the real point of (Johnstone 83), which was that the theory of locales needs to be developed constructively in order to be applied in topos-theoretic contexts (and so to yield results about, for example, the category of spaces over a base space): all of them assumed classical logic from the outset.

Of course, they had good reason for doing so. Somewhat to my chagrin, it has emerged that there is a substantial divergence between the constructive and classical theories of locales at the uniform level, which was not apparent at the "topological" level: many results which appear fundamental to the classical theory (including, for example, the result that uniformizable locales are (completely) regular) are not constructively valid. At the time of writing (83), I had hopes that restricting one's attention to open locales (cf. Johnstone 84) might enable one to "salvage" constructive proofs of apperently non-constructive results; but these hopes turned out to be larrgely illusory. It turned out that a further ingredient was needed; and this ingredient did not become available until very recently.

The new ingredient emerged from my attempts (88, 89) to understand and simplify the "Closed Subgroup Theorem" of (Isbėll et al. 88): on analysing the non-constructive elements of its proof, I found (89) that there is a "fibrewise" notion of closedness, classically equivalent to the usual notion but constructively weaker, and that if one substitutes this notion for the usual one the theorem becomes constructively valid (and indeed has a natural extension to localic groupoids, and to even more general structures). Subsequent work on this fibrewise notion of closedness (Jibladze-Johnstone 90, Johnstone 90, Vermeulen 90) has led to the introduction of weak "fibrewise" versions of the basic separation axioms for locales, which (inter alia) provide a way round the problem that uniformizability does not constructively imply regularity.

The time thus seems ripe at last for developing a constructive theory of uniform locales — even though it still cannot be claimed that the theory is as simple as one would have wished. (For example, although it is constructively true that uniformizability implies weak regularity, our only valid results in the opposite direction require the strong "classical" notion of regularity, as we shall see below.) In the present article, whose length is subject to restrictions imposed by the editors of this volume, we can do no more than take a first step along the road, by investigating the various "Tukey-style" definitions of uniformities interms of open covers, and showing how they interact with notions of regularity. In subsequent work, we plan to show how these definitions relate to the "Weil-style" ones in terms of entourages, and to the various definitions of fibrewise uniformity, for spaces over a base space, studied by (Dauns and Hofmann 68) and by (James 85, 86, 89); we shall also investigate such topics as completion, and the theory of quasi-uniform locales, in our constructive context.

Before concluding this Introduction, I should mention that J.J.C. Vermeulen has indepndently been investigating the constructive theory of uniform locales; full details of his results have not reached me at the time of writing, but I understand that he has covered at least part of the same ground as myself.

1 OPEN LOCALES AND FIBREWISE CLOSURE

Our terminology regarding frames and locales will be that of (Johnstone 82), but we shall follow the notational conventions introduced in (Johnstone 90). Thus locales will generally be denoted by letters such as X, Y, Z; the frame corresponding to a locale X will be denoted by $\mathcal{O}(X)$. Elements of $\mathcal{O}(X)$ will be denoted $U, V, W, \ldots$, and will (whenever possible) be identified with the corresponding open sublocales of X. If $f : X \to Y$ is a continuous map of locales, we write $f^* : \mathcal{O}(X) \to \mathcal{O}(Y)$ for the corresponding frame homomorphism, and f_* for the right adjoint of f^*.

It will be convenient, before proceeding further, to recall a few facts about open locales (Joyal-Tierney 84, Johnstone 84). A map $f : X \to Y$ of locales is called <u>open</u> if, for each $U \in \mathcal{O}(X)$, the image of the composite $U \to X \to Y$ is an open sublocale $f_!(U)$ of Y. This defines a mapping $f_! : \mathcal{O}(X) \to \mathcal{O}(Y)$ which is easily seen to be left adjoint to f^*; and indeed f is open iff f^* has a left adjoint $f_!$ satisfying the "Frobenius reciprocity" condition

$$f_!(U) \wedge V = f_!(U \wedge f^*(V))$$

for all $U \in \mathcal{O}(X), V \in \mathcal{O}(Y)$. If Y is a fit locale in the sense of (Isbell 72) (i.e. if every sublocale of Y is a meet of open sublocales), then the latter condition is redundant: the mere existence of the left adjoint $f_!$ ensures that the image of $U \to X \to Y$ is contained in a unique smallest open sublocale $f_!(U)$, and in the presence of fitness these two sublocales of Y must coincide. Regular locales, and in particular the terminal locale 1, are fit; we say X is an <u>open locale</u> if the unique map $X \to 1$ is open. (This terminology is slightly awkward, in that one has to distinguish carefully between open sublocales of a given locale X and sublocales which are open locales in their own right; but there does not seem to be any simple way of avoiding this difficulty.)

We write Ω for $\mathcal{O}(1)$, and think of its elements as "truth-values". We define a locale X to be <u>positive</u> if every open cover of X is inhabited, i.e. if

$$(\forall S \subseteq \mathcal{O}(X))\,((\vee S = X) \Rightarrow (\exists U)(U \in S))\,.$$

In (Johnstone 84) we showed that the left adjoint of the unique frame map $\lambda^* : \Omega \to \mathcal{O}(X)$, if it exists, necessarily sends $U \in \mathcal{O}(X)$ to the truth-value of the assertion "U is positive"; we therefore denote it by Pos rather than $\lambda_!$. A particularly useful characterization of open locales, proved in (Johnstone 84), is the "Positive Covering Lemma":

LEMMA 1.1 A locale X is open iff every open cover of an element U of $\mathcal{O}(X)$ can be refined to a cover by positive opens, i.e. iff, whenever we have $U \leq \vee S$ for some $S \in \mathcal{O}(X)$, we also have

$$U \leq \bigvee\{V \in S \mid \mathrm{Pos}(V)\}.$$

Open maps of locales are stable under pullback; in particular, if X is an open locale, the product projections $X \times X \to X$ are open. Also, if an open locale G carries a localic group structure, then the multiplication $m : G \times G \to G$ must be an open map; for we can factor it as

$$G \times G \xrightarrow{(m,\pi_2)} G \times G \xrightarrow{\pi_1} G$$

and the first factor is an isomorphism (cf. Pultr 88).

Fibrewise closedness was introduced in (Johnstone 89), where we showed that for any locale B there are notions of B-fibrewise denseness and B-fibrewise closedness for sublocales of locales over B, such that any inclusion $Y \to X$ of locales over B factors uniquely as $Y \to \overline{Y} \to X$ with $Y \to \overline{Y}$ fibrewise dense and $\overline{Y} \to X$ fibrewise closed; moreover, this factrization (the B-fibrewise closure of $Y \to X$) is functorial, and stable under pullback along open maps. When B is the terminal locale 1, we use the terms "strongly dense", "weakly closed" and "weak closure"; if we assume classical logic (i.e. that Ω is Boolean), then these concepts are equivalent to the familiar ones (as defined, for example, in (Johnstone 82, II 2.4), but constructively they are different.

We conclude this section with two lemmas, whose proofs may be found in (Johnstone 89), and which we shall need in what follows.

LEMMA 1.2 Let $X \to B$ be a locale over B, and $Y \to X$ a B-fibrewise dense sublocale. Then

(i) $X \to B$ is open iff the composite $Y \to X \to B$ is open.

(ii) $X \to B$ is surjective iff $Y \to X \to B$ is surjective; more generally, the image of $X \to B$ coincides with that of $Y \to X \to B$.

LEMMA 1.3 Let

$$\begin{array}{ccc} X' & \xrightarrow{f'} & X \\ \downarrow{\scriptstyle p'} & & \downarrow{\scriptstyle p} \\ B' & \xrightarrow{f} & B \end{array}$$

be a pullback where p is open. Then the pullback along f of any B-fibrewise dense sublocale of X is B'-fibrewise dense in X'.

2 OPEN COVERS AND STAR-REFINEMENTS

It is well known that there are two basic approaches to the definition of a uniformity on a set (or on a topological space): that using entourages, which was introduced by (Weil 37) and popularized by (Bourbaki 40), and that using uniform covers, which was introduced by (Tukey 40) and whose advantages were forcefully argued by (Isbell 64). (Incidentally, it is worth noting that, as the title of Weil's monograph indicates, a uniformity was originally seen as an additional structure to be imposed on space with a pre-existing topology. It happens that, in both the above approaches, the uniformity contains enough information to specify the topology uniquely, and so most modern textbooks begin by defining a uniformity on a set and then talk about the topology induced by the uniformity. However, this seems to be an accidental state of affairs; we shall see later that it does not persist for fibrewise uniformities, as we define them.) Both of these approaches can be copied in the locale-theoretic context: we shall begin with the uniform-cover approach.

By a cover of a locale X, we of course mean a subset $\mathcal{U}$ of $\mathcal{O}(X)$ whose join is X. We say that $\mathcal{U}$ refines a cover $\mathcal{V}$, and write $\mathcal{U} \leq \mathcal{V}$, if

$$(\forall U \in \mathcal{U})(\exists V \in \mathcal{V})(U \leq V);$$

$\leq$ is a pre-order, but not a partial order, on the set of covers of X. However, every cover $\mathcal{U}$ is $\leq$-equivalent to its downward closure $\downarrow \mathcal{U}$, and on downward-closed covers $\leq$ coincides with the inclusion partial ordering. If $\mathcal{U}$ and $\mathcal{V}$ are covers of X, we write $\mathcal{U} \wedge \mathcal{V}$ for the set

$$\{U \wedge V \mid U \in \mathcal{U}, V \in \mathcal{V}\};$$

this is a cover by the infinite distributive law, and is a greatest lower bound for $\{\mathcal{U}, \mathcal{V}\}$ in the pre-ordered set of covers. If $\mathcal{U}$ and $\mathcal{V}$ are downward-closed, then $\mathcal{U} \wedge \mathcal{V} = \mathcal{U} \cap \mathcal{V}$.

When dealing with covers of an open locale, we shall frequently wish to restrict our attention to proper covers, which are those $\mathcal{U}$ satisfying $(\forall U \in \mathcal{U})\mathrm{Pos}(U)$. By the Positive Covering Lemma, every cover $\mathcal{U}$ has a proper refinement $p\mathcal{U} = \{U \in \mathcal{U} \mid \mathrm{Pos}(U)\}$; classically, we have $\mathcal{U} \leq p\mathcal{U}$ (except for coverings of the degenerate locale !) as well as $p\mathcal{U} \leq \mathcal{U}$, but constructively things are very different. Note, however, that $\mathcal{U} \leq \mathcal{V}$ implies $p\mathcal{U} \leq p\mathcal{V}$, since $\mathrm{Pos}(U)$ and $U \leq V$ imply $\mathrm{Pos}(V)$; in fact $\mathcal{U} \mapsto p\mathcal{U}$ is right adjoint to the inclusion from proper covers to arbitrary covers.

If $\mathcal{U}$ is a cover of X, the entourage associated with $\mathcal{U}$ is the open sublocale $E(\mathcal{U})$ of $X \times X$ which is the join of $\{U \times U \mid U \in \mathcal{U}\}$. [Warning: even if $\mathcal{U}$ is downward-closed, we cannot deduce that $U \in \mathcal{U}$ from the information that $U \times U \leq E(\mathcal{U})$ — consider the covering of $\mathbf{R}$ by all open sets of measure < 1.] If V is a sublocale of X (not necessarily open), we write $\mathcal{U} * V$ for the sublocale $E(\mathcal{U}) \wedge (X \times V)$ of $X \times X$ (equivalently, for the join of $\{U \times (U \wedge V) \mid U \in \mathcal{U}\}$). And we write $\mathrm{St}_{\mathcal{U}}(V)$, the <u>star</u> of V with respect to $\mathcal{U}$, for the sublocale of X which is the image of

$$\mathcal{U} * V \longrightarrow X \times X \xrightarrow{\pi_1} X.$$

If X is an open locale and $V \in \mathcal{O}(X)$ (more generally, if V is an open locale in its own right), then $\mathrm{St}_{\mathcal{U}}(V)$ is open in X, and may be identified with the element

$$\bigvee\{U \in \mathcal{U} \mid \mathrm{Pos}(U \wedge V)\}$$

of $\mathcal{O}(X)$; but in general we have to consider $\mathrm{St}_{\mathcal{U}}(-)$ as an operation on sublocales rather than opens. Note that if $\mathcal{U} \leq \mathcal{V}$ then $E(\mathcal{U}) \leq E(\mathcal{V})$ and so $\mathrm{St}_{\mathcal{U}}(V) \leq \mathrm{St}_{\mathcal{V}}(V)$; in particular $\mathrm{St}_{\mathcal{U}}(V) = \mathrm{St}_{\downarrow\mathcal{U}}(V)$ for any $\mathcal{U}$ and V. Similarly, for any cover $\mathcal{U}$ of an open locale we have $E(\mathcal{U}) = E(p\mathcal{U})$ (by the Positive Covering Lemma applied in $X \times X$; note that $\mathrm{Pos}(U \times U)$ iff $\mathrm{Pos}(U)$), and so $\mathrm{St}_{\mathcal{U}}(V) = \mathrm{St}_{p\mathcal{U}}(V)$ for any V.

LEMMA 2.1 Let $\mathcal{U}$ be a cover of a locale X, and V a sublocale of X. Then

$$V \leq \mathrm{St}_{\mathcal{U}}(V).$$

Proof: For any $U \in \mathcal{U}$, the square $(U \wedge V) \times (U \wedge V)$ is contained in $\mathcal{U} * V$; but its image under π_1 is exactly $U \wedge V$, since the projection is split by the diagonal. So $\mathrm{St}_{\mathcal{U}}(V) \geq \bigvee\{U \wedge V \mid U \in \mathcal{U}\} = V$.

Classically, Lemma 2.1 has an immediate sterngthening which says that the closure of V is contained in $\mathrm{St}_{\mathcal{U}}(V)$. Constructively, the best we can do is

LEMMA 2.2 In the situation of Lemma 2.1, suppose additionally that V is an open locale in its own right. Then the weak closure of V is contained in $\mathrm{St}_{\mathcal{U}}(V)$.
Proof: Let $\overline{V}$ denote the weak closure of V. By Lemma 1.3, the inclusion $X \times V \to X \times \overline{V}$ is fibrewise dense over $\pi_1 : X \times \overline{V} \to X$; hence so is the inclusion $\mathcal{U} * V \to \mathcal{U} * \overline{V}$, since $E(\mathcal{U})$ is open in $X \times X$. So by Lemma 1.2(ii) we have $\mathrm{St}_{\mathcal{U}}(V) = \mathrm{St}_{\mathcal{U}}(\overline{V})$, and the result follows from Lemma 2.1.

However, we do not have a converse to Lemma 2.2: even for open sublocales V, W of an open locale X, the fact that the weak closure of V is contained in W does not imply the existence of a cover $\mathcal{U}$ with $\mathrm{St}_{\mathcal{U}}(V) \leq W$. (If the "classical" closure of V is contained in W, we do of course have such a cover, namely $\{W, \neg V\}$.) We give a fibrewise example:

EXAMPLE 2.3 Let B be the Sierpiński space $\{0,1\}$ (where the singleton $\{1\}$ is open but $\{0\}$ is not); let X be a space of the form $Y \cup \{\infty\}$, where Y is open in X (and has properties to be determined later) and the only neighbourhood of ∞ is the whole of X, and let $p : X \to B$ send ∞ to 0 and all the points of Y to 1. It is clear that p is an open continuous map, so $(X \to B)$ corresponds to an open internal locale in $\mathrm{Sh}(B)$. Now suppose Y contains open subsets V, W such that $V \neq \emptyset, W \neq Y$ and the Y-closure of V is contained in W. Then the B-fibrewise closure of V (regarded as an open set in X) is also contained in W; but the star of V with respect to any open cover of X, even in the B-fibrewise sense, is the whole of Y, since any such cover must contain X itself as a member. (For those who prefer all spaces to be Hausdorff, let us remark that "the same" counterexample can be constructed over any non-discrete Hausdorff base space B, by taking a space over B of the form $B \times Y$ and collapsing the fibre over a non-open point of B to a single point.)

Next, we need to formulate a notion of star-refinement for covers of a locale. There are (at least) three possibilities, the first of which was introduced by (Isbell 72) and also used by (Kirwan 81) and (Frith 87):

$$\mathcal{U} <^* \mathcal{V} \text{ if } (\forall U \in \mathcal{U})(\exists V \in \mathcal{V})(\mathrm{St}_{\mathcal{U}}(U) \leq V).$$

A weaker notion, which is closer to the notion of barycentric refinement in topology (Willard 70), will be denoted $<_1^*$:

$$\mathcal{U} <_1^* \mathcal{V} \text{ if there exists a cover } \mathcal{W} \text{ such that } (\forall W \in \mathcal{W})(\exists V \in \mathcal{V})(\mathrm{St}_{\mathcal{U}}(W) \leq V)$$

$$(\text{equivalently, if } \{W \in \mathcal{O}(X) \mid (\exists V \in \mathcal{V})(\mathrm{St}_{\mathcal{U}}(W) \leq V)\} \text{ covers } X).$$

Unfortunately this relation (unlike the strong notion $\mathcal{U} <^* \mathcal{V}$) does not imply $\mathcal{U} \leq \mathcal{V}$ — although, for open locales X, it does imply $p\mathcal{U} \leq \mathcal{V}$; for if $U \in \mathcal{U}$ and $\mathrm{Pos}(U)$, then from $U = \vee\{U \wedge W \mid W \in \mathcal{W}\}$ we deduce $(\exists W \in \mathcal{W})\mathrm{Pos}(U \wedge W)$, whence $(\exists W \in \mathcal{W})(U \leq \mathrm{St}_{\mathcal{U}}(W))$.

The third notion of star-refinement, which seems to make sense only for open locales, was introduced by (Pultr 84) and also used in (Kříž 86) and (Isbell et al. 88):

$$\mathcal{U} <_2^* \mathcal{V} \text{ if } \{\vee S \mid S \subseteq \mathcal{U}, S \text{ connected}\} \leq \mathcal{V},$$

where "S is connected" means that S is inhabited and $(\forall U, V \in S)\mathrm{Pos}(U \wedge V)$. Since the join of a connected subset S of $\mathcal{U}$ is contained in the $\mathcal{U}$-star of any member of S, we see that $\mathcal{U} <^* \mathcal{V}$ implies $\mathcal{U} <_2^* \mathcal{V}$. Once again, $\mathcal{U} <_2^* \mathcal{V}$ implies $p\mathcal{U} \leq \mathcal{V}$ (since if $U \in \mathcal{U}$ and $\mathrm{Pos}(U)$, then $\{U\}$ is a connected set), but not $\mathcal{U} \leq \mathcal{V}$ in general; note also that

$\mathcal{U} <_2^* \mathcal{V}$ iff $p\mathcal{U} <_2^* p\mathcal{V}$, since any connected subset of $\mathcal{U}$ is contained in $p\mathcal{U}$ and the cover $\{\vee S \mid S \subseteq \mathcal{U}, S \text{ connected}\}$ is proper.

Even classically, there is no implication in either direction between the relations $<_1^*$ and $<_2^*$; but each of them has the property that if you iterate it (i.e. take its relation product with itself) you get a relation which implies $<^*$. Constructively, the best we can do in this direction is the following:

LEMMA 2.4 Let $\mathcal{U}, \mathcal{V}, \mathcal{W}$ be three covers of an open locale X.
(i) If $\mathcal{U} <_1^* \mathcal{V} <_1^* \mathcal{W}$, then $p\mathcal{U} <^* \mathcal{W}$.
(ii) If $\mathcal{U} <_2^* \mathcal{V} <_2^* \mathcal{W}$, then $p\mathcal{U} <^* \mathcal{W}$.
Proof: (i) Let $\mathcal{Y} = \{Y \in \mathcal{O}(X) \mid (\exists V \in \mathcal{V})(\mathrm{St}_{\mathcal{U}}(Y) \leq V)\}$ and $\mathcal{Z} = \{Z \in \mathcal{O}(X) \mid (\exists W \in \mathcal{W})(\mathrm{St}_{\mathcal{V}}(Z) \leq W)\}$; by assumption, both $\mathcal{Y}$ and $\mathcal{Z}$ are covers of X. Now suppose $U \in \mathcal{U}$ and $\mathrm{Pos}(U)$. Then we have

$$U = \vee\{U \wedge Y \mid Y \in \mathcal{Y} \text{ and } \mathrm{Pos}(U \wedge Y)\}$$

and so $\mathrm{St}_{\mathcal{U}}(U) = \vee\{\mathrm{St}_{\mathcal{U}}(U \wedge Y) \mid Y \in \mathcal{Y} \text{ and } \mathrm{Pos}(U \wedge Y)\}$. Also, from $\mathrm{Pos}(U)$ we deduce $(\exists Z \in \mathcal{Z})\mathrm{Pos}(U \wedge Z)$. But if $Y \in \mathcal{Y}$ and $\mathrm{Pos}(U \wedge Y)$, then for some $V \in \mathcal{V}$ we have $U \leq \mathrm{St}_{\mathcal{U}}(U \wedge Y) \leq \mathrm{St}_{\mathcal{U}}(Y) \leq V$, and hence

$$\begin{aligned}\mathrm{St}_{\mathcal{U}}(U) &= \vee\{\mathrm{St}_{\mathcal{U}}(U \wedge Y) \mid Y \in \mathcal{Y} \text{ and } \mathrm{Pos}(U \wedge Y)\} \\ &\leq \vee\{V \in \mathcal{V} \text{ and } U \leq V\} \\ &\leq \mathrm{St}_{\mathcal{V}}(U \wedge Z) \leq \mathrm{St}_{\mathcal{V}}(Z).\end{aligned}$$

But there exists $W \in \mathcal{W}$ such that $\mathrm{St}_{\mathcal{V}}(Z) \leq W$, so we are done.

(ii) Assume $U \in \mathcal{U}$ and $\mathrm{Pos}(U)$. We have

$$\mathrm{St}_{\mathcal{U}}(U) = \vee\{U \vee U' \mid U' \in \mathcal{U} \text{ and } \mathrm{Pos}(U \vee U')\};$$

but $\mathrm{Pos}(U \vee U')$ implies that $\{U, U'\}$ is a connected subset of $\mathcal{U}$, and so $U \vee U' \in \downarrow \mathcal{V}$. Now the set

$$\{V \in \mathcal{V} \mid (\exists U' \in \mathcal{U})(\mathrm{Pos}(U \wedge U') \text{ and } U \vee U' \leq V)\}$$

is connected, since the meet of any two of its members contains U; so its join is contained in some $W \in \mathcal{W}$, and therefore $\mathrm{St}_{\mathcal{U}}(U) \leq W$.

We conclude this section with a simple example to show that the appearance of proper refinements in the statement of Lemma 2.4 (and in the discussion before it) cannot be avoided.

EXAMPLE 2.5 Let $(p : X \to B)$ be the product projection $(\pi_1 : \mathbf{R} \times \mathbf{R} \to \mathbf{R})$, regarded as an open internal locale in $\mathrm{Sh}(B)$. Given a real number $\varepsilon > 0$, let $\mathcal{U}(\varepsilon)$ be the sub-sheaf of the power-sheaf of $\mathcal{O}(X)$ generated by setting

$$[(U \times V \in \mathcal{U}(\varepsilon))] = B$$

whenever U and V are open intervals in $\mathbf{R}$ of length ε and $0 \notin U$, and also

$$[W(y_0, \varepsilon) \in \mathcal{U}(\varepsilon)] = B$$

for all $y_o \in \mathbf{R}$, where $W(y_0, \varepsilon) = \{(x, y) \mid |y - y_0| < \varepsilon/(2 + |x|)\}$. (Thus, for an arbitrary $Z \in \mathcal{O}(X)$, the truth-value $[Z \in \mathcal{U}(\varepsilon)]$ is the union of all open subsets of B over which the restriction of Z coincides with that of one of the particular opens $U \times V$ or $W(y_0, \varepsilon)$ above.) It is clear that $\mathcal{U}(\delta) \leq \mathcal{U}(\varepsilon)$ whenever $\delta < \varepsilon$. However, $\mathcal{U}(\varepsilon)$ is a proper cover, since the truth-value $\mathrm{Pos}(U \times V)$ is U, not B; in fact it is not hard to see that the only global elements of $p\mathcal{U}(\varepsilon)$ (i.e. the only Z satisfying $[Z \in p\mathcal{U}(\varepsilon)] = B$) are the sets $W(y_0, \varepsilon)$. From this it follows easily that we cannot have $\mathcal{U}(\delta) \leq p\mathcal{U}(\varepsilon)$ for any $\delta > 0$, and hence that we do not have $\mathcal{U}(\delta) <^* p\mathcal{U}(\varepsilon)$ either (although we do have $\mathcal{U}(\delta) <^* \mathcal{U}(\varepsilon)$ whenever $\delta < \varepsilon/3$). But if $\delta < \varepsilon/2$, it may be verified that we do have both $\mathcal{U}(\delta) <_1^* p\mathcal{U}(\varepsilon)$ and $\mathcal{U}(\delta) <_2^* p\mathcal{U}(\varepsilon)$.

3 T-UNIFORMITIES

We are now ready to formulate the central definition of this paper:

DEFINITION 3.1 Let X be a locale. A T-preuniformity on X (T stands for "Tukey", not "Trennungsaxiom") is a set U of covers of X such that

(1) $\mathcal{U} \in \mathsf{U}$, $\mathcal{U} \leq \mathcal{V} \Rightarrow \mathcal{V} \in \mathsf{U}$.

(2) $\mathcal{U}, \mathcal{V} \in \mathsf{U} \Rightarrow \mathcal{U} \wedge \mathcal{V} \in \mathsf{U}$.

(3) $\mathcal{U} \in \mathsf{U} \Rightarrow (\exists \mathcal{V} \in \mathsf{U})(\mathcal{V} <^* \mathcal{U})$.

A T-preuniformity is called a T-uniformity if it also satisfies

(4) For all $U \in \mathcal{O}(X)$, $U = \bigvee\{V \in \mathcal{O}(X) \mid (\exists \mathcal{U} \in \mathsf{U})(\mathrm{St}_{\mathcal{U}}(V) \leq U)\}$.

If X is open, a (pre)uniformity on X is called proper if it satisfies

(5) $\mathcal{U} \in \mathsf{U} \Rightarrow p\mathcal{U} \in \mathsf{U}$.

By a T-uniform locale we mean a pair (X, U) where X is a locale and U is a T-uniformity on X; a locale is called T-uniformizable if it admits some T-uniformity. If (X, U) and (Y, V) are T-uniform locales, a locale map $f : X \to Y$ is said to be uniform (or uniformly continuous) if $\mathcal{V} \in \mathsf{V}$ implies $\{f^*V \mid V \in \mathcal{V}\} \in \mathsf{U}$.

The (pre)uniformities on a given locale are ordered by inclusion; as usual, we shall say U is <u>finer</u> than V if $\mathsf{V} \subseteq \mathsf{U}$. Note that any preuniformity which is finer than a uniformity is itself a uniformity. We shall use the term "T_1-(pre)uniformity" and "T_2-(pre)uniformity" for the concepts obtained on replacing the relation $<^*$ by $<^*_1$ or $<^*_2$ in condition (3) of Definition 3.1 (and, in the second case, requiring that X be an open locale). However, we shall not pursue these concepts very far, since they seem to be unsatisfactory in the constructive context. We merely note that any T-(pre)uniformity is both a T_1-(pre)uniformity and a T_2-(pre)uniformity, and that (by Lemma 2.4) the converse implications hold for proper (pre)uniformities on open locales. Moreover, the analogues for T_1 and T_2 of the following Lemma are easily verified, from which it follows easily that the three notions of uniformizability are equivalent for open locales.

LEMMA 3.2 Let U be a set of covers of an open locale X, and define $\hat{\mathsf{U}}$ to be the set of all covers which are refined by some $p\mathcal{U}, \mathcal{U} \in \mathsf{U}$. If U is a T-(pre)uniformity on X, then so is $\hat{\mathsf{U}}$. Moreover, it is the coarsest proper (pre)uniformity containing U.

Proof: Conditions (1) and (5) of Definition 3.1 are immediate from the definition of $\hat{\mathsf{U}}$, and (2) follows from the fact that $p(\mathcal{U} \wedge \mathcal{V}) \leq p\mathcal{U} \wedge p\mathcal{V}$, since $\mathrm{Pos}(U \wedge V)$ implies $\mathrm{Pos}(U)$ and $\mathrm{Pos}(V)$. Moreover, $\hat{\mathsf{U}}$ inherits condition (4) from U, since it is finer than U, and the last assertion above is immediate; so it remains to verify (3). For this, we need to know that $\mathcal{U} <^* \mathcal{V}$ implies $p\mathcal{U} <^* p\mathcal{V}$; but this follows easily from that fact that $\mathrm{Pos}(U)$ implies $\mathrm{Pos}(\mathrm{St}_{\mathcal{U}}(U))$.

LEMMA 3.3 An open uniformizable locale is weakly regular.

Proof: By definition, X is weakly regular if every $U \in \mathcal{O}(X)$ can be expressed as a join of opens whose weak closures it contains. So this is immediate from condition (4) of the definition and Lemma 2.2.

However, we cannot strengthen the conclusion of 3.3 to "classical" regularity. Let us call a locale X <u>pre-discrete</u> if the diagonal $\Delta : X \to X \times X$ is open; by (Joyal-Tierney 84), Theorem V 5.1, a locale is discrete iff it is pre-discrete and open. Defining a <u>singleton</u> in X to be an element S of $\mathcal{O}(X)$ such that $S \times S$ is contained in Δ, we see that X is pre-discrete iff it can be covered by singletons, and discrete iff it can be covered by positive singletons (called "atoms" in (Joyal-Tierney 84)). Let $\mathcal{S}$ be the set of singletons in a pre-discrete locale X; then $E(\mathcal{S}) = \Delta$, whence we have $\mathrm{St}_{\mathcal{S}}(V) = V$ for any sublocale V of X. So the set S of all covers of X which are refined by $\mathcal{S}$ is a T-uniformity (note that $\mathcal{S} <^* \mathcal{S}$), and in particular we have

REMARK 3.4 Every discrete locale is uniformizable.

Thus, any non-Boolean topos, we have examples of uniformizable locales which are not regular.

For our next result, however, we shall have to use the classical notion of regularity, because of the problem presented by Example 2.3. Nevertheless, it is possible that the result remains true with "regular" replaced by "weakly regular", and/or with "compact" replaced by "weakly compact" in the sense introduced by (Vermeulen 90). (Example 2.3 does not forbid this; in fact it is regular in the strong sense provided Y is a regular space, and compact in the strong sense if Y is compact.)

LEMMA 3.5 Let X be a compact regular locale. Then the set U of all covers of X is a T-uniformity on X. If X is also open, then U is the only proper uniformity on X.
Proof: The fact that U satisfies (1) and (2) of 3.1 is immediate, and (4) follows from regularity of X: if the (classical) closure of V is contained in W, then $\mathrm{St}_{\mathcal{U}}(V) \leq W$, where $\mathcal{U}$ is the cover $\{W, \neg V\}$. To verify (3), let $\mathcal{U}$ be any cover of X. By compactness, we may refine it to a finite cover $\{U_1, U_2, \ldots, U_m\}$ (indexed by a decidable finite set $\{1, 2, \ldots, m\}$; there may be coincidences among the U_i, but this does not matter). We may then find a closure-refinement $\mathcal{V} = \{V_1, \ldots, V_m\}$ of $\{U_1, \ldots, U_m\}$, i.e. a cover such that $\overline{V}_i \leq U_i$ for each i, using the fact that the set

$$\{V_1 \vee \cdots \vee V_m \mid \overline{V}_i \leq U_i \text{ for each } i\}$$

is directed and has join $U_1 \vee \cdots \vee U_m = X$. Now we have m two-element covers $\{U_i, \neg V_i\}$ of X; we form their greatest common refinement $\mathcal{W}$, whose elements are indexed by complemented subsets S of $\{1, 2, \ldots, m\}$ and have the form

$$\bigwedge\{U_i \mid i \in S\} \wedge \bigwedge\{\neg V_j \mid j \notin S\}.$$

Writing W_S for this element of $\mathcal{W}$, we observe that $W_s \wedge V_i = 0$ whenever $i \notin S$, and hence $\mathrm{St}_{\mathcal{W}}(V_i) \leq \vee\{W_S \mid i \in S\} \leq U_i$ for each i. So $\mathcal{W} <_1^* \mathcal{U}$; and by taking $\mathcal{W}'$ to be a common refinement of $\mathcal{W}$ and $\{V_1, \ldots, V_m\}$ we can achieve $\mathcal{W}' <^* \mathcal{U}$.

For the uniqueness, let V be any proper uniformity on X and $\mathcal{U}$ an arbitrary cover of X. For each $U \in \mathcal{U}$, we have

$$U = \vee\{W \in \mathcal{O}(X) \mid (\exists \mathcal{V} \in \mathsf{V})(\mathrm{St}_{\mathcal{V}}(W) \leq U)\},$$

and hence the set

$$\mathcal{W} = \{W \in \mathcal{O}(X) \mid (\exists \mathcal{V} \in \mathsf{V})(\exists \mathcal{U} \in \mathsf{U})(\mathrm{St}_{\mathcal{V}}(W) \leq U)\}$$

is a cover of X. As before, we may find a finite (decidably-indexed) refinement $\{W_!, \ldots, W_m\}$ of $\mathcal{W}$; then for each $i \leq m$ we may choose $\mathcal{V}_i \in \mathsf{V}$ and $\mathcal{U}_i \in \mathsf{U}$ such that $\mathrm{St}_{\mathcal{V}_i}(W_i) \leq U_i$. Let $\mathcal{V}$ be the greatest common refinement of $\{\mathcal{V}_1, \mathcal{V}_2, \ldots, \mathcal{V}_m\}$; then we have $\mathrm{St}_{\mathcal{V}}(W_i) \leq U_i$ for each i, and so $\mathcal{V} <_1^* \mathcal{U}$. But this implies $p\mathcal{V} \leq \mathcal{U}$, and $p\mathcal{V} \in \mathsf{V}$ by clauses (2) and (5) of Definition 3.1, so $\mathcal{U} \in \mathsf{V}$ by 3.1 (1).

It is known that, constructively, compact regular locales need not be completely regular, so Lemma 3.5 makes it unlikely that we can strengthen the conclusion of Lemma 3.3 to any form of complete regularity. (However, it must be admitted that what is essentially the only known counter-example, the construction in (Henriksen-Isbell 58) of a proper quotient map $X \to B$ where B is completely regular but X is not, does not yield an open internal locale in $\mathrm{Sh}(B)$; I have been unable to determine whether it is weakly completely regular in the sense introduced in (Johnstone 90).) On the other hand, (classical) complete regularity does imply uniformizability, even without compactness: this is most easily proved by embedding an arbitrary complete regular locale as a sublocale of a power of the localic real line (cf. Johnstone 82), IV 1.7), and using the uniformizability of the latter (which is easily seen to be constructively valid) together with the following lemma.

LEMMA 3.6 (i) A sublocale of a T-uniformizable locale is uniformizable.
(ii) Suppose given a family of locales $(X_\gamma \mid \gamma \in \Gamma)$ and T-uniformities U_γ on X_γ for each γ. Then the product locale $\prod_{\gamma\in\Gamma} X_\gamma$ is T-uniformizable.
Note that part (ii) of the Lemma does not simply assert "A product of uniformizable locales is uniformizable"; the point is that we must make a particular choice of uniformities on the factors before we can construct one on the product.
Proof: (i) Let Y be a sublocale of X, and for each cover $\mathcal{U}$ of X let $\mathcal{U}|Y$ denote the cover $\{U \wedge Y \mid U \in \mathcal{U}\}$ of Y. It is straightforward to verify that for any sublocale V of X we have

$$\mathrm{St}_{\mathcal{U}|Y}(V \wedge Y) \leq \mathrm{St}_{\mathcal{U}}(V) \wedge Y;$$

from this it follows easily that $\{\mathcal{U}|Y \mid U \in \mathcal{U}\}$ is a T-uniformity on Y whenever $\mathcal{U}$ is a T-uniformity on X.

(ii) The argument in this case is very similar, though the actual construction of the product uniformity requires a little more care. Recalling that $\mathcal{O}\left(\prod_{\gamma\in\Gamma} X_\gamma\right)$ is generated by "open rectangles" which are finite intersections

$$\pi^*_{\gamma_1}(U_{\gamma_1}) \wedge \pi^*_{\gamma_2}(U_{\gamma_2}) \wedge \cdots \wedge \pi^*_{\gamma_m}(U_{\gamma_m})$$

where $\gamma_i \in \Gamma$ and $U_{\gamma_i} \in \mathcal{O}(X_{\gamma_i})$ for each i (note: since the equality on the index set Γ may not be decidable, we do not require that $\gamma_i \neq \gamma_j$ whenever $i \neq j$), we define U to consist of all covers which are refined by the greatest common refinement of a finite family of covers of the form $\{\pi^*_{\gamma_i}(U) \mid U \in \mathcal{U}_{\gamma_i}\}$, where $\gamma_i \in \Gamma$ and $\mathcal{U}_{\gamma_i} \in \mathsf{U}_{\gamma_i}$. The remaining details are as straightforward as before.

We remark that the analogues of Lemma 3.6 hold for T_1-uniformities and T_2-uniformities; the proofs are similar.

In both parts of the proof of Lemma 3.6, the uniformity we have constructed is an initial structure: in other words, it is the coarsest uniformity making each of the family of continuous maps (the inclusion $Y \to X$ in the first case, and the projections $\pi_\gamma : \prod_{\gamma\in\Gamma} X_\gamma \to X_\gamma$ in the second) uniformly continuous. More generally, if we are given a family of locale maps $(f_\gamma : X \to X_\gamma \mid \gamma \in \Gamma)$ with common domain, and uniformities U_γ on X_γ for each γ, we can construct the coarsest preuniformity on X making all the f_γ uniform — but it will not be a uniformity unless the f_γ are jointly regular monic, i.e. unless they embed X as a sublocale of the product of the X_γ. In any event, we may usefully note

COROLLARY 3.7 Let T-UnifLoc denote the category of T-uniform locales and uniformly continuous maps. Then the forgetful functor T-UnifLoc $\to$ Loc preserves limits.

Proof: For products, this is immediate from Lemma 3.6(ii) (plus the initiality of the uniformity constructed on the product). For equalizers, if f and g are uniform maps $(X, U) \to (Y, V)$, then their equalizer in Loc is a sublocale of X, and equipping this sublocale with the initial uniformity produces the equalizer in T-UnifLoc.

One final result which we should mention in this section concerns the uniformizability of open localic groups. It is possible to prove this directly from our definition in terms of open covers, but much easier to do so from the definition interms of entourages; so we shall postpone the proof until Part II of this paper, in which we shall establish the equivalence of the two definitions.

REFERENCES

Bourbaki, N. (1940). *Topologie Générale* (Eléments de Mathématique, Livre III), Actualités Sci. Ind. 858, Hermann, Paris.

Dauns, J. and Hofmann, K. H. (1968). *Representations of Rings by Sections.* Mem. Amer. Math. Soc. 83.

Frith, J. L. (1987). Structured frames. Ph.D. thesis, University of Cape Town.

Henriksen, M. and Isbell, J. R. (1958). Some properties of compactifications, *Duke Math. J., 25: 83-105.*

Isbell, J. R. (1964). *Uniform Spaces.* A. M. S. Surveys 12, American Mathematical Society.

Isbell, J. R. (1972). Atomless parts of spaces. *Math. Scand., 31: 5-32.*

Isbell, J. R., Kříž, I., Pultr, A., and Rosický, J. (1988). Remarks on localic groups, *Categorical Algebra and its Applications*, Lecture Notes in Math. 1348, Springer-Verlag , pp. 154-172.

James, I. M. (1985). Uniform spaces over a base. *J. Lond. Math. Soc., (2) 32: 328-336.*

James, I. M. (1986). Spaces. *Bull. Lond. Math. Soc., 18: 529-559.*

James, I. M. (1989). *Fibrewise Topology.* Cambridge Tracts in Math 91, Cambridge University Press.

Jibladze, M., and Johnstone, P. T. (1990). The frame of fibrewise closed nuclei. *Cahiers Top. Géom. Diff. Catégoriques*, to appear.

Johnstone, P. T. (1982). *Stone Spaces.* Cambridge Studies in Advanced Math. 3, Cambridge University Press.

Johnstone, P. T. (1983). The point of pointless topology. *Bull. Amer. Math. Soc. (N.S.), 8: 41-53.*

Johnstone, P. T. (1984). Open locales and exponentiation, *Mathematical Applications of Category Theory*, Contemp. Math. 30, Amer. Math. Soc., pp. 84-116.

Johnstone, P. T. (1988). A simple proof that localic subgroups are closed. *Cahiers Top. Géom. Diff. Catégoriques, 29: 157-161.*

Johnstone, P. T. (1989). A constructive "closed subgroup theorem" for localic groups and groupoids. *Cahiers Top. Géom. Diff. Catégoriques, 30: 3-23.*

Johnstone, P. T. (1990). Fibrewise separation axioms. *Math. Proc. Camb. Philos. Soc.*, to appear.

Joyal, A., and Tierney, M. (1984). *An Extension of the Galois Theory of Grothendieck.* Mem. Amer. Math. Soc. 309.

Kirwan, F. C. (1981). Uniform locales. Part III dissertation, University of Cambridge.

Kříž, I. (1986). A direct description of uniform completion in locales and a characterization of LT-groups. *Cahiers Top. Géom. Diff. Catégoriques, 27: 19-34.*

Pultr, A. (1984). Pointless uniformities I, II. *Comment. Math. Univ. Carolinae., 25: 91-120.*

Pultr, A. (1988). Some recent topological results in locale theory, *General Topology and its Relations to Modern Analysis and Algebra*, Res. Exp. Math. 16, Heldermann-Verlag, pp. 451-468.

Tukey, J. M. (1940). *Convergence and Uniformity in Topology.* Ann. of Math. Studies 2, Princeton University Press.

Vermeulen, J. J. C. (1990). Weak compactness in constructive spaces. *Math. Proc. Camb. Philos. Soc.*, to appear.

Weil, A. (1937). *Sur les Espaces à Structure Uniforme et sur la Topologie Generale.* Act. Sci. Ind. 551, Hermann, Paris.

Willard, S. (1970). *General Topology.* Addison-Wesley.

Wraith, G. C. (1981). Localic groups. *Cahiers Top. Géom. Diff., 22: 61-66.*

Wraith, G. C. (1987). Unsurprising results on localic groups. Preprint.

On P-Regular Completions of Cauchy Spaces

Darrell C. Kent and Raquel Ruiz de Eguino* Washington State University, Pullman, Washington

INTRODUCTION.

If (X, q) is a convergence space and p is a second convergence structure on X, the space (X, q) is defined to be *p-regular* if the p-closure of a filter $\mathcal{F}$ q-converges to x whenever $\mathcal{F}$ q-converges to x. The study of

* Supported by a grant from the Basque Government.

p-regularity was initiated in [9]. There it was shown that p-regularity is well-behaved relative to various structural and limit properties, and that a variety of important convergence concepts (e.g., local compactness, local boundedness, θ-continuity, and w-regularity) can be characterized in terms of p-regularity.

The versatility of p-regularity is further demonstrated when this notion is extended to Cauchy spaces and their completions (see [10]). Regular Cauchy spaces have been studied extensively because they retain many of the desirable properties of uniform space completions. Most of the attractive features of regular completion theory extend to the more general theory of p-regular completions. Indeed, the latter theory has one significant advantage: every Cauchy space which allows a p-regular completion also allows a strict p-regular completion.

In the first section, we review Cauchy spaces. Section 2 outlines the theory of p-regular completions and briefly mentions its application in the study of convergence space compactifications. Section 3 deals with the question : When is the completion of a product equivalent to the product of the completions? This question is completely answered for the "fine p-regular completion" and partially answered for the "fine regular completion".

1. CAUCHY SPACES.

The term *filter* will always mean a proper set filter. For each $x \in X$, let $\dot{x}$ denote the fixed ultrafilter containing $\{x\}$. If two filters $\mathcal{F}$ and $\mathcal{G}$ on a set X contain disjoint sets, we say that "$\mathcal{F} \vee \mathcal{G}$ fails to exist"; otherwise, these filters are said to be *linked*, and $\mathcal{F} \vee \mathcal{G}$ designates the filter generated by $\{F \cap G : F \in \mathcal{F}, G \in \mathcal{G}\}$.

A *Cauchy space* (X, C) is a set X with a collection C of filters on X satisfying:

(1) For each $x \in X, \dot{x} \in C$.

(2) If $\mathcal{G} \in C$ and $\mathcal{G} \leq \mathcal{F}$, then $\mathcal{F} \in C$.

(3) If $\mathcal{F}, \mathcal{G} \in C$ and $\mathcal{F}, \mathcal{G}$ are linked, then $\mathcal{F} \cap \mathcal{G} \in C$.

With each Cauchy space (X, C) there is associated a convergence structure q_C on X, defined by : $\mathcal{F}$ q_C-converges to x if $\mathcal{F} \cap \dot{x} \in C$. (X, C) is *complete* if each member of C is q_C-convergent. Indeed, it is often convenient to regard a convergence space as a complete Cauchy space.

For a convergence space (X, q), let "$c\ell_q$" denote the closure operator. A Cauchy space (X, C) is *regular* if $c\ell_q \mathcal{F} \in C$ whenever $\mathcal{F} \in C$. A Cauchy space is *Hausdorff* if $\dot{x} \cap \dot{y} \in C$ implies $x = y$, or, equivalently, if each q_C-convergent filter has a unique limit. *In this paper we make the standing assumption that all convergence and Cauchy spaces are Hausdorff unless otherwise indicated.*

Let (X, C) be a Cauchy space; then C is a *Cauchy structure* on X, and the members of C are called *Cauchy filters* (relative to C). A Cauchy space is *uniformizable* if there is a uniformity on the same set which has the same Cauchy filters, and *totally bounded* if every ultrafilter is a Cauchy filter. If C_1 and C_2 are Cauchy structures on X, then $C_1 \leq C_2$ (C_2 is *finer* than C_1, or C_1 is *coarser* than C_2) if $C_2 \subseteq C_1$.

A function between Cauchy spaces which preserves Cauchy filters is said to be *Cauchy-continuous*. *Cauchy-embeddings* and *Cauchy homeomorphisms* are defined in the obvious way. A *completion* $((Y, D), \varphi)$ of a Cauchy space (X, C) is a complete Cauchy space (Y, D) along with a Cauchy-embedding $\varphi : (X, C) \to (Y, D)$ such that $\varphi(X)$ is dense in Y. A completion $((Y, D), \varphi)$ of (X, C) is *strict* if, for each $\mathcal{G} \in D$, there is $\mathcal{F} \in C$ such that $c\ell_{q_D} \varphi(\mathcal{F}) \leq \mathcal{G}$.

Given a Cauchy space (X, C), an equivalence relation on C is defined as follows: $\mathcal{F} \sim \mathcal{G}$ if $\mathcal{F} \cap \mathcal{G} \in C$. For each $\mathcal{F} \in C$, let $[\mathcal{F}] = \{\mathcal{G} \in C : \mathcal{F} \sim \mathcal{G}\}$, and let $X^* = \{[\mathcal{F}] : \mathcal{F} \in C\}$ be the set of Cauchy equivalence classes. The natural injection $j : X \to X^*$ is defined by $j(x) = [\dot{x}]$, for all $x \in X$. A completion $((Y, D), \varphi)$ of (X, C) is said to be in *standard form* if $Y = X^*$

and $\varphi = \dot{j}$. Reed [11], has shown that every completion is equivalent to one in standard form. Thus, for the remainder of this paper, we shall restrict our attention to Cauchy completions in standard form.

For the purpose of constructing Cauchy spaces, it is convenient to have a characterization of the finest Cauchy structure $C_{\mathcal{A}}$ containing a given set $\mathcal{A}$ of filters. A finite set $\{\mathcal{F}_1, \cdots, \mathcal{F}_n\}$ of filters is *linked* if these filters can be arranged (by renumbering, if necessary) so that the pairs $(\mathcal{F}_1, \mathcal{F}_2), (\mathcal{F}_2, \mathcal{F}_3), \cdots,$
$(\mathcal{F}_{n-1}, \mathcal{F}_n)$ are all linked. Let $\mathcal{A}$ be a collection of filters on X which includes the fixed ultrafilters, and let $C_{\mathcal{A}}$ be the set of all filters $\mathcal{G}$ such that $\mathcal{F}_1 \cap \cdots \cap \mathcal{F}_k \leq \mathcal{G}$ for some linked family $\{\mathcal{F}_1, \cdots, \mathcal{F}_k\}$ in $\mathcal{A}$. Then $C_{\mathcal{A}}$ is called the *Cauchy structure generated by* $\mathcal{A}$; indeed, $C_{\mathcal{A}}$ is the finest Cauchy structure on X that contains $\mathcal{A}$. Of course, $C_{\mathcal{A}}$ will not generally be Hausdorff unless appropriate conditions are imposed on $\mathcal{A}$.

The Wyler completion $((X^*, C^*), \dot{j})$ of a Cauchy space (X, C) (introduced in [14]) is obtained by taking C^* to be the complete Cauchy structure on X generated by $\{\dot{j}(\mathcal{F}) \cap [\dot{\mathcal{F}}] : \mathcal{F} \in C\}$. The Wyler completion is strict and is "functorial" in the following sense: If $f : (X, C) \to (Y, D)$ is Cauchy-continuous, and $f^* : X^* \to Y^*$ is defined by $f^*([\mathcal{F}]) = [f(\mathcal{F})]$, then f^* is also Cauchy-continuous and the following diagram commutes:

$$\begin{array}{ccc} (X, C) & \xrightarrow{f} & (Y, D) \\ \dot{j}_X \downarrow & & \downarrow \dot{j}_Y \\ (X^*, C^*) & \xrightarrow{f^*} & (Y^*, D^*). \end{array}$$

Indeed, this completion has the universal lifting property relative to Cauchy-continuous functions into complete Cauchy spaces.

Results pertaining to regular completions of Cauchy spaces may be found in references [2], [6], [7], and [8]. In [7], a Cauchy space which allows a regular completion (respectively, a strict regular completion) is called a C_3 (respectively, SC_3) Cauchy space. It is shown in [8] that a regular Cauchy space is C_3 iff, whenever $\mathcal{F} \notin C$, there is a complete, regular Cauchy space (Y, D) and a Cauchy-continuous map $f : (X, C) \to (Y, D)$

such that $f(\mathcal{F}) \notin \mathcal{D}$. In [2], it is shown that each C_3 Cauchy space has a finest regular completion $((X^*, \mathcal{C}_r^*), j)$, where $\mathcal{C}_r^*$ denotes the regular modification of $\mathcal{C}^*$; this completion is called the *fine regular completion.* In [6] it is shown that a strict regular completion of a C_3 Cauchy space (if it exists) is unique (up to equivalence) and coincides with the fine regular completion. However, an example in [7] shows that the fine regular completion is not always strict. Necessary and sufficient conditions for the existence of a strict regular completion are given in Proposition 2.3, [7]; a simpler characterization of SC_3 spaces is given in Section 3.

2. P-REGULAR COMPLETIONS.

Let $(X, \mathcal{C})$ be a Cauchy space and p a convergence structure on X. $(X, \mathcal{C})$ is defined to be p-regular if $cl_p \mathcal{F} \in \mathcal{C}$ whenever $\mathcal{F} \in \mathcal{C}$. Note that $(X, \mathcal{C})$ is regular iff it is q_c-regular. If $(X, \mathcal{C})$ is a p-regular Cauchy space, then (X, q_c) is a *p-regular convergence space* as defined in [9].

It is shown in [9] that if (X, q) is a convergence space (not necessarily Hausdorff) and p is a T_1 convergence structure on X, then there is a finest p-regular convergence structure $r_p q$ coarser than q, and a coarsest p-regular convergence structure $r^p q$ finer than q. Similar results hold for Cauchy spaces.

PROPOSITION 2.1. *Let $(X, \mathcal{C})$ be a Cauchy space and let p be a convergence structure on X. Let $r_p \mathcal{C}$ be the Cauchy structure on X generated by $\mathcal{A} = \{cl_p^n \mathcal{F} : \mathcal{F} \in \mathcal{C}\}$, and let $r^p \mathcal{C} = \{\mathcal{F} : cl_p^n \mathcal{F} \in \mathcal{C}$, for all $n \in N\}$.*

(a) $r_p \mathcal{C}$ is the finest p-regular Cauchy structure on X coarser than $\mathcal{C}$. (However, $r_p \mathcal{C}$ is generally not Hausdorff).

(b) $r^p \mathcal{C}$ is the coarsest p-regular Cauchy structure on X finer than $\mathcal{C}$.

(c) If $\mathcal{C}$ is complete, then $r^p \mathcal{C}$ and $r_p \mathcal{C}$ are also complete.

In our discussion of p-regular completions, no further use will be made of the *upper p-regular modification* r^pC, but the *lower p-regular modification* r_pC plays an important role. In case $p = q_c$, it is of interest to compare $r_{q_c}C$ with the regular modification C_r. In general, we can say that $C_r \leq r_{q_c}C \leq C$; indeed, $r_{q_c}C$ is the first term in the *regularity series* for C (see [5]). If $r_{q_c}C = C_r$, the regularity series for C has length 1, and in this case we say that (X, C) is *quasi-regular.*

DEFINITION 2.2. *A completion $((X^*, \mathcal{D}), j)$ of a p-regular Cauchy space (X, C) is called a p-regular completion if there is a convergence structure p' on X^* such that:*

(i) $j : (X, p) \to (j(X), p' \mid_{j(X)})$ is a homeomorphism;

(ii) If $\mathcal{F} \in C$ is a non-q_c-convergent ultrafilter, then $j(\mathcal{F})$ p'-converges to $[\mathcal{F}]$ in X^*;

(iii) $(X^*, \mathcal{D})$ is p'-regular.

Note that the preceding definition involves the extension of both the Cauchy structure C and the convergence structure p to X^*; conditions (i) and (iii) are obvious requirements for the extension p' of p, while (ii) is a "quality-control" condition which leads to a completion theory resembling that for uniform spaces.

Let (X, C) be a p-regular Cauchy space, and let p^* denote the finest convergence structure on X^* which satisfies conditions (i) and (ii) of Definition 2.2. We shall call p^* the *standard extension* of p to X^*. Note that a completion $((X^*, \mathcal{D}), j)$ of a p-regular Cauchy space (X, C) is a p-regular completion iff $(X^*, \mathcal{D})$ is p^*-regular.

The remaining theorems of this section summarize the properties of p-regular completions; for proofs of these theorems, see [10].

THEOREM 2.3. *The following statements about a p-regular space (X, C) are equivalent:*

(a) $((X^*, r_p.C^*), j)$ *is a p-regular completion of* (X, C);

(b) (X, C) *has a p-regular completion;*

(c) If $\mathcal{F}, \mathcal{G} \in C$ *and* $\mathcal{F} \cap \mathcal{G} \notin C$, *then* $(\theta_C \mathcal{F}) \vee (\theta_C \mathcal{G})$ *fails to exist.*

For a space (X, C) which allows a p-regular completion, $((X^*, r_p.C^*), j)$ is called the *fine p-regular completion*. This completion is functorial relative to an appropriate class of morphisms. If (X, C) is p-regular and (Y, D) is s-regular, then $f : (X, C) \to (Y, D)$ is called an *admissible map* if it is Cauchy-continuous and has the further property that $f^* : (X^*, p^*) \to (Y^*, s^*)$ is continuous.

THEOREM 2.4. *Let* (X, C) *be a p-regular Cauchy space with a p-regular completion, and let* (Y, D) *be an s-regular Cauchy space with a s-regular completion. Let* $f : (X, C) \to (Y, D)$ *be an admissible map. If we regard* $(X^*, r_p.C^*)$ *as* p^**-regular and* $(Y, r_s.D^*)$ *as* s^**-regular, then the following diagram commutes, and all maps involved are admissible maps:*

$$\begin{array}{ccc} (X, C) & \xrightarrow{f} & (Y, D) \\ j_X \downarrow & & \downarrow j_Y \\ (X^*, r_p.C^*) & \xrightarrow{f^*} & (Y^*, r_s.D^*). \end{array}$$

If, in Theorem 2.4, we assume that (Y, D) is a complete s-regular Cauchy space, then we can identify (Y, D) with $(Y^*, r_s.D^*)$, and conclude that any admissible map $f : (X, C) \to (Y, D)$ can be lifted to an admissible map $f^* : (X^*, r_p.C^*) \to (Y, D)$ so that the diagram

$$\begin{array}{ccc} (X, C) & \overset{f}{\searrow} & \\ j \downarrow & & (Y, D) \\ (X^*, r_p.C^*) & \underset{f^*}{\nearrow} & \end{array} \qquad \text{commutes.}$$

THEOREM 2.5. *Let* (X, C) *be a Cauchy space which allows a p-regular completion. Then* $((X^*, r_p.C^*), j)$ *is a strict, p-regular completion of* (X, C), *and any other strict, p-regular completion of* (X, C) *is equivalent* *to the fine p-regular completion.*

THEOREM 2.6. *Let (X, C) be a Cauchy space which allows a p-regular completion. If (X, C) is uniformizable, then $((X^*, r_p \cdot C^*), j)$ is also uniformizable. If (X, C) is totally bounded, then $((X^*, r_p \cdot C^*), j)$ is also totally bounded, and consequently is a p-regular compactification of (X, q_C).*

A Cauchy space which is totally bounded and allows a p-regular completion is said to be *p-precompact.* A convergence space which allows a p-regular compactification is said to be *p-completely regular.* A subset B of a convergence space is *bounded* if each ultrafilter containing B is convergent; the space is *locally bounded* if each convergent filter contains a bounded set.

THEOREM 2.7. *Let (X, q) be a p-completely regular convergence space. There is a one-to-one, order-preserving correspondence between the p-precompact Cauchy structures compatible with q and the equivalence classes of strict, p-regular compactifications of (X, q). (X, q) has a largest p-regular compactification (obtained by taking the fine p-regular completion of the finest p-precompact Cauchy structure compatible with q); (X, q) has a smallest p-regular compactification (with one compactification point) iff (X, q) is locally bounded.*

We could conclude this section by considering a few special cases. Let (X, q) be an arbitrary convergence space, let δ be the discrete topology on X, and let C be the Cauchy structure on X consisting of all q-convergent filters and all non-q-convergent ultrafilters. Then the fine δ-regular completion of (X, C) exists and coincides with the Richardson compactification, [12], of (X, q).

Let (X, q) be a completely regular topological space, and let C be the Cauchy structure associated with the finest compatible totally bounded uniformity for q. Then the fine q-regular completion of (X, C) exists and is the Stone-Čech compactification of (X, q). The latter example can be generalized by assuming that (X, q) is a completely regular convergence space (as defined in [13]); let $\mathcal{C}$ be the finest totally bounded q-regular Cauchy structure compatible with q. In this case, the fine q-regular completion of $(X, \mathcal{C})$ exists and is the largest regular convergence space compactification constructed in [13].

3. PRODUCT THEOREMS.

A fundamental question in the study of any "functorial" type of topological extension concerns its behavior under products : When is the extension of a product equivalent to the product of the extensions? If the answer to this question is affirmative for a product involving a given set of spaces, we say that the given extension is *productive* relative to the specified set of spaces.

In 1959, Glicksberg [4] showed that the Stone-Čech compactification is productive relative to a set of completely regular spaces iff the product of these spaces is pseudo-compact. In 1987, Fric [1] showed that Novak's sequential envelope is productive relative to an arbitrary finite set of sequentially regular convergence space, and this result has recently been extended (see [3]) to a larger class of extensions. Uniform space completions are productive relative to an arbitrary set of uniform spaces, and the fine regular completion is productive relative to any set of SC_3 Cauchy spaces. In this section, we study the behavior of the fine p-regular completion under products.

We shall next introduce some notation pertaining to products which will be used in the propositions and theorems that follow. Let $\{(X_i, \mathcal{C}_i) : i \in I\}$ be a set of Cauchy spaces, and let $(X, \mathcal{C}) = \prod\{(X_i, \mathcal{C}_i) : i \in I\}$. For each $i \in I$, let $((X_i^*, \mathcal{D}_i), j_i)$ be a completion of $(X_i, \mathcal{C}_i)$, and let $(Y, \mathcal{D}) = \prod\{(X_i^*, \mathcal{D}_i) : i \in I\}$. Let $P_i : X \to X_i$ and $P_i^* : Y \to X_i^*$ be the canonical projection maps. Define $\sigma : X \to Y$ by $P_i^*(\sigma(x)) = j_i(P_i(x))$

and $\sigma^* : X^* \to X$ by $P_i^*(\sigma^*([\mathcal{F}])) = [P_i(\mathcal{F})]$; obviously the diagram

$$\begin{array}{ccc} X & \overset{\sigma}{\searrow} & \\ j\downarrow & & Y \\ X^* & \overset{\sigma^*}{\nearrow} & \end{array}$$

commutes.

PROPOSITION 3.1. *σ^* is a bijection.*

PROPOSITION 3.2. *If each (X_i, C_i) is p-regular, then (X, C) is p-regular, where $(X, p) = \prod\{(X_i, p_i) : i \in I\}$.*

PROPOSITION 3.3. *$((Y, D), \sigma)$ is a completion of (X, C) which is regular if each component completion is regular and strict if each component completion is strict.*

As a consequence of Proposition 3.3, we see that the classes of C_3 and SC_3 spaces are both closed under arbitrary products.

THEOREM 3.4. *Assume that $((X_i^*, D_i), j_i)$ is a p_i-regular completion of (X_i, C_i) for all $i \in I$, and let $(X, p) = \prod\{X_i, p_i) : i \in I\}$. Then $((Y, D), \sigma)$ is a p-regular completion of (X, C) iff exactly one of the following holds:*

(a) (X_i, C_i) is complete, for all $i \in I$;

(b) There is $i_0 \in I$ such that (X_{i_0}, C_{i_0}) is not complete, (X_i, C_i) is complete for all $i \neq i_0$ in I, and $p_i \leq q_{C_i}$ for all $i \neq i_0$;

(c) At least two members of $\{(X_i, C_i) : i \in I\}$ are not complete, and $q_{C_i} \geq p_i$, for all $i \in I$.

PROOF. It follows by Proposition 3.3 that $((Y, D), \sigma)$ is a completion of (X, C). Although this completion is not in standard form, we can consider it to be so by regarding the bijection σ^* as an identity map (in other words, by identifying Y with X^* and σ with j). Since these assumptions entail no loss of generality, we shall use them in this proof.

Let $(Y, p') = \prod\{(X_i^*, p_i^*) : i \in I\}$, where p_i^* is the standard extension of p_i to X_i^*. Let p^* be the standard extension of p to Y.

Assume that $((Y, \mathcal{D}), \sigma)$ is a p-regular completion of $(X, \mathcal{C})$. If $(X_i, \mathcal{C}_i)$ is complete for $i \neq i_0$ and $(X_{i_0}, \mathcal{C}_{i_0})$ is not complete, then there is $\mathcal{F}_0 \in \mathcal{C}_{i_0}$ such that $\mathcal{F}_0$ is non-convergent in $(X_{i_0}, \mathcal{C}_{i_0})$. For $i \neq i_0$, choose arbitrary $\mathcal{G}_i \in \mathcal{C}_i$; by assumption there is $x_i \in X_i$ such that $\mathcal{G}_i$ $q_{\mathcal{C}_i}$-converges to x_i. Let $\mathcal{H}$ be a filter on Y such that $P_i^*(\mathcal{H}) = j_i(\mathcal{G}_i)$ for $i \neq i_0$, and $P_{i_0}^*(\mathcal{H}) = j_{i_0}(\mathcal{F}_0)$. Since $\sigma^{-1}(\mathcal{H}) \in \mathcal{C}$ is not $q_{\mathcal{C}}$-convergent, $\mathcal{H}$ p^*-converges to $[\sigma^{-1}(\mathcal{H})]$ by Condition (ii) of Definition 2.2. It follows that $\mathcal{G}_i$ p_i-converges to x_i, and consequently $p_i \leq q_{\mathcal{C}_i}$, for all $i \neq i_0$.

Next, suppose $(X_{i_0}, \mathcal{C}_{i_0})$ and $(X_{i_1}, \mathcal{C}_{i_1})$ are not complete, where $i_0 \neq i_1$. The argument used in the preceding paragraph shows that $p_i \leq q_{\mathcal{C}_i}$ holds for $i \neq i_0$. Interchanging the roles of i_0 and i_1 leads to the conclusion that $p_i \leq q_{\mathcal{C}_i}$ for all $i \neq i_1$. Thus Condition (c) is established.

Conversely, assume that exactly one of conditions (a), (b), (c) holds. To establish that $((Y, \mathcal{D}), \sigma)$ is a p-regular completion of $(X, \mathcal{C})$, it is sufficient to show that p' satisfies condition (i), (ii), and (iii) of Definition 2.2. Indeed, p' is obviously an extension of p to Y, and so (i) holds. By Proposition 3.2, (iii) holds. To show that (ii) is satisfied, it suffices to show that $p^* \geq p'$.

Let $\mathcal{H}$ p^*-converge to α in Y, where $P_i^*(\alpha) = [\mathcal{F}_i]$ for all $i \in I$. If, for each $i \in I$, there is $x_i \in X_i$ such that $[\mathcal{F}_i] = [\dot{x}_i]$, then $P_i^*(\mathcal{H})$ p_i^*-converges to $[\mathcal{F}_i]$ for all $i \in I$, and hence $\mathcal{H}$ p'-converges to α. Next, suppose $\mathcal{F}_{i_0}$ is not convergent in $(X_{i_0}, \mathcal{C}_{i_0})$ for some $i_0 \in I$. Under Condition (a), this situation cannot occur. Under (b), $P_{i_0}^*(\mathcal{H})$ $p_{i_0}^*$-converges to $[\mathcal{F}_{i_0}]$, and $P_i(\sigma^{-1}(\mathcal{H}))$ $q_{\mathcal{C}_i}$-converges to x_i for $i \neq i_0$, which (because $p_i \leq q_{\mathcal{C}_i}$ for $i \neq i_0$) implies $P_i^*(\mathcal{H})$ p_i^*-converges to $[\dot{x}_i]$ for $i \neq i_0$; thus, again, $\mathcal{H}$ p'-converges to α. Under Condition (c), we can assume $I = I' \cup I''$, where $\mathcal{F}_i$ is non-$q_{\mathcal{C}_i}$-convergent for $i \in I'$, and $[\mathcal{F}_i] = [\dot{x}_i]$ for $i \in I''$. Because

$p_i \leq q_{C_i}$ for all $i \in I$, it again follows that $P_i^*(\mathcal{H})$ p_i^*-converges to $[\mathcal{F}_i]$ for all $i \in I$, and hence that $\mathcal{H}$ p'-converges to α. Thus $p' \leq p^*$, and the proof is complete. ▮

COROLLARY 3.5. *Let $\Omega = \{(X_i, C_i) : i \in I\}$ be a set of Cauchy spaces which allow p_i-regular completions. The fine p-regular completion is productive relative to Ω iff Ω satisfies exactly one of the conditions (a), (b), (c) of Theorem 3.4.*

PROOF. In Theorem 3.4, let $D_i = r_{p_i^*} C_i^*$ be the fine p_i-regular completion structure, and let $((Y, D), \sigma)$ be the product completion of $(X, C) = \prod\{(X_i, C_i) : i \in I\}$. If one of (a), (b), (c) is satisfied, then $((Y, D), \sigma)$ is a p-regular completion by Theorem 3.4, and also strict by Proposition 3.3 and Theorem 2.5. It follows by Theorem 2.5 that $((Y, D), \sigma)$ is equivalent to the fine p-regular completion of (X, C).

Conversely, if none of conditions (a), (b), (c) hold, then by Theorem 3.4, $((Y, D), \sigma)$ is not p-regular, and consequently the fine p-regular completion is not productive. ▮

COROLLARY 3.6. *If, under the assumptions of Corollary 3.5, $p_i \leq q_{C_i}$ for all $i \in I$, then the fine p-regular completion is productive with respect to Ω.*

PROPOSITION 3.7. *A C_3 Cauchy space (X, C) has a strict regular completion iff the Wyler completion space (X^*, C^*) is quasi-regular.*

PROOF. If (X, C) has a strict regular completion, then it follows from the results of [2] that $((X^*, C_r^*), j)$ is a strict regular completion. But the latter completion is also a q_C-regular completion of (X, C), and by Theorem 2.5, $C_r^* = r_{p^*} C^*$, where $p = q_C$. Thus, by definition, (X^*, C^*) is quasi-regular. The converse argument is obvious. ▮

By Proposition 3.3, a product of strict regular completions is a strict regular completion of the product. Since strict regular completions are unique up to equivalence, we obtain

COROLLARY 3.8 *If* $\Omega = \{(X_i, C_i) : i \in I\}$ *is a collection of* C_3 *Cauchy spaces whose Wyler completions are quasi-regular, then the fine regular completion is productive relative to* Ω.

REFERENCES

1. R. Frič, "Remarks on Sequential Envelopes," *Quaderni Mathematici 128,* Universitá degli Studi di Trieste, Trieste, 1987.
2. R. Frič and D. C. Kent, "Completion Functors for Cauchy Spaces," *Internat. J. Math & Math Sci 4* (1979), 589-604.
3. ____________, "The Finite Product Theorem for Certain Epireflections," *Math Nachr.* (To appear)
4. I. Glicksberg, "Stone-Čech Compactifications of Products," *Trans. Amer. Math. Soc. 90* (1959), 369-382.
5. D. C. Kent, "The Regularity Series of a Cauchy Space," *Internat. J. Math & Math Sci. 7* (1984), 1-13.
6. D. C. Kent and G. D. Richardson, "Regular Completions of Cauchy Spaces," *Pacific J. Math. 51* (1974), 483-490.
7. ____________, "Cauchy Spaces with Regular Completions," *Pacific J. Math. 111* (1984), 105-116.
8. ____________, "Cauchy Completion Categories" *Canad. Math Bull. 32* (1989), 78-84.
9. ____________, "P-Regular Convergence Spaces," *Math. Nachr.* (To appear).
10. ____________, "P-Regular Completions and Compactifications, *Math. Nachr.* (To appear).

11. E. E. Reed, "Completions of Uniform Convergence Spaces," *Math. Ann. 194* (1971), 83-108.

12. G. D. Richardson, "A "Stone-Čech Compactification for Limit Spaces," *Proc. Amer. Math, Soc. 25* (1970), 403-404.

13. G. D. Richardson and D. C. Kent, "Regular Compactifications of Convergence Spaces," *Proc. Amer. Math. Soc. 31* (1972), 571-573.

14. O. Wyler, "Ein Komplettierungsfunktor für Uniforme Limersräume," *Math. Nachr. 46* (1970), 1-12.

Which Spaces Have Metric Analogs?

T. Y. Kong Queens College, The City University of New York, Flushing, New York

Ralph Kopperman The City College of New York, The City University of New York, New York, New York

Paul R. Meyer Lehman College, The City University of New York, Bronx, New York

Metric analogs, introduced by Kong and Khalimsky in [KK], are important tools in the topological space approach to digital topology. For example, they can be used to prove a digital Jordan surface theorem for $\mathbb{Z}^n$ (where n is an arbitrary integer ≥ 2). Here we describe what we know about their existence, and finish with open questions.

Below, all spaces are assumed to have a base point, and maps are all continuous and base point preserving. If $F : X \times Y \to Z$ and $x \in X$, $y \in Y$ then $F(x, \cdot) : Y \to Z$ and $F(\cdot, y) : X \to Z$ are defined by $F(x, \cdot)(y) \equiv F(\cdot, y)(x) \equiv F(x, y)$.

DEFINITION 1 *A* (pseudo)metric analog *of a topological space X is a (pseudo)metric space M together with an open quotient map $q : M \to X$, such that whenever A is a (pseudo)metric space:*

1. *for any map $f : A \to X$ there is a map $g : A \to M$ such that $f = qg$.*
2. *for any two maps $f, g : A \to M$ such that $qf = qg$ there is a homotopy $F : A \times [0,1] \to M$ such that: $F(\cdot, 0) = f$; $F(\cdot, 1) = g$; $F(a_0, \cdot) \equiv m_0$,*

where a_0 and m_0 are the base points of A and M; and, for each $a \in A$, $qF(a, \cdot)$ is constant.

By (1), q must be onto. Also note that unique factorization cannot be expected in (1), except in trivial cases. For if $q(m) = q(p)$ where $m \neq p$, and $A = \{a, b\}$ with the discrete metric and base point b, then the (base point preserving) map $f : A \to X$ defined by $f(a) = q(m)$ is not uniquely factorable through M.

LEMMA 2 *Suppose X is T_0 and (M, q) is a pseudometric analog of X. Then (M^*, q^*) is a metric analog of X, where $M^* = M/\{(a, b) \mid d(a, b) = 0\}$ with metric d^* given by $d^*(a^*, b^*) = d(a, b)$, and where $q^*(a^*) = q(a)$. Here d is the pseudometric on M and, for $m \in M$, m^* denotes the equivalence class in M^* that contains m.*

Also, each metric analog of a T_0-space is a pseudometric analog of that space.

PROOF q^* is well-defined. To see this, suppose $a, b \in M$ and $a^* = b^*$. Then $d(a, b) = 0$. Hence $b \in \mathrm{Cl}(\{a\})$, so $q(b) \in \mathrm{Cl}(\{q(a)\})$; similarly $q(a) \in \mathrm{Cl}(\{q(b)\})$. But since X is T_0, $q(b) \in \mathrm{Cl}(\{q(a)\})$ and $q(a) \in \mathrm{Cl}(\{q(b)\})$ imply $q(a) = q(b)$.

Since q and $(a \mapsto a^*) : M \to M^*$ are open quotient maps, so is q^*. (M^*, q^*) satisfies condition (1) of a metric analog of X, for if A is a metric space, and $g : A \to M$, $f = qg$, then $f = q^*g^*$ where $g^*(c) = g(c)^*$ for each $c \in A$.

That (M^*, q^*) also satisfies condition (2) of a metric analog of X can be shown by similar reasoning, as can the second assertion of the Lemma. □

THEOREM 3 (Assume GCH.) *If X is not T_0 then X has no metric analog. Thus (by Lemma 2) a space with a pseudometric analog has a metric analog iff it is T_0.*

PROOF Suppose, for the purpose of getting a contradiction, that (M, q) is a metric analog of the non-T_0 space X. Let $\mu \geq |M| + \omega$ have cofinality ω, and pick $\beta \in \mu$.

Let H be the real Hilbert space with μ as an orthonormal base. Let A be the metric subspace of H whose elements are β (its base point), and the points in the Hilbert subspace generated by the elements of $\mu - \{\beta\}$. Then $|A| \geq \mu^\omega$, because $I(f) = \sum_{n\in\omega} f(n)/3^n$ defines a 1-1 function $I : (\mu - \{\beta\})^\omega \to A$. [1]

Since μ has cofinality ω, there exist $\mu_j < \mu$ such that $\mu = \sum_{j\in\omega} \mu_j$, and so by König's Theorem $\mu = \sum_{j\in\omega} \mu_j < \prod_{j\in\omega} \mu = \mu^\omega$. Therefore, $|A| > \mu$ and so, assuming GCH, $|A| \geq 2^\mu$.

However, $D = \{\beta\} \cup \{\sum_{j=1}^n r_j\alpha_j \mid n \in \omega, r_j \text{ rational}, \alpha_j \in \mu - \{\beta\}\}$ is a dense subspace of A, and $|D| = \sum_{n\in\omega} |\omega \times (\mu - \{\beta\})|^n = \mu$.

Since X is not T_0, it contains a two-point indiscrete set $\{x, y\}$. Any function $f : A \to X$ such that $f(\beta)$ is the base point of X and $f[A - \{\beta\}] \subseteq \{x, y\}$ is continuous and base point preserving (thus one of our maps); there are $2^{|A|} > 2^\mu$ such maps. However, every map from A into M is determined by its values on the dense subspace D, so there are no more than $|M|^{|D|} \leq \mu^\mu \leq (2^\mu)^\mu = 2^\mu$ of these maps — too few to allow every map from A into X to factor through M via q. □

LEMMA 4 *If (M_j, q_j) is a pseudometric analog of X_j for each integer $j \geq 0$, then $(\prod_{j\in\omega} M_j, \prod_{j\in\omega} q_j)$ is a pseudometric analog of $\prod_{j\in\omega} X_j$. Also, if (M, q) is a pseudometric analog of X, and $Y \subseteq X$ has the same base point as X, then $(q^{-1}[Y], q|q^{-1}[Y])$ is a pseudometric analog of Y. Thus subspaces and countable products of spaces with pseudometric analogs have pseudometric analogs, and so countable joins of topologies (on a fixed set) with pseudometric analogs have pseudometric analogs.*

The proofs are routine. □

LEMMA 5 *Any space with a pseudometric analog is first countable.*

PROOF Suppose (M, q) is a pseudometric analog of X. Given any $x \in X$, pick $a \in M$ such that $q(a) = x$; then $\{q[B_{1/n}(a)] \mid n \in \omega, n > 0\}$ is a neighborhood base about x. □

---

[1] In fact $|A| = \mu^\omega$, since there also exists a function J which maps a subset of $\mathbf{R}^\omega \times \mu^\omega$ onto A, namely $J(g, h) = \sum_{n\in\omega} g(n)h(n)$.

DEFINITION 6 *A topology is* Alexandroff *if each point is contained in a minimal neighborhood. A topology is* σ-Alexandroff *if it is a countable join of Alexandroff topologies.*

Alexandroff topologies were introduced by Alexandroff [A] (who called them "discrete"); they are also discussed by Johnstone [J]. Notice that a join of two Alexandroff topologies T, T' is Alexandroff; in fact the minimal neighborhood of x is $N(x) \cap N'(x)$, where $N(x)$ and $N'(x)$ are the minimal neighborhoods of x in T and T' respectively. Thus each σ-Alexandroff topology is an ascending join of Alexandroff topologies.

PROPOSITION 7 *Every Alexandroff space has a pseudometric analog.*

PROOF Let $\mathcal{B}$ be the base of minimal neighborhoods for an Alexandroff topology T on X. Let $M = \{(x,t) \mid x \in X,\ t : \mathcal{B} \to [0,1],\ (t(B) = 0 \Leftrightarrow x \notin B)\}$ with pseudometric $d((x,t),(y,u)) = |t-u|_\infty$, and with base point (b, b') where b is the base point of X and $b'(B) = 0$ if $b \notin B$, $b'(B) = 1$ if $b \in B$. Let $q : M \to X$ be the map $q((x,t)) = x$. We now show that (M, q) is a pseudometric analog of T.

To see that q is continuous, let $B \in \mathcal{B}$ and $(x,t) \in q^{-1}[B]$. Then $t(B) \neq 0$. If $d((x,t),(y,u)) < t(B)$ then $u(B) \neq 0$, so $y \in B$ and $(y,u) \in q^{-1}[B]$. Thus $q^{-1}[B]$ is open.

To see that q is open, let $(x,t) \in M$, and let C be the minimal neighborhood of x. If $r > 0$ then $C \subseteq q[B_r((x,t))]$ is shown as follows: if $y \in C$ define $u : \mathcal{B} \to [0,1]$ by $u(B) = r/2$ if $y \in B$ and $x \notin B$ and $u(B) = t(B)$ otherwise. Then $(y,u) \in M$ (because $y \notin B \in \mathcal{B} \Rightarrow x \notin B$), $d((x,t),(y,u)) \leq r/2$ and $y = q((y,u))$.

Now let A be a pseudometric space with pseudometric ρ and base point a_0. Given a (continuous and base point preserving) map $f : A \to X$, let $g : A \to M$ be defined by $g(a) = (f(a), t)$, where $t(B) = \min(k\rho(a, f^{-1}[X - B]), 1)$, k being a positive constant which is sufficiently large for g to be base point preserving. (It is readily confirmed that if B^* is the minimal neighborhood of b in T then any $k \geq 1/\rho(a_0, f^{-1}[X - B^*])$ can be used here.) Plainly $qg = f$. Also, $d(g(v), g(w)) \leq k\rho(v,w)$, so g is continuous.

Finally, suppose $qf = qg$, where $f, g : A \to M$ are continuous and base point preserving. Notice that if $f(a) = (x,t)$ and $g(a) = (y,u)$, then

$x = y$, so $t(B) = 0$ iff $u(B) = 0$. Define $F : A \times [0,1] \to M$ by $F(a,s) = (x(a), su(a)+(1-s)t(a))$, where $(x(a),t(a)) \equiv f(a)$ and $(x(a),u(a)) \equiv g(a)$. Then $F(\cdot,0) = f$, $F(\cdot,1) = g$, $qF(a,\cdot)$ is constant, and $F(a_0,s) = (b,b')$ for all $s \in [0,1]$. The continuity of F is shown routinely. □

THEOREM 8 *Each T_0 σ-Alexandroff space has a metric analog.*

PROOF Apply Proposition 7, Lemma 4 and Lemma 2. □

σ-Alexandroff spaces were introduced by Ceder [C] as spaces with a σ-interior-preserving base, and their history is discussed in Fletcher and Lindgren [FL].

DEFINITION 9 *A* quasimetric *on X is a map $d : X \times X \to \mathbb{R}$ such that $d(x,y) \geq 0$, $d(x,x) = 0$ and $d(x,z) \leq d(x,y) + d(y,z)$. The quasimetric d is said to be C_0 if $d(x,y) + d(y,x) = 0 \Rightarrow x = y$, and is said to be* non-Archimedean *if $d(x,z) \leq \max(d(x,y), d(y,z))$.*

If d is a quasimetric then the open balls $B_r(x) = \{y \mid d(x,y) < r\}$ form a base for a topology $T(d)$, the topology arising from d.

Notice that the topology arising from a quasimetric d is T_0 iff d is C_0.

PROPOSITION 10 *A topology is σ-Alexandroff iff it arises from a non-Archimedean quasimetric.*

PROOF ($\Rightarrow$) Let T be the ascending join of the Alexandroff topologies $T_1, T_2, T_3 \ldots$ on the same set, where T_1 is the indiscrete topology. Let $N_i(x)$ denote the minimal neighborhood of x in T_i. Notice that if $i < j$ then $N_j(x) \subseteq N_i(x)$. Let $d(x,y) = \inf\{1/n \mid y \in N_n(x)\}$. Clearly $d(x,x) = 0$. If $d(a,b) \leq 1/n$ then $b \in N_n(a)$ and, since $N_n(a)$ is an open set in T_n containing b, $N_n(b) \subseteq N_n(a)$. Thus if $d(x,y), d(y,z) \leq 1/n$ then $z \in N_n(z) \subseteq N_n(y) \subseteq N_n(x)$ which implies $d(x,z) \leq 1/n$; the arbitrary nature of n now shows that $d(x,z) \leq \max(d(x,y), d(y,z))$. So d is a non-Archimedean quasimetric. Also, $T = T(d)$ because the sets in T are just the sets P which satisfy the condition that for all $x \in P$ there is some n such that $N_n(x) = B_{1/(n-1)}(x) \subseteq P$.

($\Leftarrow$) Let $T = T(d)$, where d is a non-Archimedean quasimetric on a set X. Define $T_n = \{P \subseteq X \mid$ if $y \in P$ then $C_{1/n}(y) \subseteq P\}$ where $C_r(x) = \{y \mid d(x,y) \leq r\}$. T_n is easily seen to be a topology. We show it is Alexandroff by showing that $C_{1/n}(x) \in T_n$, which implies that $C_{1/n}(x)$ is the minimal neighborhood of x in T_n. Specifically, if $y \in C_{1/n}(x)$ and $z \in C_{1/n}(y)$ then $z \in C_{1/n}(x)$ because $d(x,z) \leq \max(d(x,y), d(y,z)) \leq 1/n$; so for all $y \in C_{1/n}(x)$, $C_{1/n}(y) \subseteq C_{1/n}(x)$. The fact that $C_{1/n}(x) \in T_n$ also implies that T is the join of the T_i, since the sets $C_{1/n}(x)$ form a local base for T at x. □

The T_1 case of Proposition 10 is essentially shown in [FL]. Nedev [N] attributes the realization that spaces with a σ-interior-preserving base are quasimetrizable to Doitchinov.

We close with two open questions:

QUESTION 1 *For T_0-spaces, here is a summary of the inclusion relations which follow from the above results:*

σ-Alexandroff $\subseteq$ Has a Metric Analog $\subseteq$ First Countable

σ-Alexandroff $\subseteq$ Quasimetrizable $\subseteq$ First Countable

Just where do the spaces that have a metric analog fit along the second line?

QUESTION 2 *By Theorem 3 it is consistent with ZFC that only T_0-spaces have metric analogs. Is it also consistent with ZFC that there are non-T_0-spaces with metric analogs? If so, to what extent can the GCH hypothesis of Theorem 3 be weakened?*

References

[A] P. Alexandroff, Diskrete Räume, *Mat. Sbornik* **2** (1937), 501 - 518.

[C] J. G. Ceder, Some generalizations of metric spaces, *Pacific J. Math.* **11** (1961), 105 - 125.

[FL] P. Fletcher and W. F. Lindgren, *Quasi-uniform Spaces*, M. Dekker, NY, 1982.

[J] P. T. Johnstone, *Stone Spaces*, Cambridge U. Press, Cambridge, 1982.

[KK] T. Y. Kong and E. Khalimsky, Polyhedral analogs of locally finite topological spaces, in: R. M. Shortt, Ed., *General Topology and Applications: Proceedings of the 1988 Northeast Conference*, M. Dekker, 1990, 153 - 164.

[N] S. Nedev, On generalized-metrizable spaces, *C. R. Acad. Bulg. Sci.* **20** (1967), 513 - 516 (in Russian); *MR* **35**, no. 7292.

New Invariant Differential Operators on Supermanifolds and Pseudo-(co)homology

Dimitry Leites, Yuri Kochetkov*, and Arkady Weintrob Stockholm University, Stockholm, Sweden

Introduction. This talk is a direct continuation of [BL1] and [L1], where there are listed, among other things, unary differential operators invariant with respect to a supergroup of diffeomorphisms. A rival exposition [Ki] is also recommended. This talk provides also with examples of calculations lacking in [LP1].

The topology of differentiable manifolds has always been related with various geometric objects on them and in particular with operators invariant with respect to the group of diffeomorphisms of the manifold, operators which act in natural bundles. For example, an important invariant of the manifold, its cohomology, stems from the de Rham complex whose terms are connected by an invariant differential operator -- the exterior differential.

The role of invariance had been appreciated already in XIX century in physics (differential operators invariant with respect to the group of diffeomorphisms preserving a geometric structure are essential both in Maxwell's laws of electricity and magnetism and in Einstein's (and Hilbert's) formulation of general relativity).

Simultaneously invariance became a topic of conscious interest for mathematicians (the representation theory flourished in works of F.Klein, followed by Lie, Levi-Civita, É.Cartan; it provided with the language and technique adequate to study geometric structures). Still it was not untill O.Veblen's talk in 1928 at Math. Congress in Bologna that invariant operators (such as, say, Lie derivative, the exterior differential or integral) became the main object of the study.

Schouten and Nijenhuis tackled Veblen's problem: they rewrote it in modern terms and found several new bilinear invariant differential operators. Schouten conjectured that there is essentially one *unary* invariant differential operator: the exterior differential of differential forms.

This conjecture had been proved in particular cases by a number of people, and in full generality in 1977-78 by A.A.Kirillov and, independently, C.L.Terng [Ki]. Thanks to the usual clarity and an enthusiastic way of Kirillov's presentation he drew new attention to this problem. Under the light of this attention it became clear that in 1973 A.Rudakov ([R1]) also proved this conjecture by a simple and powerful method which, owing to Theorem 0.2 below, reduces Veblen's problem for differential operators to a "computerizable" one for any given "arity" and order of the operator.

**Current Affiliation*: University of Twente, Enschede, The Netherlands

Later there had been classified **unary** differential operators on manifolds invariant with respect to any of the remaining three pseudogroups of transformations corresponding to a simple *infinite* dimensional $\mathbb{Z}$-graded Lie algebra L. (Rudakov's method is poorly applicable to the finite-dimensional groups; therefore it does not apply, say, to isometries of a Riemannian manifold or the group preserving the Laplace operator, see [Ki], [L1].)

The next step in solution of Veblen's problem was due to Grozman, who in 1978 described all **binary** invariant differential operators [G1, G2]. It turned out that there are plenty (but not too many) of them. Miraculously, the 1st order differential operators determine (for a few exceptions) a Lie *super*algebra structure on their domain. Here superalgebras timidly indicated their usefulness in a seemingly nonsuper problem. Other examples (such as Quillen's proof of the index theorem) followed suit.

Our study of invariant differential operators on **super**manifolds began in 1976 as a byproduct of attempts to construct an integration theory on supermanifolds similar to the integration theory of differential forms on manifolds. In this way new objects which it is possible to integrate (integrable forms) were discovered, see [BL1, BL2].

The unary invariant differential operators acting in the superspaces of tensor fields with finite-dimensional fibers are classified by now for any known simple Lie superalgebra L of vector fields in their standard realization (see the review [L1] and subsequent papers [Ko5]-[Ko7]). Here we continue the account of new results, ideas and conjectures, and study the extent to which the results presented in [L1] and [Ko5-7] are final.

The main results of this paper are as follows:

First, we describe *new operators* invariant with respect to the already considered (super)groups but acting in the superspaces of sections of vector bundles with *infinite dimensional fibers*: These operators of high order have no counterparts on manifolds. (A year after this talk was delivered, I.Penkov and V.Serganova interpreted these new operators as acting in the superspaces of certain tensor fields on "curved" superflag and supergrassmann supervarieties [PS]).

Second, we propose to relate with some of these operators *new topological invariants* (or perhaps old, like cobordisms, but from a new viewpoint). Recall that since the de Rham cohomology of a supermanifold are the same as those of its underlying manifold, the "old type" operators are inadequate to study "topological" invariants of supermanifolds. The operators described here and related to vector bundles of infinite rank lead to new (co)homology theories (we loosely call them pseudo(co)homology) that provide us with invariants different from de Rham ones. We will calculate these invariants elsewhere; so far see Shander's elucidations in [L2], v.1, Ch.4.

Lastly, we show more of *new, typically supermanifoldish, geometric structures* briefly mentioned in [L2] .

The approach adopted here for supermanifolds prompts us to *raise* (for manifolds, too) *some problems* previously not discussed, and it is clear that their solution will give us more of invariant operators in addition to those listed here.

Acknowledgements. We are grateful to J.Bernstein, E.Galaktionova, P.Grozman and A.Rudakov for help; to P.Deligne, B.Feigin and D.Fuchs for helpful discussions; D.L. is also thankful to I.Bendixson (1986-88) and NSF (1987 and 1989) grants, MPI, Bonn (1987) and IHES (1986) for financial support and to the organisers of the conference, especially to Prof. P.Misra, for the encouragement, patience and help.

0. Some background and formulation of Veblen's problem. In what follows the ground field is $\mathbb{R}$ and everything is smooth, i.e. of C^∞-class if we speak about manifolds and groups; passing to algebras we consider everything over $\mathbb{C}$.

0.1. Veblen's problem. Let V-->E-->B be a rank n vector bundle over an m-dimensional (connected) manifold B with the action of the group Diff(B) of diffeomorphisms of B extended to an action on the total space E so that the stabilizer St_b of any $b \in B$ linearly acts on the fiber over b (isomorphic to the vector space V) and the action only depends on k-jets of elements from St_b. The sections of this bundle constitute the space denoted by $\Gamma_k(B, V, E)$, or just $\Gamma_k(V)$ if B is fixed; it is called the space of *geometric objects of differential order* $\leq k$. The precise formulation of Veblen's problem for smooth invariant operators (SIO):

describe Diff(B)*-invariant operators* $D: \Gamma_k(V_1) \otimes \ldots \otimes \Gamma_k(V_r) \to \Gamma_k(V)$ (SIO)

The number r here is called *arity* of D (i.e. D is un*ary* for r = 1, bin*ary* for r = 2, etc.), k the *differential order* of D (do not confuse with order if D is a differential operator, for example the integral is a (nondifferential) operator of differential order 1); in what follows we will only consider operators homogeneous with respect to differential order.

Particular case: tensor fields. For k = 1 geometric objects are called *tensor fields.* Tensor fields are sometimes defined in the following, equivalent, way. Let ρ be a representation of GL(m) in a (finite-dimensional, traditionally) space $V \subset T^p(id) \otimes T^q(id^*)$, where id is the (space of) the *identity*, or *standard*, representation of GL(m). A *tensor field of type* ρ (sometimes called a tensor field of *type V*) is a section of the vector bundle with fiber V associated with the frame bundle over B. A tensor field t is given by a vector-valued function t: B --> V such that the passage from local coordinates x to another coordinates y is given by the formula

$$t(y(x)) = (\rho(\partial y/\partial x)(t(x)),$$

or, in still other words,

$$(gt)(x) = (\rho(J_g(x)(t))(g^{-1}x), \text{ where } g \in \mathrm{Diff}(B),\ J_g = \partial y/\partial x \text{ is the Jacobi}$$

matrix of g.

It was a **tacit convention:** to consider in the above problem only *finite dimensional* fibers. Since GL(m) is reductive, it suffices, assuming the convention, to confine ourselves to irreducible GL(m)-modules.

For superalgebras complete reducibility is *rara avis* and the above convention is absolutely unjustified. But even on manifolds it is natural (in field theory) to consider *infinite dimensional* fibers. Besides, as we will see, it is natural to consider not only algebraic representations of GL(m); it pays to pass to "all" (say, with highest (lowest) weight vector) representatons of $\mathfrak{gl}(m)$. However, even though we should study the problem (SIO) for indecomposable modules, the first describable case is that of irreducible ones.

0.2. Rudakov' s method. Let $\mathcal{L} = \mathfrak{vect}(m/n) = \mathfrak{der}\ \mathbb{C}[[x]]$, where $x = (u_1,...,u_m, \xi_1, ..., \xi_n)$, be the Lie superalgebra of formal vector fields, $\mathcal{L} = \mathcal{L}_{-1} \supset \mathcal{L}_0 \supset ...$ its standard filtration corresponding to (x)-adic topology of $\mathbb{C}[[x]]$. Let $T_k(V) = \mathrm{Hom}_{U(\mathcal{L}_0)}(U(\mathcal{L}),V)$ for an $(\mathcal{L}_0/\mathcal{L}_k)$-module V be a formal analogue of $\Gamma_k(V)$.

An $\mathcal{L}$-module which is also a $\mathbb{C}[[x]]$-module so that both structures are compatible will be called a *continuous* $\mathcal{L}$-module. Let Tens_k be the category of continuous $\mathcal{L}$-modules M such that $\mathcal{L}_k M \subset (x)^k M$.

Theorem (J.Bernstein, [BL1]). *In $Tens_k$, for* ***diferential*** *operators acting in the superspaces of tensor fields on supermanifolds Veblen's problem (SIO) is equivalent to the following formal one:*

describe $\mathcal{L}$-invariant operators $D: T_k(V_1)\otimes ... \otimes T_k(V_r) \dashrightarrow T_k(V)$. (FIO)

Convention. In what follows we will only consider *differential* operators D.

As is explained in [BL1], [Ko5, 6], A.Rudakov's approach enables one

1) to consider in the above problem (FIO) not only $\mathfrak{vect}$ but any filtered Lie (super) algebra of finite *depth* d (i.e. of the form $\mathcal{L} = \mathcal{L}_{-d} \supset ... \supset \mathcal{L}_0 \supset \mathcal{L}_1 \supset ...$);

2) to pass from $\mathcal{L}$ to the associated graded Lie (super) algebra $L = \oplus_i L_i$ with $L_i = \mathcal{L}_i/\mathcal{L}_{i+1}$ and from Tens_k to its dual category Ind_k whose objects are discreetly topologized $\mathcal{L}$-modules, the main examples beeng the induced modules $I_k(V) = U(L)\otimes_{U(\mathcal{L}_0)} V$ for a $\mathcal{L}_0/\mathcal{L}_k$-module V. (The answer should be formulated in lucid geometric terms of the category Tens; but proof goes easier in Ind.)

From our point of view the main results of [R1] are the following ones (in [R1] they are disguised as Corollary 13.8 or some such):

1) **Main Lemma**. *Any discretely topologized irreducible $\mathcal{L}$-module with highest weight is a quotient of some $I_k(V)$.*

The importance of this Lemma is due to the duality between $I_k(V)$ and $T_k(V^*\otimes \mathrm{tr})$, cf. [BL1]. The counterpart of Main Lemma for superalgebras see in [Ko5, 6].

2) **There are no invariant operators for k > 1 and r = 1**: if V is an irreducible $\mathcal{L}_0/\mathcal{L}_k$-module with lowest weight, then the L-module $T_k(V)$ is irreducible for a simple infinite dimensional vectory Lie algebra $\mathcal{L}$.

Problem. *Write explicitely invariant operators of arity r>1 acting in the spaces of geometric objects depending on k-jets of diffeomerphisms for k>1.*

3) **Unary invariant operators exist for finite-dimensional fibers only.** More exactly, denote for brevity $T(\phi) = T(V_\phi)$ for an irreducible L_0-module V_ϕ with the lowest weight ϕ; then the only $\mathfrak{vect}(n)$-invariant operators are 1-st order operators d: $T(-\phi_i) \to T(-\phi_{i+1})$ for $0 \le i \le n$, where ϕ_i is the highest weight of the i-th fundamental representation of L_0 and $\phi_0 = (0, ..., 0)$. All these operators are, essentially, one operator -- the exterior differential in the de Rham complex. (For $\mathcal{L}$ other than $\mathfrak{vect}$ there are *seemingly* other differential operators; but all of them are realized in the space of differential forms and are compositions of d and the structure tensor that $\mathcal{L}$ preserves, at least if $\mathcal{L}$ is considered in the standard realisation, see [R2], [L2], [Ko5-7]. For a nonstandard realization this paper provides with NEW examples, see sec.5.)

0.3. Passage to supermanifolds. Let V be a superspace of an irreducible $\mathfrak{gl}(m/n)$-module with an even highest (lowest) vector; we will identify V by the labels of its highest (lowest) weight with respect to a Borel subalgebra. Let Vol = Σ_0 = $\mathbb{C}[[x]]v$ be the superspace of volume forms. Let $\mathrm{Vol}^\mu = T(\mu, ..., \mu; -\mu, ..., -\mu) = \mathbb{C}[[x]]v^\mu$ for $\mu \in \mathbb{C}$ be the space of *μ-densities*, the "μ-th tensor power (over functions)" of Vol -- the natural extention of the notion well-defined for $\mu \in \mathbb{Z}$ as a GL(m/n)-module by the formula

$$L_X(fv^\mu) = (X(f) + (-1)^{p(f)p(X)} \mu f \operatorname{div} X)\, v^\mu$$

Set F = T(0, ..., 0), the superspace of functions; $\Omega_\mu{}^r = \Omega^r \otimes_F \mathrm{Vol}^\mu$, where $\Omega^r = T(E^r(\mathrm{id}^*))$, the superspace of differential forms; $L_\mu{}^r = L^r \otimes_F \mathrm{Vol}^\mu$, where $L^r = T(E^r(\mathrm{id}))$, the superspace of *r-vector fields*; (the elements of $\oplus_{r\in\mathbb{N}} L^r$ are called *polyvector* fields); in particular, $L_1{}^r = \Sigma_{-r}$.

0th order operators: Given irreducible representations ρ_1 and ρ_2 of $\mathfrak{gl}(p/q)$ and a projection $p_{123}: \rho_1 \otimes \rho_2 \to \rho_3$ onto an irreducible component, we define $Z_{123} : T(\rho_1) \otimes T(\rho_2) \to T(\rho_3)$ as extention of p_{123} via F–linearity.

Duality. Let ρ be a representation of $\mathcal{L}_0/\mathcal{L}_k$, str the (superspace of the) representation of $\mathfrak{gl}(m/n)$ determined by the supertrace, and $\rho^+ = \rho^* \otimes \mathrm{str}$. Obviously, for any m|n-dimensoional supermanifold M there is a 0th order operator Z_{ρ,ρ^+}: $T_k(\rho) \otimes T_k(\rho^+) \to \mathrm{Vol}(M)$ which determines a pairing of the spaces of functions with compact support (or for a compact M, or when residue is defined, etc.):

$$t, t^+ \dashrightarrow \int Z_{\rho,\rho+}(t, t^+) \quad \text{for } t \in T_k(\rho),\ t^+ \in T_k(\rho^+).$$

Thus, to every operator B: $T_k(\rho_1) \otimes T_k(\rho_2) \dashrightarrow T_k(\rho_3)$ we assign its left and right dual operators

$$B^{*1}: T_k(\rho_3{}^+) \otimes T_k(\rho_2) \dashrightarrow T_k(\rho_1{}^+) \quad \text{and} \quad B^{*2}: T_k(\rho_1) \otimes T_k(\rho_3{}^+) \dashrightarrow T_k(\rho_2{}^+)$$

These operators induce operators (that we will denote by the same symbols) in the superspaces of *formal* λ-densities and tensor fields.

Clearly, the duality and the operators Z_{123} are applicable to operators of any arity and therefore classification to follow is performed up to duality, and 0th order operators are disregarded unless they are crucial.

0.4. Specifics of supermanifolds. In 1975-76 J.Bernstein and Leites tried to figure out how to formulate an integration theory on supermanifolds. The idea was that if the theory is to possess an analogue of Stockes formula and reduce to the usual one in the absence of odd parameters we have to superize Rudakov's result to describe all analogues of d. It turned out ([BL1]) that for supermanifolds there are at least two analogues of the de Rham complex.

0.4.1. Several de Rham complexes. The complex of *differential forms:*

$0 \dashrightarrow \mathbb{C} \dashrightarrow \Omega^0 \dashrightarrow \Omega^1 \dashrightarrow \dots$ (ad infinum), where $\Omega^i = T(E^i(id^*))$ and E^i is the functor of raising to the ith exterior power

and the (dual of the above) complex of *integrable forms* (called so since the elements with compact support from the smooth version of Σ_{-i} can be integrated over a subsupermanifold of codim i; in particular, Σ_0 = Vol, is the superspace of volume forms):

$$\dots \dashrightarrow \Sigma_{-1} \dashrightarrow \Sigma_0 \dashrightarrow 0, \qquad \text{where } \Sigma_{-i} = T(E^i(id) \otimes str)$$

For m = 1 there are many more complexes:

$$\dots \dashrightarrow \Phi_{\lambda-1} \dashrightarrow \Phi_\lambda \dashrightarrow \Phi_{\lambda+1} \dashrightarrow \dots, \quad \text{where } \Phi_\lambda = T(E^\lambda(id^*)) \text{ for } \lambda \in \mathbb{C}$$

In [BL1] it is shown that these complexes are "mixtures" of Ω^* and Σ_*. What is so special in m = 1, wherefrom do these extra complexes appear? The answer: only for m = 1 we have dim $E^\lambda(id) < \infty$.

The objects corresponding to "$\oplus_{\lambda \in \mathbb{C}}$"$T(E^\lambda(id))$ on an arbitrary supermanifold are called pseudodifferential forms, see [BL2]. More precisely, recall, that *pseudodifferential forms* on a supermanifold X are functions on the supermanifold X' associated with the bundle τ^*X obtained from the cotangent one by fiber-wise change of parity. *Differential forms* on X are fiber-wise *polynomial* functions on X'. The *exterior differential* on X is now considered as an odd vector field d on X'. Let $x = (u_1, \dots, u_p, \xi_1, \dots, \xi_q)$ be local coordinates on X, $x_i' = \pi(x_i)$. Then $d = \Sigma x_i' \pi(x_i)$ is a familiar coordinate expression of d.

The Lie superalgebra $\mathfrak{C}(d) \subset \mathfrak{vect}(m+n/m+n)$, where $(m/n) = \dim X$, -- the Lie superalgebra of vector fields preserving the field d on X' (see definition of the Nijenhuis operator P_4 in 3.1) -- is neither simple nor transitive and therefore did not draw much atention so far. Still, an interesting problem was associated with it:

Calculate structure functions (see [LP2]) *of the corresponding G-structure.*

While this talk was being made into paper D.L. has calculated these structure functions: they are zero! Therefore the extention of d from X to X' is always well-defined.

In sec. 1 we will investigate the possibility rejected in [BL1], that of *infinite dimensional fibers*, and, using the calculations already carried out in [BL1], find new invariant operators. Some of these operators are essentially the good old d, now acting in the superspaces of certain pseudodifferential forms. This suggests a possibility of a new approach to cohomology theory, see sec. 2 below.

Similarly, making use of calculations by Leites and A.Shapovalov who classified finite-dimensional irreducible representations of the remaining simple Lie superalgebras of vector fields: divergence-free, Poisson and their various derived algebras, see review [L1], we can get new operators corresponding to infinite dimensional fibers. For lack of space we leave these answers out. (These answers were recently confirmed by different techniques of "odd reflections" and via BWB-theorem in [PS].)

We will also indicate how to get new operators assosiated with *odd parameters* of representations. The operators themselves will be listed elsewhere.

0.4.2. Odd parameters of representations. If we define Lie superalgebras in terms of functor of points, see [L2], we can view the collection of (say, irreducible) representations of a Lie superalgebra as a supervariety. The Zariski tangent space to this supervariety at the point corresponding to a representation V of a Lie superalgebra $\mathfrak{g}$ is isomorphic to $H^1(\mathfrak{g}; \mathrm{End}\ V)$. In order to verify whether or not there exists a global representation corresponding to a cocycle $c \in H^1(\mathfrak{g}; \mathrm{End}\ V)$ one has to compute Massey powers of c as explained in [F].

Example. Let $\mathfrak{g}$ be a Lie superalgebra of dimention $q\varepsilon$ (see definition of superdimension in [L1]). By Schur's lemma its only (up to change of parity functor Π) irreducible representation is the trivial one, $\mathbf{1}$. Obviously, $\mathrm{End}\ (\mathbf{1}) = \mathrm{End}\ (\Pi(\mathbf{1})) = \mathbf{1}$ and $H^1(\mathfrak{g}) = \mathfrak{g}$. Since $\mathfrak{g}$ is clearly supercommutative, the supervariety of irreducible representations of $\mathfrak{g}$ is two copies of the $(0, \dim \mathfrak{g})$-dimensional supermanifold associated with the vector superspace $\mathfrak{g}$. The representations corresponding to odd parameters are constructed as follows.

Consider the case $q = 1$; the general case is obtained by tensoring q copies of the representation constructed for $q = 1$.

The representation of $\mathfrak{g}$ corresponding to an odd parameter τ is given on C-points, where C is a supercommutative superalgebra with the nonzero odd part C_1, by the formula

$$\rho_C(g) = \tau \,\mathrm{antidiag}(\lambda, \lambda) \text{ for } \lambda \in C_1 \text{ and a nonzero } g \in \mathfrak{g}.$$

0.5. A homological generalisation. From homology point of view to describe $\mathcal{L}$-invariant differential operators (FIO) is the same as to describe

$$H_0(\mathcal{L};(T_k(V_1))^* \otimes \ldots \otimes (T_k(V_r))^* \otimes T_k(V)) = \mathrm{Hom}_{U(\mathcal{L})}((T_k(V_1)) \otimes \ldots \otimes (T_k(V_r)), T_k(V))$$

It is interesting to consider the following
Generalized Veblen's problem:

Describe $\mathrm{Ext}^*_{U(L)}(T_k(V_1) \otimes \ldots \otimes T_k(V_r)\, ,\, T_k(V))$.

Until recently this problem with many interesting byproducts was only studied for $k = r = 1$ and $\mathcal{L} = \mathfrak{vect}(1|0)$, see [F] and references therein. Recently (spring 1990, unpublished) V.Serganova solved it for $\mathfrak{vect}(0/2) = \mathfrak{sl}(1/2)$ and for $\mathfrak{sl}(1/n)$.

0.6. A misterious operator. Since in the absence of even coordinates, all operators in superspaces of tensors or jets (with finite dimensional fiber) are differential ones (even the integral), one of us (D.L.) hoped that having worked out this finite-dimensional model we could, by analogy, find new nonlocal invariant operators for manifolds as well. Their arity is, clearly, >1. So far, no such operator is written except an example of a symbol of such a binary operator acting in the spaces of certain tensors on the line; see Kirillov's review [Ki].

1. Invariant unary operators. Rudakov's method described in Introduction allows one to list invariant differential operators acting between modules of tensor fields of type T(V) for an irreducible V with lowest weight (resp. between modules I(V), where V is irreducible with highest weight). Rudakov's answer showed that there are no invariant differential operators for $\dim V = \infty$. The calculations from [BL1] show that on supermanifolds the situation is different; however, these operators had never been written.

Theorem. *The differential operators listed in 1.1-1.3 below exaust all unary* $\mathfrak{vect}(m/n)$*-invariant operators acting in the modules T(V) with an irreducible* $\mathfrak{gl}(m/n)$*-module V with lowest weight.*

1.1. $\mathfrak{vect}(0/2)$-invariant operators:

1st order operators -- extensions of the exterior differential:

$\ldots \to I(0, n) \to I(0, n + 1) \to \ldots \quad , \qquad n \in \mathbb{C}$

$\ldots \to I(n, 1) \to I(n + 1, 1) \to \ldots \quad , \qquad n \in \mathbb{C}$

In primed coordinates ξ, ξ' on $\mathbb{C}^{0,2'}$ the above spaces are explicitely described as

$I(0, n) \quad = \mathbb{C}[\xi]\, (\xi_1'\xi_2')^{-1} <\xi_2'^{-n}, ..., \xi_2'^{-n} (\xi_2'/\xi_1')^k, ...>, \qquad k \in \mathbb{Z}_+$

$I(n + 1, 0) = \mathbb{C}[\xi] <\xi_1'^{n}, ..., \xi_1'^{n}(\xi_2'/\xi_1')^k, ...>, \qquad k \in \mathbb{Z}_+$

For $n \in \mathbb{Z}_+$ and $k \le n$ we have $I(n + 1, 1) = \Omega^n$ and $I(0, n + 1) = \Sigma_{-n}$; for a general k the spaces Ω^n and Σ_{-n} we get subsuperspaces of formal pseudodifferential and pseudointegrable n-forms, respectively.

2nd order operator -- Berezin integral: $\int = \xi_1'\xi_2'\partial_1\partial_2: \Sigma_0 \dashrightarrow \Omega^0$.

1.2. $\mathfrak{vect}$(0/n)-invariant operators, n ≥ 2. They are given by singular (definition see in [R], [BL1]) vectors in $\mathfrak{vect}$(0/n)-modules I(V) and, as follows from calculations in [BL1], these singular vectors are (cf. [PS]):

deg = -1: $i_1 = \Sigma\partial_j v_j$, where if $v_{j_0} = 0$ then $v_{j_0 - 1} = ... = v_1 = 0$ and if $v_{j_0+1} \neq 0$ it is the highest vector in V of weight $(0, ..., 0, \lambda, 1, 1, ..., 1)$ with j_0-many 0s. All the corresponding operators are, essentially, one exterior differential.

deg = -k: $i_k = \partial_n \dots \partial_{n-k+1} v$, the highest weight of V is (0, ..., 0,1, ..., 1) with k-many 1s. The corresponding operators are Berezin integration with respect to the last k primed variables over an ambient supermanifold.

1.3. $\mathfrak{vect}$(m/n)-invariant operators for mn ≠ 0. They are only 1st order ones; but in addition to the cases listed in [BL1, Lemma 5.2] there are operators corresponding to singular vectors of the form

$$\Sigma_{j\le r} \text{ with weight of } v_r = \begin{cases} (0, ..., 0, -1, ..., -1; 1, ..., 1) & \text{with r-many 0s, } r \le m \\ (0, ..., 0, ; p,1, ..., 1) & \text{for } r = m + 1 \\ (0, ..., 0, ; 0, ..., 0, p,1, ..., 1) & \text{for } r > m \text{ with any } p \in \mathbb{C}. \end{cases}$$

(Note that there is a misprint in [BL1]: the case bγ of Lemma 5.2 should read as $\lambda = (0, ..., 0, ; 0, ..., 0,-p; 0)$ which corresponds to $r = m + n$ above.)

2. Homological digressions.

2.1. Massey operations. It is tempting to consider Berezin's integral as composition of two operators acting in infinite dimensional superspaces with indicated highest weights:

I(0, 0) --> I(0, 1) --> I(1, 1)

This composition, however, is 0 since the both operators are actualy the same operator, $d = \Sigma\, \xi_i'\, \partial/\partial\xi_i$, and $d^2 = 0$. Still, the Berezin integral (and some other differential operators, e.g. those invariant with respect to the nonstandard realization of $\mathfrak{vect}$(m/n)) is a composition of sorts: in 1976 J.Bernstein verified for $\mathfrak{vect}$(0/2) that the polarization of the integral is a Massey power (definition see in [F]) of d.

Problem. *Which of the invariant operations listed in [L1] and in this paper (especially, in 1.3 and 5.2) are Massey products of other operations?*

2.2. Pseudo(co)homology of supermanifolds and of Lie superalgebras. It is well known that de Rham cohomology of a supermanifold is the same as that of the underlying manifold. Here we continue the train of thought initiated in [BL1] to introduce more adequate (co)homology of supermanifolds and as a byproduct those of Lie superalgebras. Another approach to homology of supermanifolds is reviewed in [MPV]; conventional definition of (co)homology of Lie superalgebras, see in [F].

Let us consider the space of pseudoforms on a supermanifold X. Since $d^2 = [d, d] = 0$, the homologies of d in diferent superspaces of tensor fields lifted on X' from X are well-defined and will be called *pseudocohomology* of X (a prudent one would have called them "cohomology of pseudoforms").

The simplest of such tensors are just functions. Even their space is, however, too broad; we should confine ourselves to subspaces for which homology of d are computable: see various candidates for such subspaces in V.Shander's *very lucid and inspiring* exposition of cohomology theories and integration in #31 of [L2]. Take, say, homogeneous functions on $X'_0 = X' \setminus \{\text{zero section}\}$; the *degree of a cochain* is the homogeneity degree $\lambda \in k$ of the cochain, where k is the ground field.

For example, let X = G be a Lie supergroup. A *pseudocochain of degree* λ on the Lie superalgebra $\mathfrak{g}$ = Lie G is a left-invariant homogeneous function on G'_0 of homogeneity degree λ. The *pseudocohomology of* $\mathfrak{g}$ *with coefficients in a* $\mathfrak{g}$*-module M* is homology $PH^*(\mathfrak{g}; M)$ of d in homogeneous sections of the vector bundle over G' with fiber M. Selecting a basis of $\mathfrak{g}$ we can confine ourselves to *different* theories corresponding to homogeneous functions with "highest" or "lowest" vector, for which we can take different vectors from $\mathfrak{g}$. Thus we get functions in even variables $x = d\theta$ and odd variables $\xi = dq$, where q and θ are even and odd coordinates on G, of the form (where + corresponds to the case with highest vector , - to that with lowest vector):

$$f_0(q,\theta)\,(x_1)^{\lambda} + \ldots + f_K(q,\theta)(x_1)^{\lambda \pm k}(x_2)^{k_1}\ldots(\xi_1)^{\kappa_1}\ldots + \ldots ,$$
where $K = (k, k_1, \ldots, \kappa_1, \ldots)$, $k, k_1, \ldots, \kappa_1, \ldots \in \mathbb{Z}$, and $k = \Sigma k_i + \Sigma \kappa_j$).

Examples, where the answer is easy to see: 1)supercommutative Lie superalgebras; 2) Lie superalgebra $\mathfrak{q}(n)$, cf. [F].

The *pseudohomology* of X is defined as the homology of the fiber-wise Fourier transform of d, the operator of the inner multiplication by the bivector Q that determines the Buttin (a.k.a. Schouten) bracket (see Remark 3.1 below) in sections of the supermanifold $'X = \tau X$ similarly to the above associated with the tangent bundle on X. If X is a supergroup G the *pseudohomology* $PH_*(\mathfrak{g}; M)$ of $\mathfrak{g}$ = Lie G with cooefficients in a $\mathfrak{g}$-module M are connected with invariant homogeneous sections of the vector bundle over τG with fiber being M.

If dim X = (n, 1) pseudoforms are *tensors* on X' or 'X. As is shown in [BL1], the de Rham complex and its dual "homology" complex of integrable forms are, respectively, the subcomplex and quotient of the complex of pseudoforms of integer degrees.

3. Invariant binary operators. In this section we make a conjecture that extends Grozman's list [G1] to supermanifolds of dimension m|n. The proof of it is known for n = 0 and m = n = 1.

3.1. *1st order operators*. (Since there are 3 types of analogues of differential forms on supermanifolds, there are about 3 times more bilinear operators on supermanifolds than there are on manifolds):

$^{\Omega}P_1: \Omega^r, T(\rho_2) \dashrightarrow T(\rho_3) \quad (\omega,t) \dashrightarrow Z(d\omega, t); \quad {}^{\Omega}P_1{}^{*1}: T(\rho_3{}^+), T(\rho_2) \dashrightarrow \Sigma_{-r}$

$^{\Sigma}P_1: \Sigma_{-r}, T(\rho_2) \dashrightarrow T(\rho_3) \quad (\sigma,t) \dashrightarrow Z(d\sigma, t); \quad {}^{\Sigma}P_1{}^{*1}: T(\rho_3{}^+), T(\rho_2) \dashrightarrow \Omega^r$

$^{\Phi}P_1: \Phi^{\lambda}, T(\rho_2) \dashrightarrow T(\rho_3) \quad (\phi,t) \dashrightarrow Z(d\phi, t); \quad {}^{\Phi}P_1{}^{*1}: T(\rho_3{}^+), T(\rho_2) \dashrightarrow \Phi^{n-\lambda-1}$

P_2: Vect, $T(\rho) \dashrightarrow T(\rho)$ *Lie derivative* ; $P_2{}^{*1}: T(\rho^+), T(\rho) \dashrightarrow \Omega^1 \otimes_F \mathrm{Vol}$

$P_3 = \mathrm{PB}: T(S^p(id^*)), T(S^q(id^*)) \dashrightarrow T(S^{p+q-1}(id^*))$ *Poisson bracket* ;

$P_3{}^{*1} = \mathrm{PB}^{*1}: \quad T(S^{p+q-1}(id)) \otimes_F \mathrm{Vol}, T(S^q(id^*)) \dashrightarrow \quad T(S^p(id)) \otimes_F \mathrm{Vol},$

To describe P_4, let us identify $\Omega^* \otimes_F \mathfrak{vect}(M)$, where F = F(M) is the superspace of functions on a supermanifold M, with the Lie subsuperalgebra of $\mathfrak{vect}(M')$:

$$\mathfrak{C}(d) = \{D \in \mathfrak{vect}\ M': [D, d] = 0\}, \text{ the centralizer of the exterior differential,}$$

setting

$$(\omega \otimes X)(f) = \omega X(f), \quad (\omega \otimes X)(df) = (-1)^{p(\omega)+p(X)} d(\omega X(f)).$$

In other words,

$$f(x, x')\partial/\partial x_i \dashrightarrow f(x, x')\partial/\partial x_i + (-1)^{p(f)+p(x_i)} f(x, x')\partial/\partial x'_i$$

The Lie bracket in $\mathfrak{C}(d)$ induces new operators NB^{Ω}, NB^{Φ}, and NB^{Σ}, respectively, in the superspaces

$$\Omega^r \otimes_F \mathrm{Vect} = \Omega^{r-1} \oplus \begin{cases} T(1, \ldots, 1, 0, \ldots, 0, -1) & (r\text{-many } 1\text{s}) \text{ if } r \leq n \\ T(1, \ldots, 1; r-n, 0, \ldots, 0, -1) & \text{if } r \geq n \end{cases}$$

and

$$\Phi^{\lambda} \otimes_F \mathrm{Vect} = \Phi^{\lambda-1} \oplus T(1, \ldots, 1, 0; \lambda - n - 1)$$

and

$$\Sigma_{-r}\otimes_F \text{Vect} = \Sigma_{-r-1}\otimes T(1, ..., 1,0; -1, ..., -1, -r-2)$$

On manifolds, the bracket in $\mathfrak{C}$ (d), is called the *Nijenhuis bracket*. This bracket is a linear combination of operators P_1, P_1^{*1}, their composition with the permutation operator $T(V)\otimes T(W) = T(W)\otimes T(V)$, and entirely new operators NB. More precisely, speaking in terms of irreducible fibers, we have to consider several operators for different particular cases:

$$NB^{\Omega} \begin{array}{l} T(1, ..., 1,0, ..., 0, -1), T(1, ..., 1,0, ..., 0, -1) \to T(1, ..., 1,0, ..., 0, -1) \\ T(1, ..., 1,0, ..., 0,-1), T(1, ..., 1;s-n,0, ..., 0,-1) \to T(1, ..., 1;s+r-n,0, ...,0,-1) \\ T(1, ..., 1;r-n,0,...,0,-1), T(1, ..., 1;s-n,0,...,0,-1) \to T(1, ..., 1;s+r-n,0,...,0,-1) \end{array}$$

and similarly for NB^{Σ}:

$$T(1, ..., 1,0, ...,0, -1), T(1, ..., 1,0;-1 ..., -1,-s-2) \to T(1, ..., 1,0;-1, ..., -1,-s+r-2)$$
$$T(1, ..., 1;r-n,0,...,0,-1), T(1, ..., 1,0;-1,...,-1, -s-2) \to T(1, ..., 1,0;-1, ..., -1,s+r-2)$$

and for NB^{Φ}:

$$T(1, ..., 1, 0;\lambda-n-1), T(1, ..., 1, 0;\mu-n-1) \to T(1, ..., 1,0;\lambda+\mu-n-1)$$

The dualization gives new operators NB^{*1}.

We could not trace who was the first to discover the following family of operators:

$$P_5^{\Omega}:\Omega^p, \Omega^q \to \Omega^{p+q+1};\ \omega_1, \omega_2 \to a(d\omega_1\omega_2) + (-1)^{p(\omega_1)}b(\omega_1 d\omega_2),\ \text{where } a,b\in\mathbb{C}. \quad (*)$$

The same formula (*) determines an extension of P_5^{Ω} to Φ^*:

$$P_5^{\Phi}:\quad \Phi^{\pi}, \Phi^{\theta} \to \Phi^{\pi+\theta+1}$$

and, with $\omega_2 \in \Sigma$, the same formula (*) determines, for $p < q$, the operator

$$P_5^{\Omega\Sigma}:\ \Omega^p, \Sigma_q \to \Sigma_{q-p+1}$$

The dualization gives new operators P_5^{*1}.

The operators P_5 may be generalized as follows. Let $|\mu|^2 + |\nu|^2 \neq 0$; in addition, on maifolds, let $p + q < n$. Define

$$P_6^{\Omega}:\ \Omega_{\mu}^{p}, \Omega_{\nu}^{q} \to \Omega_{\mu+\nu}^{p+q+1} \quad\text{and}\quad P_6^{\Omega L}:\ \Omega_{\mu}^{p}, L_{\nu}^{q} \to L_{\mu+\nu}^{p+q+1}$$

setting

ω_1, ω_2 --> $(\nu d\omega_1\omega_2 - (-1)^{p(\omega_1)}\mu\omega_1 d\omega_2)v^{\mu+\nu}$

The dualization gives new operators $P_6{*}^1$.

Denote:

P_7: L^p, L^q --> L^{p+q-1} the *Schouten bracket.*

(It was C.Buttin who first published the proof of the fact that the Scouten bracket determines a *Lie superalgebra* structure on $\Pi(\oplus_{p\in\mathbb{N}} L^p)$; in her memory then we call the bracket in this Lie superalgebra the *Buttin bracket.*)

The dualization gives new operator $P_7{*}^1$.

Define a generalization of the Schouten bracket (on manifolds, for $p + q < n$; on supermanifolds of dimension n|1, for $p,q \in \mathbb{C}$) the operator

P_8: L_μ^p, L_ν^q --> $L_{\mu+\nu}^{p+q-1}$

setting

Xv^μ, Yv^ν --> $[(\nu-1)(\mu+\nu-1)\ \mathrm{div}X\ Y + (-1)^{p(X)}(\mu-1)(\mu+\nu-1)X\ \mathrm{div}\ Y -$
$-(\mu-1)(\nu-1)\ \mathrm{div}(XY)]v^{\mu+\nu}$,

where the *divergence of a polyvector field* is best described in local coordinates (x, 'x) on the supermanifold 'M = (M, $L^* = \oplus_i L^i$) associated to any manifold M. In other words, to coordinates x = (u, ξ) on a supermanifold assign coordinates 'u, 'ξ of opposite parities whose rule of transformation is induced by that of the x according to the formila $'x_i = \pi(\partial/\partial x_i)$ and set

$$\mathrm{div} = \Sigma\, \partial^2/\partial x_i \partial' x_i$$

Remark. Physicists ([BV]) rediscovered (the first to discover was S.Belkin, see Kirillov's Correction, Russian Math. Surveys, 32, #1, 1977) the operator div while studying quantization of gauge fields and BRST-symmetries. They called this operator Q. A prosaic explanation of Q: div is the Fourier transform with respect to primed (hatted in notations of [BL1]) variables of the exterior differential

$$d = \Sigma\, x'_i \partial/\partial x_i$$

Since $d^2 = 0$, then $Q^2 = 0$; physicists called the homology of Q *BRST-cohomology*; but nowadays everything is called BRST-cohomology. An analogue of this transformation in mechanics is the Legendre transformation: $C^\infty(T^*M)$ --> $C^\infty(TM)$; in the above example everything is the same as in classical mechanics only the fibers of these bundles have shifted parity.

The dualization gives new operator $P_8{*}^1$.

3.2. *2nd order operators* :

$S_1 = Z(d\ .\ \ , d\ .\ )$, where Z corresponds to the projection on the highest of the irreducible components; S_1 is defined on (Ω^p, Ω^q) and for $p \le q$ on (Ω^p, Σ_q) (recall that on a manifold M of dimension m, $\Sigma_0 = \Omega^m$)

$S_2 = P_8(d \otimes id)$: $\Sigma_{-1}, L_\nu{}^q \dashrightarrow L_{\mu+2}{}^{q-1}$

$S_3 = dP_5$: $\quad \Omega^p, \Omega^q \dashrightarrow \Omega^{p+q+2}$, $\phi^\pi, \Phi^\theta \dashrightarrow \Phi^{\pi+\theta+2}$

and, for $p < q-1$, $\quad \Omega^p, \Sigma_q \dashrightarrow \Sigma_{q-p+2}$

and their ^{*1}-duals.

3.3. *3rd order operators*:

$T_1 = P_8(d\ .\ \ , d.\ )$: $\Sigma_{-1}, \Sigma_{-1} \dashrightarrow L_2{}^{-1}$ and its ^{*1}-dual (recall again that on a manifold M of dimension m, $\Sigma_0 = \Omega^m$).

Special cases :

dim M = 1/0 : see the list of $\mathfrak{k}(1/1)$-invariant differential operators for an entirely new 3rd order operator.

dim M = 0/m:

$\int_1 : \Sigma_0, T(\rho) \dashrightarrow T(\rho) \quad \omega, t \dashrightarrow (\int \omega)t$ (order m)

$\int_1{}^{*1} : T(\rho+), T(\rho) \dashrightarrow \Omega^0 \quad t+, t \dashrightarrow \int t^+(t)$ (order m)

$\int_{11} : \Sigma_0, \Sigma_0 \dashrightarrow \Omega^0 \quad \omega_1, \omega_2 \dashrightarrow (\int\omega_1)(\int\omega_2)$ (order 2m)

$\int \circ S_2 : \Sigma_{-1}, L_1{}^0 \dashrightarrow \Omega^0$ (order m + 2)

Conjecture. *The above list exhausts all* $\mathfrak{vect}(m/n)$*-invariant binary differential operators in tensor superspaces whose fibers are irreducible finite-dimensional* $\mathfrak{gl}(m/n)$*-modules.*

Problem. *List the* $\mathfrak{vect}(m/n)$*-invariant binary differential operators in tensor superspaces whose fibers are irreducible* $\mathfrak{gl}(m/n)$*-modules with lowest weight.*

Remark. The computations required to prove this conjecture are voluminous: for n = 0 and arbitrary m, about 100 pages ([G1, 2]) and about 75 pages ([Kol]) for n = m = 1. These computations are labour consuming but very simple (calculating ranks of rectangular matrices); we will illustrate that with the case slightly more difficult than m = 0, n = 1, see sec. 4. The corresponding **problem** *is a challendge for a computer scientist*, see [LP1].

4. Operators invariant with respect to Neveu-Schwarz superalgebras.

The notations we use in this sec. are borrowed from [L1].

Let $\alpha = dt + \Sigma\, \theta_i d\theta_i$ be a contact form on (1/n)-dimensional supermanifold, $\mathfrak{k}(1/n)$ the Lie superalgebra of contact vector fields that preserve α up to a functional factor. It can be shown that $\mathfrak{k}(1/n)$ is spanned by vector fields

$K_f = (2-E)f\partial/\partial t + H_f + \partial f/\partial t E$, where $E = \Sigma\theta_i\partial/\partial\theta_i$, $H = \Sigma\partial f/\partial\theta_i\partial/\partial\theta_i$

The most popular $\mathfrak{k}(1/n)$-modules are modules F_λ of "conformal weight λ":

$F_\lambda = \mathbb{C}[t^{-1}, t, \theta]\alpha^{\lambda/2}$ if $n \neq 0$ and $\mathbb{C}[t^{-1}, t]\alpha^\lambda$ otherwise.

For n = 2 generalise the definition of F_λ setting $F_{\lambda,\mu} = \mathbb{C}[[t, \xi, \eta]]\alpha^{\lambda/2}(\xi'/\eta')^{\mu/2}$.

4.1. Unary case ([L1]). Denote by d' the composition of the exterior differential (applied first) and the projection onto the quotient superspace modulo the subspace of forms generated by α and $d\alpha$. The structure of the superspace of differential forms as $\mathfrak{k}(2m+1/n)$-module is investigated in [L1]. Let us concentrate on m = 0.

n = 1. The space T(V) for an irreducible V with lowest weight λ is just F_λ and there is only one operator $d' = \mathrm{pr}\circ d$, the composition of the exterior differential and the projection: $F_0 = \Omega_0 \to F_1 = \Omega^1/\Omega^0\alpha = \Sigma_0$. Explicitely, $d = \alpha K_1 + \xi' K_\xi$, hence $d' = \xi' K_\xi$.

n = 2. Since in this case $\Sigma_0 = F_0$ (see [L1]) and $L_0/L_1 = \mathfrak{co}(2)$, the irreducible L_0/L_1-modules are 1- (or ε-)dimensional and d' splits into $d_\xi' = \xi' K_\xi$ and $d_\eta' = \eta' K_\eta$. These operators act on the spaces of pseudoforms:

$$d_\xi' : F_{\lambda,-\lambda} \to F_{\lambda+1,-\lambda-1}, \qquad d_\eta' : F_{\lambda,\lambda} \to F_{\lambda+1,\lambda+1}.$$

There are also the following 2nd order invariant differential operators. Denote by $pr_{1,1}$ and $pr_{1,-1}$ the projections of $\Omega^1 = F_{1,1} \oplus F_{1,-1} \oplus \Omega^0\alpha$ onto $F_{1,1}$ and $F_{1,-1}$, respectively. The composition of

$$d\circ v^{-1}\circ d : \Sigma_{-1} \to \mathrm{Vol} = \Omega_0 \to \Omega^1$$

and $pr_{1,1}$ or $pr_{1,-1}$ are the only invariant 2nd order operators.

There are no more $\mathfrak{k}(1/2)$-invariant differential binary operators acting in the superspaces of tensors corresponding to linear bundles, see [L1].

4.2. Binary case.

n = 0: **Theorem** (P.Grozman, BA diploma work, 1977). *The following differential operators exhaust, together with their duals, all $\mathfrak{vect}(1/0)$-invariant bilinear differential operators acting in the spaces of tensors whose fibers are irreducible $\mathfrak{gl}(1)$-modules.*

$Z: F_\lambda \otimes F_\mu \to F_{\lambda+\mu}$ $\qquad f dx^\lambda, g dx^\mu \to fg dx^{\lambda+\mu}$

$P: F_\lambda \otimes F_\mu \to F_{\lambda+\mu+1}$ $\quad f dx^\lambda,\ g dx^\mu \to (\lambda f g' - \mu f' g) dx^{\lambda+\mu+1}$

$a(1\otimes d') + b(d'\otimes 1): F_0 \otimes F_0 \to F_1$ (for $a,b \in \mathbb{C}$) $\quad f, g \to (afg' + bf'g)dx$

$S_1 = Z(d\otimes d): F_0 \otimes F_0 \to F_2$ $\quad f, g \to f'g'dx^2$

$S_2 = P(d\otimes id): F_0 \otimes F_\mu \to F_{\mu+2}$ (similarly for $F_\mu \otimes F_0$) $\quad f, g \to (fg'' - \mu f''g)dx^2$

$S_3 = dP: F_\lambda \otimes F_\mu \to F_1$ (for $\lambda + \mu = -1$) $\quad f, g \to [\lambda f g'' - \mu f'' g + (\lambda - \mu) f'g']dx^1$

$T_1 = P(d\otimes d): F_0 \otimes F_0 \to F_3$ $\quad f, g \to (fg'' - f'g')dx^3$

$T: F_{-2/3} \otimes F_{-2/3} \to F_{5/3}$ $\quad f dx^{-2/3},\ g dx^{-2/3} \to [2(f'''g - fg''') + 3(f''g' - f'g'')]dx^{5/3}$

Comments. 1) Denote: $p = (dx)^{-1}$. The operation P defines the bracket in the space $\oplus_{\lambda \in \mathbb{Z}} F_\lambda$. This is just the Poisson bracket extended onto the space of (not only polynomial in p, as here) functions in x,p.

2) The operator T, discovered by Grozman, has an interesting generalization to skewsymmetric invariant operators of higher arity, cf. [F]. Strange to say, it is difficult to guess what are supercounterparts of these operators.

Problem. *Superize this result of* [F].

n = 1: **Theorem.** (In what follows $f' = K_1(f)$, notations see in [L1].) *The following differential operators exhaust, together with their dual, all $\mathfrak{k}(1/1)$-invariant bilinear differential operators acting in the spaces of tensors whose fibers are irreducible $\mathfrak{gl}(1) = \mathfrak{k}(1/1)$-modules.*

$Z: F_\lambda \otimes F_\mu \to F_{\lambda+\mu}$ $\quad f\alpha^\lambda,\ g\alpha^\mu \to fg\alpha^{\lambda+\mu}$

$P: F_\lambda \otimes F_\mu \to F_{\lambda+\mu+1}$ $\quad f\alpha^\lambda,\ g\alpha^\mu \to (\mu K_\xi(f)g - (-1)^{p(f)}\lambda f K_\xi(g))\alpha^{\lambda+\mu+1}$

$a(1\otimes d') + b(d'\otimes 1): F_0 \otimes F_0 \to F_1$ (for $a,b \in \mathbb{C}$): $\quad f, g \to (aK_\xi(f)g + (-1)^{p(f)} b f K_\xi(g))\alpha$

$S: F_\lambda \otimes F_\mu \to F_{\lambda+\mu+2}$ $\quad f\alpha^\lambda,\ g\alpha^\mu \to (\mu f'g + (-1)^{p(f)} K_\xi(f) K_\xi(g) + \lambda f g')\alpha^{\lambda+\mu+2}$

$T_1: F_\lambda \otimes F_\mu \to F_{\lambda+\mu+3}$ (for $\lambda\mu = 0$):

$f\alpha^\lambda,\ g\alpha^\mu \to (\mu K_\xi(f')g + (-1)^{p(f)}\lambda f K_\xi(g') + K_\xi(f)g' + (-1)^{p(f)} f' K_\xi(g))\alpha^{\lambda+\mu+3}$

$T_2: F_\lambda \otimes F_{-\lambda-2} \to F_1$ (for $\lambda(2-\lambda) = 0$):

$f\alpha^\lambda,\ g\alpha^{2-\lambda} \to ((\lambda+2)K_\xi(f')g + (-1)^{p(f)}\lambda f K_\xi(g') + (\lambda+1)(K_\xi(f)g' + (-1)^{p(f)} f' K_\xi(g)))\alpha$

$S(d_\xi \otimes d_\xi): F_0 \otimes F_0 \to F_4$ $\quad f, g \to (K_\xi(f')K_\xi(g) + (-1)^{p(f)} f'g' + K_\xi(f)K_\xi(g'))\alpha^4$

$Q_{03}: F_0 \otimes F_{-3} \to F_1$ $\quad f, g\alpha^{-3} \to (3f''g + (-1)^{p(f)}(4K_\xi(f')K_\xi(g) + K_\xi(f)K_\xi(g')) - 2f'g')\alpha$

$Q_{30}: F_{-3} \otimes F_0 \to F_1$ $\quad f\alpha^{-3}, g \to (3fg'' + (-1)^{p(f)}(K_\xi(f')K_\xi(g) + 4K_\xi(f)K_\xi(g')) - 2f'g')\alpha$

Comments. 1) Denote: $p = \pi(\alpha)$. The operation P defines the bracket in the space $\oplus_{\lambda \in \mathbb{Z}} F_\lambda$. This is just the Buttin bracket extendable onto the space of (not only polynomial in p, as here) functions in ξ, p with t serving as a parameter.

2) $Q_{03} = S(d_\xi \otimes d_\xi)^{*2}$; $Q_{30} = S(d_\xi \otimes d_\xi)^{*1}$; $T_1 = T_2{}^{*1} = T_1{}^{*2}$.

Proof. *Existence.* Let V and W be irreducible $\mathfrak{gl}(1) = (\mathfrak{k}(1/1))_0$-modules with highest weights λ and μ, respectively. First, note that the subalgebra $\mathfrak{k}(1/1)_+$ of elements of positive degree (with respect to the standard grading) is generated by $K_{t\xi}$ and $K_{t^2\xi}$. Let f_i be a singular vector of degree -i. Let us write down consequences of singularity starting with i = 1:

$$f_1 = aK_\xi v \otimes w + bv \otimes K_\xi w, \qquad \text{where } a,b \in \mathbb{C}$$

We have

$$K_{t\xi} f_1 = -(a\lambda + (-1)^{p(v)} b\mu) v \otimes w$$

Hence $a\lambda = -(-1)^{p(v)} b\mu$.

Now let

$$f_2 = aK_1 v \otimes w + bv \otimes K_1 w + cK_\xi v \otimes K_\xi w, \qquad \text{where } a,b,c \in \mathbb{C}$$

Then $K_{t\xi} f_2 = 0$ implies

$$-2aK_\xi v \otimes w - (-1)^{p(v)} 2bv \otimes K_\xi w - c(\lambda v \otimes K_\xi w - (-1)^{p(v)} \mu K_\xi v \otimes w) = 0$$

Hence,

$$2a = (-1)^{p(v)} \mu c, \qquad 2b = -(-1)^{p(v)} \lambda c$$

Let

$$f_3 = aK_1 K_\xi v \otimes w + bv \otimes K_1 K_\xi w + cK_1 v \otimes K_\xi w + dK_\xi v \otimes K_1 w, \qquad \text{where } a,b,c,d \in \mathbb{C}$$

Then

$$K_{t\xi} f_3 = -a(1 + \lambda) K_1 v \otimes w - (-1)^{p(v)} (1 + \mu) bv \otimes K_1 w -$$

$$c(2K_\xi v \otimes K_\xi w + (-1)^{p(v)} \mu v \otimes K_1 w) - d(\lambda v \otimes K_1 w - (-1)^{p(v)} K_\xi v \otimes K_\xi w)$$

$$K_{t^2\xi} f_3 = 4(a\lambda + (-1)^{p(v)} b\mu) v \otimes w$$

Therefore,

$$a(1 + \lambda) + (-1)^{p(v)} \mu c = 0, \quad (-1)^{p(v)} (1 + \mu) b + d\lambda = 0, \quad c = (-1)^{p(v)} d;$$

$$4(a\lambda + (-1)^{p(v)} b\mu) = 0$$

The compatibility condition on a,b,c,d is:

$$\lambda\mu(1 + \lambda) + \lambda\mu(1 + \mu) = \lambda\mu(2 + \lambda + \mu) = 0$$

a) $\lambda = \mu = 0$, hence $a = b = 0$

b) $\lambda = 0$, hence $a = d\mu$, $b = 0$

c) $\mu = 0$, hence $a = 0$, $b = (-1)^{p(v)+1} d\lambda$

d) $\lambda\mu \neq 0$, $\lambda + \mu + 2 = 0$, hence

$a = (-1)^{p(v)} (2 + \lambda) b/\lambda$, $d = (-1)^{p(v)} (1 + \lambda) b/\lambda$, $c = (1 + \lambda) b/\lambda$.

Let

$f_4 = aK_1{}^2 v\otimes w + bK_1 v\otimes K_1 w + cv\otimes K_1{}^2 w + dK_1K_\xi v\otimes K_\xi w + eK_\xi v\otimes K_1K_\xi w$, where $a,b,c,d,e \in \mathbb{C}$.

Then

$$K_{t\xi}f_3 = -4aK_1K_\xi v\otimes w - 4c(-1)^{p(v)}v\otimes K_1K_\xi w - 2b(dK_\xi v\otimes K_1 w + (-1)^{p(v)}K_1 v\otimes K_\xi w) -$$
$$-d((1+\lambda)K_1 v\otimes K_\xi w - (-1)^{p(v)}\mu v\otimes K_1K_\xi w)-$$
$$-e(\lambda v\otimes K_1K_\xi w - (-1)^{p(v)}(1+\mu)K_\xi v\otimes K_1 w)$$

and

$$K_{t^2\xi}f_4 = 8aK_\xi v\otimes w + 8c(-1)^{p(v)}v\otimes K_\xi w + 4d\lambda v\otimes K_\xi w - (-1)^{p(v)}4e\mu K_\xi v\otimes w$$

implying

$$2b = -(-1)^{p(v)}(1+\mu)e,\quad 2b = -(-1)^{p(v)}(1+\lambda)d;$$
$$4a = (-1)^{p(v)}d\mu,\quad 4c = -(-1)^{p(v)}e\lambda, \qquad (*)$$
$$2a = (-1)^{p(v)}e\mu,\quad 2c = -(-1)^{p(v)}d\lambda \qquad (*)$$

Consider the following cases:

1) $\lambda = \mu = 0$. Then $a = c = 0$, $2b = (-1)^{p(v)}e$, $d = -e$;

2) $\lambda = 0$, $\mu \neq 0$. Then

$c = 0$, $2d = (-1)^{p(v)}e\mu$, $2b = -(-1)^{p(v)}(1+\mu)e$, $d = -(1+\mu)e$; hence, $\mu = -3$;

3) $\lambda \neq 0$, $\mu = 0$ is symmetric to the above;

4) $\lambda\mu \neq 0$. The system constituted by (*) is incompatible: its discriminant is $12\lambda\mu$.

Uniqueness. A priori there might have existed a singular vector in degrees -5 to -8 (but not lower than -8 thanks to Grozman's result: the restriction to $\mathfrak{vect}(1/0)$ forbids). Straightforward calculations similar to the ones above show that there is no such vector ([Ko2]).

Hint for a programmer who will solve problems of this type. So far the following empirical fact is observed: if there is no invariant operator of degree k than there is no operator of higher degree. This, together with Kirillov's restriction from above of the order of such operators for any given arity and differential order ([Ki]), is helpful to know when to stop computing.

5. Unary differential operators invariant with respect to $\mathfrak{vect}(m/n; 1)$ -- a nonstandard realization of $\mathfrak{vect}(m/n)$. Any simple Lie algebra of vector fields $\mathcal{L}$ contains a unique maximal subalgebra of finite codimension which determines its natural filtration (called the Weisfeiler filtration) and the associated $\mathbb{Z}$-grading (which for $\mathfrak{vect}$ is deg $x_i = 1$ for all i). As is explained in [L1], for Lie *super*algebras of vector fields, unlike the Lie algebra case, there might be several Weisfeiler filtrations, in other words, there is a possibility to present these superalgebras as vector fields on different supermanifolds. For $\mathfrak{vect}(m/n) = \mathfrak{der}\,\mathbb{C}[[x_1, ..., x_m, \xi_1, ..., \xi_n]]$ all these nonstandard realizatons correspond to filtrations associated with the following gradings: deg x_i = deg ξ_j = 1 for all i and $r+1 \le j \le n$ and deg $\xi_i = 0$ for $1 \le j \le r$. This realization of $\mathfrak{vect}(m/n)$ is denoted by $\mathfrak{vect}(m/n; r)$.

5.1. The Lie superalgebra $\mathfrak{vect}(m/n;\ 1)$ and modules over it. In this section we will consider the simplest case of the nonstandard realization: that of $\mathfrak{vect}(m/n;\ 1)$, cf. [W]. Clearly, $\mathfrak{vect}(m/n;\ 1)$ is a $\mathbb{Z}$-graded Lie subsuperalgebra in $\mathfrak{vect}(m+n/m+n)$; it is the Cartan prolong -- the result of Cartan prolongation -- (definition see e.g. in [L1]) of $L_0 = \mathfrak{gl}(m/n)\otimes\Lambda(1)\ltimes\mathfrak{vect}(0/1) = \mathfrak{gl}(m+n)\otimes\Lambda(1)\ltimes\mathfrak{vect}(0/1)$.

In "geometric" terms the meaning of $\mathfrak{vect}(m/n;\ 1)$ is as follows. Let M be an m/n-dimensional supermanifold with coordinates $X = (x, \xi)$, M' the supermanifold described in Introduction with even coordinates $y = (x, \xi')$ and odd ones $\eta = (\xi, x')$. Then $D = \Sigma Z'_i\otimes Z_i$, where $Z = (y, \eta)$, is the exterior differential on M', i.e. D is a vector field on $(M')'$. Clearly, D is invariant with respect to $\mathfrak{vect}(m+n/m+n)$, whereas $d = \Sigma X'_i\otimes X_i$ is not. The desired interpretation of $\mathfrak{vect}(m/n;\ 1)$: it is the formal version of the Lie superalgebra of vector fields that send d to a linear combination of d and D.

For an explicit calculation of $\mathfrak{vect}(m/n;1)$-invariant differential operators we have to know a bit about irreducible L_0-modules, which is also of interest *per se*.

Theorem. *Any irreducible finite dimensional L_0-module of dimension different from 1 (or ε) being considred as $\mathfrak{gl}(m+n)$-module splits into the direct sum of two equvivalent irreducible modules.*

In what follows we denote the numerical labels of the highest weight of a $\mathfrak{gl}(k)$-module V as $(a_1, \ldots, a_k; \lambda)$, where $(a_1, \ldots, a_k)$ are the numerical labels with respect to the standard basis of $\mathfrak{sl}(k)$, cf. [OV], and λ is the weight with respect to the unit matrix. We will denote the module with highest weight χ by V_χ.

Proof. Any irreducible finite-dimensional L_0-module is, by A.Rudakov's arguments discussed in Introduction, a quotient of I(V), the module induced from an irreducible $\mathfrak{gl}(m+n)$-module V. Since $I(V) = V\otimes\Lambda(\xi) = V\oplus V\xi$, the computation of singular vectors is very easy and the answer is: $I(V_\chi)$ is irredudible unless $\chi = (0; \lambda)$.

As above, instead of $I(V_\chi)$ we will write $I(\chi)$.

Examples of L_0-modules. The superspaces of (formal) differential or integrable forms of given degree are of the form $T(E^k(id))$ and $T(E^k(id^*)\otimes str)$, respectively. Let us consider $E^k(id)$ and $E^k(id^*)\otimes str$ as L_0-modules. It is not difficult to verify that these L_0-modules are completely reducible and of the form

$$E^k(id) = I(k, 0, \ldots, 0; k)\oplus I(k-1, 0, \ldots, 0; k-1)\oplus\ldots\oplus$$
$$I(1, \ldots, 1, 0, \ldots, 0; 1) \text{ (k-many 1's) for } k\le n$$
$$I(k-n+1, 1, \ldots, 1; k-n+1) \text{ for } k > n$$

$$E^k(id^*)\otimes str = I(0, \ldots, 0, -k; 1-k+n)\oplus I(0, \ldots, 0, -1, 1-k; k-1)\oplus \ldots \oplus$$
$$I(0, \ldots, 0, -1, \ldots, -1; -n) \text{ (k-many -1's) for } k \le n$$
$$I(-1, \ldots, -1. n-1-k; -k) \text{ for } k > n$$

5.2. $\mathfrak{vect}(m/n; 1)$-invariant unary differential operators. In the following theorem there are only described operators corresponding to *finite-dimensional* fibers. There are many more operators corresponding to infinite dimensional fibers; we will consider them elsewhere.

Theorem.([W]) *Let V_1 and V_2 be irreducible L_0-modules, c: $T(V_1)$ --> $T(V_2)$ a vect(m/n; 1)-invariant differential operator. Then (up to the change of parity) $T(V_1)$ and $T(V_2)$ are neighbouring terms in one of the following sequences which are obtained from the de Rham complexes of integro-differential forms:*

case A:

... -->I(a, b, c, ...; a+n) -->I(a-1, b, c, ...; a+n-1) --> ... --> I(b, b, c, ...; b+n)

I(b-1, b-1, c, ...; b+n-2) -->I(b-1, b-2, c, ...; b+n-3) --> ...

....

... --> I(..., d+1, d, d, ... ; d+k+1) -->I(..., d, d, d, ... ; d+k)

I(..., d-1, d-1, d-1, ... ; d+k-3) -->I(..., d-1, d-1, d-2, ... ; d+k-4) --> ...

....

..... -->I(b-1, c-1, ..., d-1, d-1,..., z, z ; z+2)

I(b-1, c-1, ..., d-1, d-1, ..., z-1, z-1; z) --> I(..., z-1, z-2; z-1) --> ....

....

case B:

I(0, ..., 0; 0) -->I(0, ..., 0, -1; 0) --> ... --> I(0, -1, ..., -1; 0) --> I(-1, -1, ..., -1; 0)

case C:

I(0, ..., 0; 1)
↓
I(0, ..., 0, -1, -1; 0)

where the horisontal arrows are restrictions of the exterior differential and the other arrows are (new) higher order differential operators.

Proof of this theorem is tedious but straightforward (about 5 pages of calculations of singular vectors).

Bibliography

[BL1] Bernstein J., Leites D. Invariant diferential operators and irreducible representations of the Lie superalgebra of vector fields. Sel. Math. Sov., v. 1, n2, 1981,143 -160

[BL2] Bernstein J., Leites D. How to integrate differential forms over a supermanifold, v. 11, n3, 1977, 70-71

[BV] Batalin I., Vilkovisky G., Quantization of gauge fields. Phys. Lett., 102 B, 1982, 273-280

[F] Fuchs D., Cohomology of infinite dimensional algebras. Consultants Bureau, NY, 1987

[G1] Grozman P. Classification of bilinear invariant differential operators on tensor fields. Funct. Anal. Appl. v.14, n2, 1984, 58-59 (Russian)

[G2] Grozman P. Area-preserving invariant bilinear differential operators on a plane. Vestnik MGU, ser. math., #6, 1980, 3-6 (Russian)

[Ki] Kirillov A. Invariant operators on geometric quantities. In: Itogi nauki, Sovr. Probl. Matem., Nov. dostizheniya, v.16, 1980, 3-29 (Russian = Engl. transl. in JOSMAR (J. Sov. Math.), 1980)

[Ko1] Kochetkov Yu. Binary operations invariant with respect to Lie superalgebra W(1|1).VINITI depositions, 1983, n6847-83-dep.(Russian)

[Ko2] Kochetkov Yu. Binary operations invariant with respect to Lie superalgebras K(1|1) and K(1|2). VINITI depositions, 1983, n68478- 83-dep. (Russian)

[Ko3] Kochetkov Yu. Binary operations invariant with respect to Lie superalgebra K(1|3). VINITI depositions, 1985, n385-85-dep. (Russian)

[Ko4] Kochetkov Yu. Irreducible representations of Lie superalgebras $Щ_1$ and $Щ_2$[#]. In: Questions of Group Theory and Homologic Algebra, Yaroslavl Univ. Press, Yaroslavl, 1985, 145-147 (Russian)

[Ko5] Kochetkov Yu. Singular vectors and discrete modules. Russian Math. Surveys, v.41, #5, 1986, 182-183

[#] Two (of the three known so far) exceptional simple infinite dimensional Lie superalgebras discovered by Shchepochkina (Щепочкина)

[Ko6] Kochetkov Yu. Singular vectors and discrete modules II. In: Questions of Group Theory and Homologic Algebra, Yaroslavl Univ. Press, Yaroslavl, 1987, 115-117 (Russian)

[L1] Leites D., Lie superalgebras, JOSMAR, 30, n6, 1984, 2481-2513

[L2] Leites D., ed. Seminar on Supermanifolds, v. 1-7, Kluwer, to appear; see preprinted version in: Reports of Dept. of Math. of Stockholm University,##1-35, 2800pp., 1988-90

[LP1] Leites D., Post G. Cohomology to compute. In: Computers and mathematics. Springer, NY ea, 1989.

[LP2] Leites D., Poletaeva E., Analogues of Riemannian structure on supermanifolds. In: Proc. Internatnl. Alg. Conf., Novosibirsk, 1989, to appear

[MPV] Manin Yu., Penkov I., Voronov A., Elements of supergeometry. In: Itogi nauki. Sovremennye problemy matematiki. v.32, VINITI, Moscow, 1988 (Russian = Engl. transl. in JOSMAR)

[OV] Onishchik A., Vinberg É. Seminar on algebraic groups and Lie groups. Springer, 1990

[PS] Penkov I., Serganova V. Borel-Weil-Bott theorem for Lie superalgebras of vector fields and cohomology of invertible sheaves on curved supergrassmannians. SFB-170 preprint. Göttingen, July, 1990

[R1] Rudakov A. Irreducible representations of infinite dimensional Lie algebras of Cartan type. Math. USSR Izvestiya. 38, n3,1974, 835-866

[R2] Rudakov A. Irreducible representations of infinite dimensional Lie algebras of types S and H. Math. USSR Izvestiya. 39, n3,1975, 496-511

[W] Weintrob (Vaintrob) A., Differential operators invariant with respect to a new series of Lie superalgebras, Diploma thesis, Department of Mechanics and Mathematics of Moscow University, 1979 (in Russian)

Monic Sometimes Means α-Irreducible

Anthony J. Macula Westfield State College, Westfield, Massachusetts

Abstract

α denotes an uncountable cardinal number or the symbol ∞. A continuous function $f: Y \longrightarrow X$ is called an α-*SpFi* morphism if $f^{-1}(G)$ is dense in Y for each dense α-cozero set G in X. Thus, we have a category, α-*SpFi* , which consists of compact Hausdorff spaces and α-*SpFi* morphisms. α-*SpFi* , like any other category, has its monics, which need not be one-to-one. What are the monics in α-*SpFi*? When $\alpha \neq \omega_1, \infty$, there is no known characterization of the monics in α-*SpFi*. However, in this paper, we show that if X is α-cozero complemented, then $f: Y \longrightarrow X$ is monic in α-*SpFi* iff f is a range α-irreducible α-*SpFi* morphism

---

AMS(MOS) Subj. Class. : Primary 18A20, 18B30, 54C10, 54E99

Secondary 06E15

Key Words : spaces with filters, α-cozero complemented, monic irreducible, α-complete Boolean algebra

§ 1 The category α-SpFi

In this section, and throughout this paper, X, Y, and Z will denote compact Hausdorff spaces. α will denote an uncountable cardinal number or the symbol ∞; the meaning of "$\alpha = \infty$" will be clear from, or explained in, the context.

1.1 Definition. Let C(X) be the real-valued continuous functions on X. Let $\mathrm{Coz}(X) = \{ f^{-1}(\mathbb{R}\setminus\{0\}) : f \in C(X) \}$ be the set of cozero sets of X. A subset $V \subset X$ is called an α-cozero set if

$$V = \bigcup \{ U_i : i \in I, |I| < \alpha, U_i \in \mathrm{Coz}(X) \}.$$

Note that an ω_1-cozero set is a cozero set. By "$|I| < \infty$" we mean that $|I|$ is unrestricted, so every open set is an ∞-cozero set. We denote the collection of α-cozero sets by $\mathrm{Coz}_\alpha X$. Let $G_\alpha X$ be the filter generated by the dense members of $\mathrm{Coz}_\alpha X$; $G_\infty X$ will denote the filter generated by the dense open sets.

1.2 Definition. A continuous function $f: X \longrightarrow Y$ is called an α-*SpFi* morphism if $f^{-1}(G) \in \mathcal{G}_\alpha X$ for each $G \in \mathcal{G}_\alpha Y$.

If f is an α-*SpFi* morphism, we often say "f is α-*SpFi*".

1.3 Definition. The category *α-SpFi* consists of compact Hausdorff spaces and *α-SpFi* morphisms.

1.4 Definition. Let E be closed in X. If $E \cap G$ is dense in E for each $G \in G_\alpha X$, then E is called an *α-SpFi* subset of X, and we write $E \overset{\alpha}{\subset} X$.

1.5 Proposition. (from § 1 in [BHM])

(a) $E \overset{\alpha}{\subset} X$ iff the embedding of E into X is *α-SpFi*.

(b) $E \overset{\infty}{\subset} X$ iff E is a regular closed set in X.

(c) $E \overset{\infty}{\subset} X$ iff $E \overset{\alpha}{\subset} X$ for all $\alpha < \infty$.

1.6 Proposition. (from 1.6 in [BHM]) A continuous function $f\colon X \longrightarrow Y$ is *α-SpFi* iff $E \overset{\alpha}{\subset} X$ implies $f[E] \overset{\alpha}{\subset} Y$.

In general, recall that, in a category, a morphism τ is monic iff it is left cancelable, i.e., $\tau \circ \varphi_1 = \tau \circ \varphi_2$ implies $\varphi_1 = \varphi_2$.

What are the monics in *α-SpFi* ?

1.7 Definition. A continuous function $f\colon X \longrightarrow Y$ is called *range α-irreducible* if for each $U \in \mathrm{Coz}_\alpha X$ there is a $V \in \mathrm{Coz}_\alpha Y$ such that $f^{-1}(V) \subset U$ and

$cl(f^{-1}(V)) = \overline{U}$. If f is surjective and range α-irreducible, then f is just called α-irreducible. It follows that f is α-range irreducible if and only if $f: X \longrightarrow f(X)$ is α-irreducible.

Note, cl( ) and $\overline{\ }$ both denote the closure operator. Also, ∞-irreducibility is equivalent to what is commonly referred to in the literature as just irreducibility. Recall, a continuous surjection $f: X \longrightarrow Y$ is said to be irreducible if $f[K] \neq Y$ for any proper closed set K in X.

1.8 Proposition. Let $f: X \longrightarrow Y$ be a continuous function .

(a) If f is α-irreducible, then f is α-*SpFi*.

(b) If f is α-*SpFi*, then $f(X) \overset{\alpha}{\subset} Y$. On the other hand, if f is range α-irreducible and $f(X) \overset{\alpha}{\subset} Y$, then f is α-*SpFi*.

(c) (from 3.12 in [BHN]) f is range α-irreducible iff whenever $E_1, E_2 \overset{\alpha}{\subset} X$ and $E_1 \neq E_2$, then $f[E_1] \neq f[E_2]$. (Thus f induces a one-to-one map from the α-*SpFi* subsets of X to the α-*SpFi* subsets of f(X).)

(d) If f is range α-irreducible and $f(X) \overset{\alpha}{\subset} Y$, then f is monic in α-*SpFi*.

Proof: (a) Suppose f is not α-*SpFi*. Then there is a $G \in \mathcal{G}_\alpha Y$ such that $f^{-1}(G)$ is not dense in X. Then there is a non-empty $U \in \mathrm{Coz}_\alpha X$ such that $U \subset X \setminus \mathrm{cl}(f^{-1}(G))$. Since $V \cap G \neq \emptyset$ for all $V \in \mathrm{Coz}_\alpha Y$, there is no $V \in \mathrm{Coz}_\alpha Y$ with $f^{-1}(V) \subset U$ and $\mathrm{cl}(f^{-1}(V)) = \overline{U}$. Hence, f is not range α-irreducible.

(b) For the first part apply 1.6. The second follows from (a), 1.6, and the easily verified fact that $K \overset{\alpha}{\subset} f(X) \overset{\alpha}{\subset} Y$ implies $K \overset{\alpha}{\subset} Y$.

(d) Suppose f is not monic. Then there are α-*SpFi* maps, $h_i\colon Z \longrightarrow X$, $i = 1,2$, for which $f \circ h_1 = f \circ h_2$, but $h_1 \neq h_2$. It follows that there is a regular closed set U such that $h_1[U] \neq h_2[U]$. By 1.5 and 1.6, $h_i[U] \overset{\alpha}{\subset} X$, and $f(h_1[U]) = f(h_2[U])$. Therefore, by (c) here, f is not range α-irreducible.

1.9 Proposition. Let $f\colon X \longrightarrow Y$ and $h\colon Y \longrightarrow Z$ be continuous.

(a) If $h \circ f$ is α-*SpFi*, and h is α-irreducible, then f is α-*SpFi*.

(b) $h \circ f$ is range α-irreducible iff h and f are range α-irreducible.

Proof: (a) Let $G \in \mathcal{G}_\alpha Y$. Then there is a $U \in \mathrm{Coz}_\alpha Z$ such that $h^{-1}(U) \subset G$, and $\mathrm{cl}(h^{-1}(V)) = \overline{G} = Y$. It follows that $U \in \mathcal{G}_\alpha Z$; thus, $f^{-1}(h^{-1}(U)) = (h \circ f)^{-1}(U)$ is in $\mathcal{G}_\alpha X$. Therefore $f^{-1}(G) \in \mathcal{G}_\alpha X$ because

$f^{-1}(h^{-1}(U)) \subset f^{-1}(G)$. Hence f is α-*SpFi*.

(b) ($\Leftarrow$) Let $U \in Coz_\alpha X$, there is a $V \in Coz_\alpha Y$ with $f^{-1}(V)$ dense in U. There also is a $W \in Coz_\alpha Z$ with $h^{-1}(W)$ dense in V. It follows from the range α-irreducibility of f that $f^{-1}(h^{-1}(W))$ is dense in U.

($\Rightarrow$) If $h \circ f$ is range α-irreducible, then for each $U \in Coz_\alpha X$ there is a $V \in Coz_\alpha Z$ such that $f^{-1}(h^{-1}(V)) \subset U$ and $cl(f^{-1}(h^{-1}(V))) = \overline{U}$. Since $h^{-1}(V) \in Coz_\alpha Y$, f is range α-irreducible.

Let $W \in Coz_\alpha Y$. Then $f^{-1}(W) \in Coz_\alpha X$. As above, there is a $V \in Coz_\alpha Z$ such that $f^{-1}(h^{-1}(V)) \subset f^{-1}(W)$ and $cl(f^{-1}(h^{-1}(V))) = cl(f^{-1}(W))$. It follows that $h^{-1}(V) \subset W$ and $cl(h^{-1}(V)) = \overline{W}$. Hence h is range α-irreducible.

From 1.8 (c), we see that if an α-*SpFi* morphism, f is range α-irreducible, then f is monic in α-*SpFi*. However, when $\alpha \neq \infty$, monics in α-*SpFi* need not be range α-irreducible. See [M_1] and [BH].

When $\alpha = \infty$ we have:

1.10 Theorem. ([BHM],[W]) In ∞-*SpFi*, a morphism f is monic in ∞-*SpFi* iff f is range ∞-irreducible.

When $\alpha = \omega_1$, monics need not be range ω_1-irreducible. Below is a characterization of the monics in ω_1-*SpFi*.

1.11 Theorem. (from 3.1 in [BH]) An ω_1-*SpFi* morphism $f: X \longrightarrow Y$ is monic in ω_1-*SpFi* iff for each pair of points $x_1 \neq x_2$, there are neighborhoods U_i of x_i and a sequence $G_1, G_2, \ldots$ in $G_{\omega_1}X$ for which

$f[U_1 \cap \bigcap_n G_n] \cap f[U_1 \cap \bigcap_n G_n] = \emptyset$.

A complete characterization of α-*SpFi* the monics when $\alpha \neq \omega_1, \infty$ is not known.

We do have some additional information about monics in general α-*SpFi*. Recall that X is said to be α-*disconnected* if $\overline{U}$ is open for each $U \in Coz_\alpha X$.

1.12 Theorem. (from 3.4 [BHM]) If $f: X \longrightarrow Y$ is monic in α-*SpFi*, and Y is α-disconnected, then f is one-to-one.

Let $\mathcal{B}$ be a category and $\mathcal{A}$ a subcategory of $\mathcal{B}$. We call $\mathcal{A}$ a *monocoreflective subcategory* of $\mathcal{B}$ if, for each $B \in |\mathcal{B}|$, there is an $A_B \in |\mathcal{A}|$ and a monic, $m_B : A_B \longrightarrow B$, in $\mathcal{B}$, such that for each $\mathcal{B}$ morphism, $f : A \longrightarrow B$, from an $\mathcal{A}$ object, there is a unique $\mathcal{A}$ morphism $\overline{f} : A \longrightarrow A_B$ such that $f = m_B \circ \overline{f}$. The pair (A_B, m_B) is called the $\mathcal{A}$ monocoreflection of B.

1.13 Theorem. (from 4.2 in $[M_3]$) In α-*SpFi*, the full subcategory of α-disconnected spaces is monocoreflective.

For each X, we write $(m_\alpha X, m_\alpha)$ to denote the α-disconnected monocoreflection of X in α-*SpFi*.

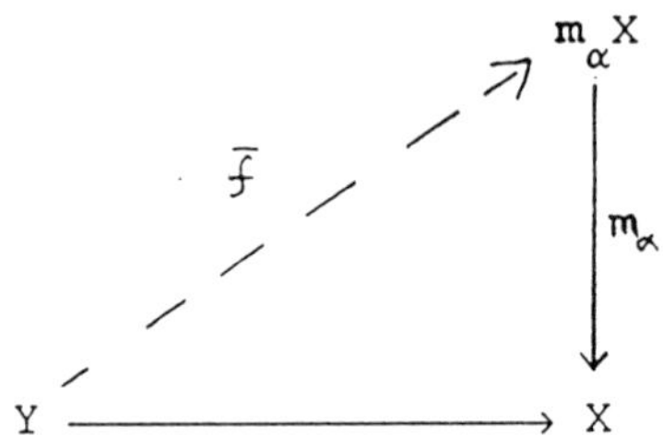

What is $(\mathbf{m}_\alpha X, \mathbf{m}_\alpha)$? In $[M_3]$ it is constructed by abstract categorical arguments, which only tell us that $(\mathbf{m}_\alpha X, \mathbf{m}_\alpha)$ has the desired universal property, $\mathbf{m}_\alpha X$ is α-disconnected, and $\mathbf{m}_\alpha$ is monic. They fail to give much insight into the structure of $\mathbf{m}_\alpha X$, and they don't give information about additional properties of $\mathbf{m}_\alpha$.

Information about $\mathbf{m}_\alpha X$ and $\mathbf{m}_\alpha$ leads to information about monics in general. In what follows, assuming that X is α-cozero complemented (i.e., for each $U \in \mathrm{Coz}_\alpha X$, there is a $V \in \mathrm{Coz}_\alpha X$ with $U \cup V \in G_\alpha X$ and $U \cap V = \emptyset$), we construct $m_\alpha X$ as the Stone space of $\overline{\mathrm{Coz}_\alpha X}$ ($\overline{\mathrm{Coz}_\alpha X} = \{ \overline{U} : U \in \mathrm{Coz}_\alpha X \}$) and we show that $\mathbf{m}_\alpha$ is α-irreducible. Note that the Stone space of $\overline{\mathrm{Coz}_\alpha X}$ need not be Hausdorff if X is not α-cozero complemented. Therefore we call this Stone space $m_\alpha X$ to distinguish it from the $\mathbf{m}_\alpha X$ above that is always compact and Hausdorff . $m_\alpha X \cong \mathbf{m}_\alpha X$ if and only if X is α-cozero complemented. Our construction is a special case of a construction in $[HVW_2]$. In the terminology of $[HVW_2]$, if X is α-cozero complemented , $m_\alpha X$ is a Wallman cover of X . Our construction also coincides with the space $T_\alpha X$ in section 6 of [BHN]. Since an α-cozero complemented X satisfies Theorem 6.2 (c) of [BHN], it follows that $m_\alpha X$ is the α-quasi-F cover in this case.

§ 2 α-complete Boolean algebras

In this section, A and B denote Boolean algebras, and φ denotes a Boolean algebra homomorphism. *BA* denotes the category of Boolean algebras and Boolean algebra homomorphisms.

2.1 Definition. B is called *α-complete* if $\bigvee^{B} S$ exists for each $S \subset B$ with $|S| < \alpha$. A morphism $\varphi: A \longrightarrow B$ is also called α-complete if $\bigvee^{A} S = s$ implies $\bigvee^{B} \varphi[S] = \varphi(s)$ whenever $|S| < \alpha$.

2.2 Proposition. (See [S]) Let $\varphi: A \longrightarrow B$. The following are equivalent:

(a) φ is α-complete.

(b) $\bigvee^{A} S = 1$ implies $\bigvee^{B} \varphi[S] = 1$ whenever $|S| < \alpha$.

Let *BS* denote the category of Boolean spaces (i.e., zero dimensional compact Hausdorff spaces) with continuous functions. We have, of course, the functors

$$BS \underset{S}{\overset{\text{clop}}{\rightleftarrows}} BA$$

of Stone duality.

For B , an element of S(B) (the Stone space of B)

is an ultrafilter μ of B, and the family $B^* = \{ b^* : b \in B \}$, where $b^* = \{ \mu \in S(B) : b \in S(B) \}$, is clop(S(B)) (the family of clopen sets of S(B)). Recall that if μ is an ultrafilter on B, then, for each $\varphi: A \longrightarrow B$, the family

$$\varphi^{-1}(\mu) = \{ a \in A : \varphi(a) \in \mu \}$$

is an ultrafilter on A. The map $s(\varphi): S(B) \longrightarrow S(A)$ is defined by $s(\varphi)(\mu) = \varphi^{-1}(\mu)$. See [S] for a more thorough discussion of the above.

The next proposition exhibits a connection between α-*SpFi* and *BA*.

2.3 Proposition.

(a) (from 22.4 of [S]) B is α-complete iff S(B) is α-disconnected.

(b) (from 22.5 of [S]) $\varphi: A \longrightarrow B$ is α-complete iff $s(\varphi): S(B) \longrightarrow S(A)$ is α-*SpFi*.

2.4 Definition. $\overline{\mathrm{Coz}_\alpha X} = \{ \overline{U} : U \in \mathrm{Coz}_\alpha X \}$.

Notice that the representation of an element of $\overline{\mathrm{Coz}_\alpha X}$ is not unique; it may happen that $\overline{U} = \overline{V}$, $U \neq V$ for $U, V \in \mathrm{Coz}_\alpha X$. Also, notice that $\overline{\mathrm{Coz}_\infty X} = R(X)$, where R(X) denotes the regular closed sets in X.

2.5 Definition. A space X is called *α-cozero complemented* (abbr. α.c.c.) if for each $U \in \mathrm{Coz}_\alpha X$ there is a $V \in \mathrm{Coz}_\alpha X$ for which $U \cup V$ is dense in X, and $U \cap V = \emptyset$.

Many spaces are α.c.c. . α-disconnected spaces are α.c.c. . Any space with cellularity $< \alpha$ is α.c.c. (e.g., dyadic spaces are ω_1.c.c.).

2.6 Proposition. If X is α.c.c., then $\overline{\mathrm{Coz}_\alpha X}$ is an α-complete Boolean algebra under the following operations: $\overline{U} \wedge \overline{V} = \overline{U \cap V}$, $\overline{U} \vee \overline{V} = \overline{U \cup V}$, and $\overline{U}' = \overline{W}$ where $W \cup U$ is dense in X, and $W \cap U = \emptyset$. Moreover, if $|I| < \alpha$, then $\bigvee_I \overline{U}_i = \mathrm{cl}(\bigcup_I U_i)$.

Proof: It is left to the reader to verify most of the above assertions. The last claim, indicating that $\overline{\mathrm{Coz}_\alpha X}$ is α-complete, follows from the fact that $\mathrm{Coz}_\alpha X$ is closed under $< \alpha$ unions. We show that the definition of $\overline{U}'$ is well-defined. If W_1, $W_2 \in \mathrm{Coz}_\alpha X$ are such that $U \cup W_i$ is dense and $U \cap W_i = \emptyset$, then $W_i \subset X \backslash \overline{U}$ and $\overline{W}_i \cup \overline{U} = X$; thus $W_i \subset \overline{W}_j$. Hence $\overline{W}_1 = \overline{W}_2$.

Note, $\bigvee \overline{U}_i = 1$ in $\overline{\mathrm{Coz}_\alpha X}$ iff $\bigcup U_i$ is dense in X, i.e., $\cup U_i \in G_\alpha X$.

2.7 Proposition. Let X be α.c.c. and $U, V \in Coz_\alpha X$. Then $\overline{U} = \overline{V}$ iff there is a $W \in Coz_\alpha X$ for which $W \cup U$ and $W \cup V$ are dense, and $W \cap U = W \cap V = \emptyset$.

Proof: (⇒) Clear.

(⇐) Suppose there is such a W and $\overline{U} \neq \overline{V}$. Without loss of generality, assume $U \backslash \overline{V} \neq \emptyset$. Then $(W \cup V) \cap U \backslash \overline{V} = \emptyset$, but this can't happen because $W \cup V$ is dense.

§ 3 α-disconnected monocoreflections of α.c.c. spaces

In this section we prove the main result of this paper.

3.1 Definition. Let $f: X \longrightarrow Y$. Define $\overline{f}: \overline{Coz_\alpha Y} \longrightarrow \overline{Coz_\alpha X}$ by $\overline{f}(\overline{U}) = cl(f^{-1}(U))$ for $\overline{U} \in \overline{Coz_\alpha Y}$.

The next proposition is similar to 2.3 (b).

3.2 Proposition. Let X and Y be α.c.c. . If

$f: X \longrightarrow Y$ is α-*SpFi*, then $\overline{f}$ is an α-complete Boolean algebra homomorphism.

Proof: It's not hard to see that $\overline{f}$ is a homomorphism once we see that it is well-defined. To this end let $\overline{U} = \overline{V}$ for $U,V \in Coz_\alpha Y$, then there is a $W \in Coz_\alpha Y$ as in 2.7. Therefore, since f is α-*SpFi*, $f^{-1}(U) \cup f^{-1}(W)$ and $f^{-1}(V) \cup f^{-1}(W)$ are dense in X, and $f^{-1}(U) \cap f^{-1}(W) = f^{-1}(V) \cap f^{-1}(W) = \emptyset$. Thus, by 2.7, $cl(f^{-1}(U)) = cl(f^{-1}(V))$.

To see that $\overline{f}$ is α-complete, let $|I| < \alpha$ and $\bigvee_I \overline{U}_i = 1$ in $\overline{Coz_\alpha Y}$. Then $\bigcup_I U_i \in G_\alpha Y$. Therefore, $\bigvee_I \overline{f}(\overline{U}_i) = cl(\bigcup_I f^{-1}(U_i)) = 1$ in $\overline{Coz_\alpha X}$ because $\bigcup f^{-1}(U_i) = f^{-1}(\bigcup U_i) \in G_\alpha X$.

3.3 Definition. Let X be α.c.c. . Define $m_\alpha X$ by setting $m_\alpha X = S(\overline{Coz_\alpha X})$. In addition, define $m^X_\alpha: m_\alpha X \twoheadrightarrow X$ by $m^X_\alpha(\mu) = \bigcap\mu = \bigcap\{ \overline{V} : \overline{V} \in \mu \}$ for $\mu \in m_\alpha X$

If μ is an ultrafilter on $\overline{Coz_\alpha X}$, then $\bigcap\mu$ is a singleton, because, if $x \neq y \in \bigcap\mu$, there are $U,V \in Coz_\alpha X$ with $x \in U$, $y \in V$, and $U \cap V = \emptyset$. From this, it follows that $\overline{U},\overline{V} \in \mu$ and $\overline{U} \wedge \overline{V} = 0$ in $\overline{Coz_\alpha X}$, which is absurd. Moreover, the family $\{ \overline{V} : x \in V \in Coz_\alpha X \}$ is easily seen to be a filter, which is contained in an some ultrafilter μ. Clearly, $x = \cap\mu$; thus m_α^X is indeed surjective.

For m_α^X , we will drop the superscript ^{X} when the context is clear.

When $\alpha = \infty$, $\overline{\mathrm{Coz}_\infty X} = R(X)$, so $m_\alpha X = S(R(X)) = EX$, where EX is the absolute of X. What follows below is, in many ways, a cardinal generalization of EX. See chapter six of [PW].

We now state the main result.

3.4 Theorem. Let X be α.c.c. .

(a) m_α is α-irreducible.

(b) $m_\alpha X$ is α-disconnected.

(c) If X is α-disconnected, then m_α is a homeomorphism.

(d) If Y is α-disconnected, and $f: Y \longrightarrow X$ is α-*SpFi*, then there is a unique α-*SpFi* morphism $f^*: Y \longrightarrow m_\alpha X$ such that $m_\alpha \circ f^* = f$. In other words, $(m_\alpha X, m_\alpha)$ is the α-disconnected monocoreflection , $(\mathbf{m}_\alpha X, \mathbf{m}_\alpha)$, of X in α-*SpFi*.

(e) $f: Y \longrightarrow X$ is monic in α-*SpFi* iff f is a range α-irreducible α-*SpFi* morphism.

Proof: (a) To see that m_α is continuous, we show that for each $V \in Coz_\alpha X$, $m_\alpha^{-1}(V) = \bigcup\{ \overline{U}^* : \overline{U} \subset V,\ U \in Coz_\alpha X \}$. If $\mu \in m_\alpha^{-1}(V)$, then $x = \bigcap\mu \in V$. Hence there is $U \in Coz_\alpha X$ such that $x \in U$, and $\overline{U} \subset V$. Clearly $\mu \in \overline{U}^*$. On the other hand, if $\mu \in \overline{U}^*$ for some $\overline{U} \subset V$, $\bigcap\mu \in \overline{U} \subset V$. Hence $\mu \in m_\alpha^{-1}(V)$.

To see that m_α is α-irreducible, it suffices to show that $cl(m_\alpha^{-1}(V)) = \overline{V}^*$ for each $V \in Coz_\alpha X$. (Clearly, $m_\alpha^{-1}(V) \subset \overline{V}^*$.)

$$\begin{aligned} cl(m_\alpha^{-1}(V)) &= \{ \mu : \mu \in \overline{W}^* \Rightarrow \overline{W}^* \cap m_\alpha^{-1}(V) \neq \emptyset \} \\ &= \{ \mu : \overline{W} \in \mu \Rightarrow \exists\ \gamma \ .\ni.\ \overline{W} \in \gamma \text{ and } \textstyle\bigcap\gamma \in V \} \\ &= \{ \mu : \overline{W} \in \mu \Rightarrow W \cap V \neq \emptyset \} \\ &= \{ \mu : \overline{W} \in \mu \Rightarrow \overline{W} \wedge \overline{V} \neq 0 \} \\ &= \{ \mu : \overline{V} \in \mu \} \\ &= \overline{V}^* \end{aligned}$$

If $U \in Coz_\alpha m_\alpha X$, then $U = \bigcup\{ \overline{V}_i^* : i \in I,\ |I| < \alpha \}$ for some family $\{ V_i \} \subset Coz_\alpha X$. Therefore $m_\alpha^{-1}(\bigcup_i V_i) \subset U$, and $cl(m_\alpha^{-1}(\bigcup_i V_i)) = \overline{U}$.

(b) Apply 2.3 and 2.6.

(c) By (a) here and 1.8 (c), m_α is a monic in α-*SpFi*. Hence, by 1.12, m_α is a homeomorphism.

(d) We have, by 3.2, that $\overline{f}: \overline{Coz_\alpha X} \longrightarrow \overline{Coz_\alpha Y}$ is an α-complete homomorphism. We claim that for $s(\overline{f}) : m_\alpha Y = S(\overline{Coz_\alpha Y}) \longrightarrow S(\overline{Coz_\alpha X}) = m_\alpha X$, $f \circ m_\alpha^Y = m_\alpha^X \circ s(\overline{f})$.

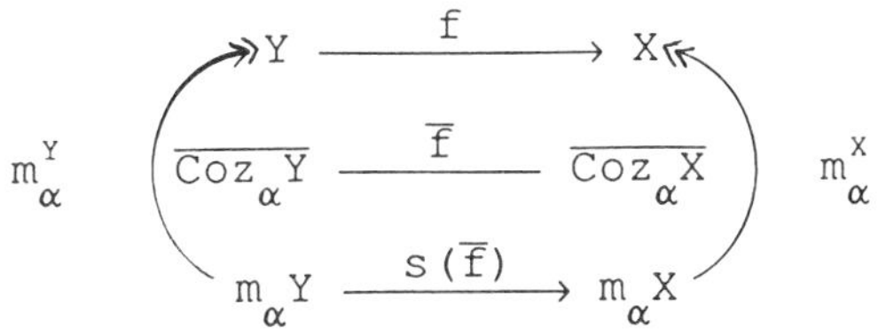

Since m_α^Y is a homeomorphism ((c) here), and $s(\overline{f})$ is α-*SpFi* (2.3), a proof of the above claim will give us (d). (Take $f^* = s(\overline{f}) \circ (m_\alpha^Y)^{-i}$.)

To this end, let $\mu \in m_\alpha Y$. Then $f \circ m_\alpha^Y(\mu) = f(\bigcap\mu)$, and $s(\overline{f})(\mu) = \overline{f}^{-1}(\mu) = \{ \overline{W} : cl(f^{-1}(W)) = \overline{V} , \overline{V} \in \mu \}$. Let $x = \bigcap\mu$. Then $f(x) \in f[\overline{V}]$ for each $\overline{V} \in \mu$. Therefore, $f(x) \in f[cl(f^{-1}(W))] = \overline{W}$ for each $\overline{W}$ with $cl(f^{-1}(W)) = \overline{V} \in \mu$, i.e., for each $\overline{W} \in \overline{f}^{-1}(\mu) = s(\overline{f})(\mu)$. Therefore, $f \circ m_\alpha^Y(\mu) = f(\bigcap\mu) = f(x) = \bigcap s(\overline{f})(\mu) = m_\alpha^X \circ s(\overline{f})(\mu)$.

(e) ($\Leftarrow$) 1.8 (c).

($\Rightarrow$) If $f: Y \longrightarrow X$ is monic in α-*SpFi*, then $f \circ m_\alpha^Y: m_\alpha Y \longrightarrow X$, being the composition of two monics, is monic. By (d) here, there is an α-*SpFi* morphism $f^*: m_\alpha Y \longrightarrow m_\alpha X$ such that $m_\alpha^X \circ f^* = m_\alpha^Y \circ f$. f^* is monic in α-*SpFi* because it's the first factor of a monic. By 1.12, f^* is one-to one, hence range α-irreducible. Therefore, by 1.9 (b), $m_\alpha^X \circ f^* = m_\alpha^Y \circ f$ is range α-irreducible. Again applying 1.9 (b), we see that f is range α-irreducible.

3.5 Proposition. Any space with an α-irreducible preimage which is α.c.c. is itself α.c.c. . Thus X has an α-irreducible preimage which is a-disconnected iff X is a.c.c. .

Proof. Suppose f: Y —— X is α-irreducible and Y is α.c.c. . Given $U \in \mathrm{Coz}_\alpha X$ there is a $V \in \mathrm{Coz}_\alpha Y$ such that $f^{-1}(U) \cup V$ is dense in Y and $f^{-1}(U) \cap V = \emptyset$. Since f is α-irreducible, there is a $W \in \mathrm{Coz}_\alpha X$ with $f^{-1}(W)$ dense in V. Then $f^{-1}(W) \cup f^{-1}(U)$ dense in Y implies $W \cup U$ dense in X, and clearly $W \cap U = \emptyset$.

The second part follows from the first and 3.4 (a) and (b).

3.6 Corollary. $(m_\alpha X, m_\alpha)$ is essentially the same as $(\mathbf{m}_\alpha X, \mathbf{m}_\alpha)$ iff X is α.c.c.

Proof. Apply 3.4 and 3.5.

§ 4 Remarks about other covers

All spaces in this section are compact.

Consider, in addition to $m_\alpha: m_\alpha X \twoheadrightarrow X$, the following preimages of X:

(1) $q_\alpha: QF_\alpha X \twoheadrightarrow X$

(2) $\lambda_\alpha: \Lambda_\alpha X \twoheadrightarrow X$

The pairs $(QF_\alpha X, q_\alpha)$ and $(\Lambda_\alpha X, \lambda_\alpha)$, described below, are examples of what is generally known as a *cover* of X. (P,f) is said to be a cover of X if $f: P \twoheadrightarrow X$ is ∞-irreducible. Two covers (P,f) and (Q,h) are said to be essentially identical if there is a homeomorphism $g: Q \longrightarrow P$ such that $h = f \circ g$. We denote this by $(P,f) \cong (Q,h)$. (P,f) is called the essentially unique * cover of X , where * denotes some properties of P and/or f (beyond ∞-irreducibility) , if for any other * cover (P′,f′), we have $(P,f) \cong (P',f')$.

(1) $(QF_\alpha X, q_\alpha)$ is called the (minimum) α-quasi-F cover of X. See [DHH],[HVW],[BHN],[Mo],[M_2]. $QF_\alpha X$ is α-quasi-F , and q_α is α-irreducible. (Y is α-quasi-F if each dense α-Lindelöf set in Y is C^*-embedded in Y.)

(i) $(QF_\alpha X, q_\alpha)$ is the essentially unique α-irreducible α-quasi-F cover of X. Therefore $(QF_\alpha X, q_\alpha)$, is the projective cover in the category of compact Hausdorff spaces and α-irreducible functions.

(ii) In α-*SpFi*, the full subcategory of α-quasi-F spaces is monocoreflective, and for each X, $(QF_\alpha X, q_\alpha)$ is the α-quasi-F monocoreflection of X.

(2) $(\Lambda_\alpha X, \lambda_\alpha)$ is called the (minimum) α-disconnected cover of X. $\Lambda_\alpha X$ is α-disconnected, and λ_α is monic in α-*SpFi*. See [V_1], [V_2], [H], and [M_1].

(i) If $f: Y \longrightarrow X$ is ∞-irreducible, and Y is α-disconnected, then there is an ∞-irreducible $f': Y \longrightarrow \Lambda_\alpha X$ such that $f = \lambda_\alpha \circ f'$.

(ii) $(\Lambda_\alpha X, \lambda_\alpha)$ is the essentially unique α-disconnected α-*SpFi* monic cover of X. Therefore $(\Lambda_\alpha X, \lambda_\alpha)$ is the projective cover of X in the category of compact spaces and ∞-irreducible α-*SpFi* monic morphisms.

In [HVW], it is shown that $(\Lambda_{\omega_1} X, \lambda_{\omega_1}) \cong (QF_{\omega_1} X, q_{\omega_1})$ iff X is α.c.c. . The result below is from [M_1].

4.1 Theorem. (a) If X is α.c.c. , then

$$(QF_\alpha X, q_\alpha) \cong (\Lambda_\alpha X, \lambda_\alpha) \cong (m_\alpha X, m_\alpha) = (S(\overline{Coz_\alpha X}), m_\alpha) \ .$$

(b) If X is not α.c.c. , then no two of the above three are essentially identical.

References

[BH] R.N. Ball and A.W. Hager, Application of spaces with filters to Archimedean ℓ-groups, Proc. Conf. on Ordered Alegebraic Structures, Curaçao, 1988, J. Martinez Editor, Kluwer Acad. Pub.

[BHM] R.N. Ball, A.W. Hager, and A.J. Macula, An α-disconnected space has no proper monic preimage, Topology and Its Appl., to appear.

[BHN] R.N. Ball, A.W. Hager, and C. Neville, The κ-ideal completion of an Archimedean ℓ-group and the κ-quasi-F cover of a compact space, Gen. Top. and Appl, Proc. Northeast Top. Conf., Wesleyan Univ., 1988, R.M.Shortt, ed., 7-50 , Marcel Dekker, New York, 1990.

[DHH] F.Dashiell, A.W. Hager, and Mel Henriksen, Order-Cauchy completions of rings and Vector lattices of continuous functions, Can. J. Math. 32(1980), 657-685.

[G] A.M. Gleason, Projective topological spaces, Illinois J. Math. 2(1958), 482-489.

[GJ] L. Gillman and M. Jerison, Rings of Continuous Functions, Van Nostrand, Princeton, 1960

[H] A.W. Hager, Minimal covers of topological spaces, Proc. Northeast Topology Conference, 1987, Annals N.Y. Acad. Sciences, 552(1989) 44-59.

[HVW] M. Henriksen, J. Vermeer, and R.G. Woods, Quasi-F covers of Tychonoff spaces, Trans. Amer. Math Soc. 303(1987), 779-803.

[HVW_2] ——————, Wallman covers of compact spaces, Diss. Math, 280(1989)

[HS] H. Herrlich and G. Strecker, Category Theory, Allyn and Bacon Co., Boston, 1973.

[M_1] A.J. Macula, Thesis, Wesleyan Univ., 1989.

[M_2] —————, α-Dedekind complete Archimedean vector lattice vs. α-quasi-F spaces, submitted for publication.

[M_3] —————, Free α-extensions of an Archimedean vector lattice and their topological duals, to appear, Trans. Amer. Math. Soc..

[Mo] A. Molitor, Quasi-F-like covers of a compact Hausdorf space, Gen. Top. and Appl., Proc. Northeast Topology Conf, Wesleyan Univ, 1988, R.M. Shortt, ed., 219-226, Marcel Dekker, New York 1990.

[PW] J.R. Porter and R.G. Woods, Extensions and Absolutes of Hausdorff Spaces, Springer Velag, New York, 1988.

[S] R. Sikorski, Boolean Algebra, (Third Ed.), Springer-Verlag , Berlin, 1969.

[V_1] J. Vermeer, The smallest basically disconnected preimage of a space, Topology and Its Appl., 17(1984), 217-232.

[V_2] —————, On perfect irreducible preimages, Topology Proceedings, 9(1984), 173-189.

[W] R.G. Woods, Covering properties and coreflective subcategories, Proc. Northeast Topology Conf. 1987, Annals N.Y. Acad. Sci., 552(1989).

Nonnormality of $C^p(M,N)$ in Whitney's Topology

Vitor Neves Universidade da Beira Interior, Covilhã, Portugal

ABSTRACT

We show that $C^P(M,N)$ $(n\in\mathbb{N})$ is not normal under Whitney's W^p-topology when M is open, by means of an adaptation of the embedding given in [4]. This answers a question of Hirsch [2; 2.4, page 65].

INTRODUCTION

Some time after having proved, in [4], that $C^\infty(M,N)$ is not normal under any topology finer than Whitney's C^∞ and coarser than Michor's $\mathcal{D}$, we noticed that a finer use of convolution allows an adaptation of the embedding given therein, so that the same ideas apply to $C^p(M,N)$, as we proceed to describe.

One of the basic ingredients in the foregoing proof is that each term of a well chosen sequence $f_\infty^n \in C^{p-1}(\mathbb{R}^\mu,\mathbb{R}^\nu)\setminus C^p(\mathbb{R}^\mu,\mathbb{R}^\nu)$ $(n=1,2,\dots)$ can be approximated by C^p-functions, uniformly on compact sets up to the derivative of order p-1, and pointwise at points of continuity, up to the pth derivative. This is so because the following equality, where * denotes convolution and ρ is a C^∞-kernel,

$$D^k(f*\rho) = (D^k f)*\rho,$$

still holds when $D^k f$ is bounded but not necessarily continuous, and

$$f*\rho_n \longrightarrow f$$

at points of continuity of f, when the diameter of $\mathrm{supp}(\rho_n)$ tends to zero ([2; 2.3] and [8; 9.8]).

Let W^p denote the Whitney C^p-topology on $C^p(M,N)$.

In section 3 below we build a set S of compactly supported C^p-series whose C^p-addends are supported in compact disjoint sets. This is done so that, on the one hand, S as a subspace of $(C^p(\mathbb{R}^\mu,\mathbb{R}^\nu),W^p)$ behaves like $\mathbb{N}^\mathbb{N}$ endowed with the product topology as a subspace of $\mathbb{R}^\mathbb{N}$ and, on the other hand, the series in S either approach elements of S or a non-C^p-function. Therefore S is W^p-closed.

Next we embed S into a pointwise closed subset of $C^p(M,\mathbb{R}^\nu)$ whose elements are supported in a fixed compact set K_0.

The other important observation is that the box topology on $\mathbb{R}^\mathbb{N}$ and the W^p topologies have a remarkably similar definition that can be explored locally via uniform continuity (section 4), so that we are also able to embedd $(\mathbb{R}^\mathbb{N},\mathrm{box})$ into a W^p-closed subset of series in $C^p(M,\mathbb{R}^\nu)$ whose addends are supported in disjoint compact sets that "tend" to infinity and are also disjoint from K_0.

Embedding the non-normal space $D=(\mathbb{N}^\mathbb{N},\mathrm{product})\times(\mathbb{R}^\mathbb{N},\mathrm{box})$ into a W^p-closed subset of $C^p(M,\mathbb{R}^\nu)$ will therefore show the non-normality of this set under the Whitney-C^p topology.

In section 5, the embedded image of D into $C^p(M,\mathbb{R}^\nu)$ will itself be embedded into a W^p-closed subset of $C^p(M,N)$, where ν is the dimension of N.

We use the theory of infinitesimals [6; Chap.8] to prove that our 1-1 map from D into $C^p(M,N)$ is continuous, has closed image and has a continuous inverse. In this context, $x\approx y[t]$ means **x is infinitely close to y for the uniformity t.**

1 THE WHITNEY C^P-TOPOLOGY

Let M and N be open (non-compact, borderless) connected (paracompact) C^p-manifolds ($p \geq 0$) modeled, respectively, on $\mathbb{R}^\mu$ and $\mathbb{R}^\nu$. The jet spaces $J^k(M,N)$ ($0 \leq k \leq p$) are metrizable, so let $d^k(.,.)$ denote a compatible metric on $J^k(M,N)$. As the projections $j^kf(m) \longmapsto j^{k-1}f(m)$ are continuous, the Whitney C^p-topology is given by the family of pseudometrics

$$\sup[\eta(m)d^p(j^pf(m), j^pg(m)): m \in M]$$

for all positive continuous functions $\eta: M \longrightarrow]0,\infty[$. Equivalently

$$f \approx g[W^p] \text{ iff } \eta(m)d^p(j^pf(m), j^pg(m)) < 1$$

for all standard positive continuous $\eta \in {}^\sigma C(M,]0,\infty[)$.

If $M = \mathbb{R}^\mu$ and $N = \mathbb{R}^\nu$, let $\|.\|$ denote the usual norms for multilinear functions on $\mathbb{R}^\mu$ and for each compact $K \subseteq \mathbb{R}^\mu$, let

$$\|f\|_K^p = \sup\{\|D^if_m\|: m \in K, 0 \leq i \leq p\} \quad (f \in C^p(\mathbb{R}^\mu, \mathbb{R}^\nu)).$$

$$\|f\|_m^p = \|f\|_{\{m\}}^p.$$

The Whitney C^p-topology is then given by the family of pseudometrics

$$\sup\{\eta(m)\|f - g\|_m^p: m \in M\}$$

for all positive continuous $\eta: \mathbb{R}^\mu \longrightarrow]0,\infty[$. Equivalently,

$$f \approx g[W^p] \text{ iff } \eta(m)\|f - g\|_m^p < 1$$

for all standard $\eta \in C^p(\mathbb{R}^\mu,]0,\infty[)$.

2 A VAN DOUWEN SPACE

The space $\mathbb{R}^{\mathbb{N}}$ with the usual Tychonoff product topology is characterized by the infinitesimal relation

$$t \approx s[\text{prod}] \text{ iff } t_n \approx s_n \textbf{ for standard } n \in {}^\sigma\mathbb{N}.$$

The space $\mathbb{R}^{\mathbb{N}}$ with the box topology is characterized by

$$t \approx s[\text{box}] \text{ iff } \varepsilon_n|t_n - s_n| < 1$$

for all $n \in {}^*\mathbb{N}$ **and all standard** $\varepsilon \in {}^\sigma\mathbb{N}^{\mathbb{N}}$.

Let P be $\mathbb{N}^{\mathbb{N}}$ as a subspace of $(\mathbb{R}^{\mathbb{N}}, \text{prod})$. Let A be $\{0\} \cup \{\frac{1}{n} : n \in \mathbb{N},\ n > 0\}$ and take $B = A^{\mathbb{N}}$ as a subspace of $(\mathbb{R}^{\mathbb{N}}, \text{box})$. Van Douwen's space $P \times B$ is a non-normal [5; 10.4] closed subspace of $(\mathbb{N}^{\mathbb{N}}, \text{prod}) \times (\mathbb{R}^{\mathbb{N}}, \text{box}) = D$, so that D is not normal.

Our embedding $E_0 : D \longrightarrow (C^p(M, \mathbb{R}^\nu), W^p)$ is the direct sum of two maps $\Phi : \mathbb{N}^{\mathbb{N}} \longrightarrow C^p(M, \mathbb{R}^\nu)$, $\Psi : \mathbb{R}^{\mathbb{N}} \longrightarrow C^p(M, \mathbb{R}^\nu)$:

$$E_0(t,s) = \Phi(t) + \Psi(s).$$

3 CONSTRUCTION OF $\Phi : \mathbb{N}^{\mathbb{N}} \longrightarrow C^p(M, \mathbb{R}^\nu)$

The map Φ is the composition $\phi \circ \Phi'$ of two functions $\Phi' : \mathbb{N}^{\mathbb{N}} \longrightarrow C^p(\mathbb{R}^\mu, \mathbb{R}^\nu)$ (eq. 6) and $\phi : \Phi'(\mathbb{N}^{\mathbb{N}}) \longrightarrow C^p(M, \mathbb{R}^\nu)$ (eq. 7).

Choose a sequence of pairwise disjoint balls $B(a_n, r_n) = \{x \in \mathbb{R}^\mu : \|x - a_n\| \le r_n\}$ with a_n converging to zero. We start by building a family of functions supported on $\bigcup_{n=0}^{\infty} B(a_n, r_n) \cup \{0\}$.

For each n choose a function f^n_∞ which is of class C^{p-1}, with support in $B(a_n, r_n)$, such that $D^p f^n_\infty$ is bounded in $\mathbb{R}^\mu$ and discontinuous only at a_n, and also such that

$$\|f^n_\infty\|^p_{B(a_n, r_n)} < \frac{1}{n+1} \tag{1}$$

Each of these functions may be separately approximated by C^p-functions f^n_k supported in $B(a_n, r_n)$ using convolution kernels , so that we have the following, with $p - 1 = 0$ if $p = 0$:

$$\|f^n_k - f^n_\infty\|^{p-1}_{B(a_n, r_n)} < \frac{1}{k+1} \tag{2}$$

$$\lim_{k \to \infty} \|f^n_k - f^n_\infty\|^p_m \longrightarrow 0, \text{ if } m \ne a_n \tag{3}$$

$$\|f^n_k\|^p_{B(a_n, r_n)} < \frac{1}{n+1} \tag{4}$$

for example see [2; 2.3] and [8; 9.8].

$$f^n_i \ne f^n_k \text{ , if } i \ne k. \tag{5}$$

Define a function $\Phi':\mathbb{N}^{\mathbb{N}}\longrightarrow C^p(\mathbb{R}^\mu,\mathbb{R}^\nu)$ by

$$\Phi'(t) = \sum_{n\in\mathbb{N}} f^n_{t_n} \tag{6}$$

The map Φ' is one-to-one by (5). Each function $f = \Phi'(t)$ is of class C^p at zero, and hence on all of $\mathbb{R}^\mu$. First, $f(0) = 0$ by definition. If $x \approx 0$, then $x \in B(a_n,r_n)$ for some infinite $n \in {}^*\mathbb{N}$ and $|f(x)| \le \|f^n_{t_n}\|^p_{B(a_n,r_n)}$; but $\|f^n_{t_n}\|^p_{B(a_n,r_n)} \approx 0$ by (4). Thus, f is continuous at zero. It follows from the mean value theorem that $D^if_0 = 0$ $(0\le i\le p)$ and D^pf is continuous at zero since $\|D^pf_x\| \le \|f^n_{t_n}\|^p_{B(a_n,r_n)} \approx 0$ when $x \approx 0$ by (4).

Let $\varphi: U\longrightarrow\mathbb{R}^\mu$ be a chart of M such that U has compact closure and $\varphi(U)=\mathbb{R}^\mu$. Define $\phi:\Phi'(\mathbb{N}^{\mathbb{N}})\longrightarrow C^p(M,\mathbb{R}^\nu)$ by

$$\phi(f)(m) = \begin{cases} f(\varphi(m)) & \text{if } m \in U \\ 0 & \text{if } m \in M\setminus U \end{cases} \tag{7}$$

Note that, for each $f\in\Phi'(\mathbb{N}^{\mathbb{N}})$, $\varphi^{-1}(\mathrm{supp}(f))$ is a compact subset of U, so that $\phi(f)$ does belong to $C^p(M,\mathbb{R}^\nu)$. Let

$$\Phi = \phi\circ\Phi'.$$

LEMMA 1 $\Phi(\mathbb{N}^{\mathbb{N}})$ is closed in the pointwise convergence topology of $C^p(M,\mathbb{R}^\nu)$, hence closed in every finer topology and W^p in particular.

PROOF: By [6;8.3.8] it suffices to show that the standard part $\mathrm{st}({}^*\Phi(\mathbb{N}^{\mathbb{N}})) = \Phi(\mathbb{N}^{\mathbb{N}})$, that is, if $t \in {}^*(\mathbb{N}^{\mathbb{N}})$ and $g \in C^p(M,\mathbb{R}^\nu)$ are such that for every standard $m \in {}^*M$ $\Phi(t)(m) \approx g(m)$, then for some standard $s \in {}^*(\mathbb{N}^{\mathbb{N}})$, $g = \Phi(s)$. Since $\Phi(t)$ is identically zero off a compact subset of U, the standard map g must be zero off the same subset. And it remains to be shown that $g\circ\varphi^{-1} = \Phi'(s)$ for some standard s, if $\Phi(t)(m) \approx g(m)$ for all $m\in{}^\sigma U$. As $\varphi: U\longrightarrow\mathbb{R}^\mu$ is a homeomorphism, it suffices to show that $\Phi'(\mathbb{N}^{\mathbb{N}})$ is pointwise closed in $C^p(\mathbb{R}^\mu,\mathbb{R}^\nu)$.

Suppose $\Phi'(t)$ has a pointwise standard part $g \in C^p(\mathbb{R}^\mu,\mathbb{R}^\nu)$. Given $x\in\mathbb{R}^\mu$, either $x \notin \bigcup_{n=0}^{\infty} B(a_n,r_n) \cup \{0\}$ and the standard $g(x)=0$, or x is in some standard $B(a_n,r_n)$ and $g(x) \approx \Phi'(t)(x) = f^n_{t_n}(x)$; t_n must be finite when n is finite. Otherwise $g(x) \approx f^n_{t_n}(x) \approx f^n_\infty(x)$, by (2), so the C^p-function g equals the non-C^p-function f^n_∞ on the standard ball $B(a_n,r_n)$. Since t_n is finite for finite n, there is a standard sequence $s_n = t_n$ for standard n. On the standard balls $B(a_n,r_n)$ we have $f^n_{t_n}(x) = f^n_{s_n}(x)$. On the nonstandard balls $f^n_{s_n}(x) \approx 0$ and since $\Phi'(t)(0) = 0$, $\Phi'(s)$ is the pointwise standard part of $\Phi'(t)$.

LEMMA 2 The map $\Phi: (\mathbb{N}^{\mathbb{N}}, \text{prod}) \longrightarrow (C^p(M,\mathbb{R}^\nu), W^p)$ is continuous.

PROOF: Note that $\varphi: U \longrightarrow \mathbb{R}^\mu$ is a homeomorphism, hence both φ and φ^{-1} are proper, so that $f \longmapsto f\circ\varphi$ is a homeomorphism from $(C^p(\mathbb{R}^\mu,\mathbb{R}^\nu), W^p)$ onto $(C^p(U,\mathbb{R}^\nu), W^p)$. Also, the supports of all the the functions in $\Phi(\mathbb{N}^{\mathbb{N}})$ are contained in the compact set $\varphi^{-1}[\bigcup_{n=0}^{\infty} B(a_n,r_n) \cup \{0\}] \subset U$, and $\phi(\Phi'(t))$ is obtained simply extending $\Phi'(t)\circ\varphi$ to $M\backslash U$ with value zero. Hence, ϕ is itself a W^p-homeomorphism and we may again limit ourselves to showing that $\Phi': (\mathbb{N}^{\mathbb{N}}, \text{prod}) \longrightarrow (C^p(\mathbb{R}^\mu,\mathbb{R}^\nu), W^p)$ is continuous.

It suffices to prove that if $t \in {}^*(\mathbb{N}^{\mathbb{N}})$ and $s \in {}^\sigma(\mathbb{N}^{\mathbb{N}})$, and if $t_n = s_n$ for standard $n \in {}^\sigma\mathbb{N}$, then $D^i\Phi'(t)(x) \approx D^i\Phi'(s)(x)$ for all $x \in {}^*(\mathbb{R}^\mu)$ and all $i = 0,1,\ldots,p$ [6; 8.3.1 and 8.4]. This condition is trivially satisfied off the support as well as on the balls $B(a_n,r_n)$ when n is standard.

When $x \approx 0$ in ${}^*\mathbb{R}^\mu$, the standard C^p-function $\Phi'(s)$ verifies $D^i\Phi'(s)_0 = 0$ for all $i = 0,\ldots,p$ as we saw above. Therefore we conclude our proof by observing that $D^i\Phi'(t)_x \approx 0$ when $x \approx 0$. The only case to test is when $x \in B(a_n,r_n)$ for infinite n, Where $\Phi'(t)(x) = f^n_{t_n}(x)$. But this is guarantied by condition (4) as $\|f^n_{t_n}\|^p_{B(a_n,r_n)} < \frac{1}{n+1} \cdot \approx 0$, So $D^i f^n_{t_n}(x) \approx 0$.

LEMMA 3 The map $\Phi^{-1}: (\Phi(\mathbb{N}^{\mathbb{N}}), \text{pointwise}) \longrightarrow (\mathbb{N}^{\mathbb{N}}, \text{prod})$ is continuous, hence Φ^{-1} is continuous in the W^p topology.

PROOF: By [6; 8.3.1] it suffices to show that if $t \in {}^*(\mathbb{N}^{\mathbb{N}})$, $s \in {}^\sigma(\mathbb{N}^{\mathbb{N}})$ and $\Phi(t)(m) \approx \Phi(s)(m)$ for all standard $m \in {}^*M$, then $t_n = s_n$ for finite n, that is, if $\Phi'(t)(x) \approx \Phi'(s)(x)$ for all standard $x \in {}^*\mathbb{R}^{\mu}$, then $t_n = s_n$ for all finite n. Let n be finite, so s_n is finite since s is standard. The number $t_n \in {}^*\mathbb{R}$ cannot be infinite, because we would have the non-C^p function $f^n_\infty(x) \approx f^n_{t_n}(x) \approx f^n_{s_n}(x)$, by (2), forcing a standard C^p function $f^n_{s_n}$ to equal a non-C^p function. The number t_n must be equal to s_n by (5). Hence $t_n = s_n$ for standard n.

Summarizing, what we proved

PROPOSITION 1 The map $\Phi: (\mathbb{N}^{\mathbb{N}}, \text{prod}) \longrightarrow (C^p(\mathbb{R}^{\mu}, \mathbb{R}^{\nu}), W^p)$ is a closed embedding.

4 CONSTRUCTION OF $\Psi: \mathbb{R}^{\mathbb{N}} \longrightarrow C^p(M, \mathbb{R}^{\nu})$

Let K_0 be the closure of the coordinate domain U considered in section 3. Recall that M is open and choose a sequence of charts $\varphi_n: U_n \subset M \longrightarrow \mathbb{R}^{\mu}$, $n=1,2,\ldots$, such that $\varphi_n(U_n) \supset B(0,1)$, the closures of the U_n, $cl(U_n)$ are compact and pairwise disjoint and $K_0 \cap \bigcup_{n=1}^{\infty} cl(U_n) = \emptyset$.

Consider the smooth bump function

$$\beta(t) = \begin{cases} e^{\frac{t^2}{t^2-1}}, & |t| < 1 \\ 0, & |t| \geq 1, \end{cases}$$

fix $u \in \mathbb{R}^{\nu} \setminus \{0\}$ and define $b_n: \mathbb{R} \times M \longrightarrow \mathbb{R}^{\nu}$ $(n=1,2,\ldots)$ by

$$b_n(t,m) = \begin{cases} t\beta(\|\varphi_n(m)\|)u & \text{if } m \in U_n \\ 0 & \text{if } m \in M \setminus U \end{cases}$$

The embedding Ψ is given by

$$\Psi(s)(m) = \sum_{n=1}^{\infty} b_n(s_n, m)$$

Since at most one term at a time of this series is nonzero and since $\Psi(s)[\varphi_n^{-1}(0)] = s_n u$, with $u \neq 0$, Ψ is 1-1 from $\mathbb{R}^{\mathbb{N}}$ into $C^p(M, \mathbb{R}^{\nu})$.

LEMMA 4 $\Psi(\mathbb{R}^{\mathbb{N}})$ is pointwise closed in $C^p(M,\mathbb{R}^\nu)$, hence closed for W^p.

PROOF: By [6;8.3.8] it suffices to show that the standard part of an element in $\Psi\{^*(\mathbb{R}^{\mathbb{N}})\}$ is in $\Psi(\mathbb{R}^{\mathbb{N}})$. Suppose $f = \Psi(s)$ and some standard $g \in {}^\sigma C^p(M,\mathbb{R}^\nu)$ verifies $g(m) \approx f(m)$ for all standard $m \in {}^\sigma M$. In particular $g(\varphi_n^{-1}(0)) \approx f(\varphi_n^{-1}(0)) = s_n u$ for all standard $n \in {}^\sigma\mathbb{N}$, therefore $g(\varphi_n^{-1}(0)) \approx s_n u$. Thus, s_n is finite and has a standard part c_n. Let $c = (c_n)$. We claim that $g = \Psi(c)$.

It suffices to show that, for each $n \in \mathbb{N}$ and each $m \in K_n$, $g(m) = s_n(c_n, m)$, because $\Psi(s)$ is supported in $\bigcup_{n=0}^{\infty} K_n$ and $g(m) \approx f(m)$ forces the standard function g also to be zero off $\bigcup_{n=0}^{\infty} K_n$.

Let n be an arbitrary standard index, $c_n = st(s_n)$, $m \in K_n$. Each b_n is a continuous (smooth) function, hence $g(m) \approx f(m) = b_n(s_n, m) \approx b_n(c_n, m)$, so that the standard $g(m)$ and $b_n(c_n, m)$ must be equal, as we wanted to show.

LEMMA 5 The map $\Psi: (\mathbb{R}^{\mathbb{N}}, \text{box}) \longrightarrow (C^p(M,\mathbb{R}^\nu), W^p)$ is continuous.

PROOF: Let $c \in {}^\sigma(\mathbb{R}^{\mathbb{N}})$ and $s \in {}^*(\mathbb{R}^{\mathbb{N}})$ satisfy $s \approx c$ [box]. We must show that for any standard positive continuous $\eta: \mathbb{R}^\mu \longrightarrow]0,\infty[$

$$\eta(m)d^p(j^p\Psi(s)(m), j^p\Psi(c)(m)) < 1 \qquad \text{for all } m \in {}^*M.$$

For each n, the function $\eta(m)j^p b_n(t,m)$ is uniformly continuous on the compact set $[c_n-1, c_n+1]\times K_n$, thus there is a standard sequence of tolerances ε_n so that if $\varepsilon_n|t-r|<1$ for $c_n-1 \le t,r \le c_n+1$, then $\eta(m)d^p(j^p b_n(t,m), j^p b_n(r,m)) < 1$ for all m in K_n. Since $s \approx c$ [box] we know that $\varepsilon_n|c_n - s_n|<1$ for all $n \in {}^*\mathbb{N}\setminus\{0\}$, thus

$$\eta(m)d^p(j^p b_n(c_n, m), j^p b_n(s_n, m)) < 1 \qquad \text{for all } m \in K_n$$

and all $n \in {}^*\mathbb{N}$. As

$$d^p(j^p\Psi(c)(m), j^p\Psi(s)(m)) = d^p(j^p b_n(c_n, m), j^p b_n(s_n, m))$$

on K_n, this proves the lemma.

LEMMA 6 The map $\Psi^{-1}: (\Psi(\mathbb{R}^{\mathbb{N}}), W^p) \longrightarrow (\mathbb{R}^{\mathbb{N}}, \text{box})$ is continuous.

PROOF: Suppose $\Psi(s) \approx \Psi(c)$ $[W^p]$ for $s \in {}^*(\mathbb{R}^{\mathbb{N}})$ and $c \in {}^{\sigma}(\mathbb{R}^{\mathbb{N}})$, that is, for any positive continuous $\eta: \mathbb{R}^{\mu} \longrightarrow]0,\infty[$

$$\eta(m)d^p(j^p\Psi(s), j^p\Psi(c)(m)) < 1 \qquad \text{for all } m \in {}^*M.$$

We must show that for any positive sequence (ε_n), $\varepsilon_n|s_n - c_n| < 1$, for all $n \in {}^*\mathbb{N}$.

We know by construction that

$$[\Psi(s) - \Psi(c)](\varphi_n^{-1}(0)) = (s_n - c_n)u$$

with standard $u \neq 0$. Given a positive sequence (ε_n), simply extend to a positive continuous $\eta: \mathbb{R}^{\mu} \longrightarrow]0,\infty[$ so that $\eta(\varphi_n^{-1}(0)) = \varepsilon_n/|u|$. Then, with $\|.\|$ a norm on $\mathbb{R}^{\nu}$,

$$\varepsilon_n|s_n - c_n| = \eta(\varphi_n^{-1}(0))\|[\Psi(s) - \Psi(c)](\varphi_n^{-1}(0))\| < 1.$$

Summarizing, we obtained

PROPOSITION 2 $\Psi: (\mathbb{R}^{\mathbb{N}}, \text{box}) \longrightarrow (C^p(M,\mathbb{R}^{\nu}), W^p)$ is a closed embedding.

Finaly, observe that $\text{supp}[\Phi(t)] \cap \text{supp}[\Psi(s)] = \emptyset$ so that the map $E_0: D \longrightarrow (C^p(M,\mathbb{R}^{\nu}), W^p)$ given by $E_0(t,s) = \Phi(t) + \Psi(s)$ is a closed embedding.

5 THE MAIN RESULT

So far we have proved

LEMMA 7 The space $(C^p(M,\mathbb{R}^{\nu}), W^p)$ is not normal.

And finally we have

THEOREM The space $(C^p(M,N), W^p)$ is not normal.

PROOF: Let ν be the dimension of N and $\psi: V \longrightarrow \mathbb{R}^{\nu}$ be a chart of N such that $\psi(V) = \mathbb{R}^{\nu}$. The map $\psi_{\bullet}: C^p(M,\mathbb{R}^{\nu}) \longrightarrow C^p(M,N)$ given by $\psi_{\bullet}(f) = \psi^{-1} \circ f$ is a W^p-closed embedding. Thus, by lemma 4.6, $(C^p(M,N), W^p)$ is not normal.

Note that we actually proved that no W^p-open subset of $C^p(M,N)$ is normal.

REFERENCES

1. Golubitzky, M., Guillemin, V., **Stable Mappings And Their Singularities**, Springer-Verlag, GTM 14, 1980.
2. Hirsch, M. W, **Differential Topology**, Springer-Verlag, GTM 33, 1976.
3. Michor, P. W., **Manifolds Of Smooth Mappings**, Shiva Mathematical Series, 1980.
4. Neves, V. M., **Non-Normality Of $C^{\infty}(M,N)$ In Whitney's And Related Topologies When M Is Open**, (to appear in Topology And Its Applications).
5. Rudin, M. E., **Lectures in Set Theoretic Topology**, CBMS-RCSM 23, Amer. Math. Soc., 1974.
6. Stroyan, K. D., Luxemburg, W. A. J., **Introduction To the Theory Of Infinitesimals**, Acad. Press 1976.
7. Willard, S., **General Topology**, Addison-Wesley 1970.
8. Wheeden, R. L., Zygmund, A., **Measure And Integral**, Marcel Dekker, Inc., 1977.

Extensions, Filters, and Proximities

Ellen E. Reed Trinity School at Greenlawn, South Bend, Indiana

0. Introduction

Thron [9] has shown that principal T_0-extensions correspond 1-1 to systems of open filters which contain all the neighborhood filters. Under this correspondence, the T_2-compactifications of a completely regular T_2- space arise from the minimal Cauchy filters of some totally bounded uniformity compatible with the topology. This correspondence between T_2-compactifications and minimal Cauchy filters can be extended to the T_1 case if we replace the Cauchy filters by <u>contigual</u> filter systems. Contigual filter systems are the duals of the contigual nearnesses of Herrlich [3]. The principal T_1-compactifications correspond 1-1 to contigual T_1-filter systems. This was essentially proved in Reed [7], using the concept of a nearness. It is proved here for filter systems, which can be regarded as duals of cluster-generated nearnesses. The first section is simply a review of these known correspondences between extensions and filter systems.

In a completely regular T_2-space the compatible uniformities can be grouped into proximity classes. Each class contains a unique totally bounded member, which is its smallest member. This uniformity produces the largest family of Cauchy filters of any uniformity in the class. The minimal Cauchy filters are the maximal round filters of the proximity, and yield the unique T_2-compactification associated with the proximity. In the T_1 case, filter systems can also be grouped into proximity classes. A filter system Θ induces a Lodato proximity π if we call two sets near whenever they are both members of the dual of some filter in Θ. Each proximity class of filter systems has a largest member, which consists of all the π-filters. The minimal π-filters yield the compactification of Gagrat and Naimpally [2]. These compactifications can be characterized as the <u>clan-complete</u> principal T_1-compactifications, and they are in 1-1 correspondence with the Lodato proximities on the space. This is essentially the content of the second section.

Systems of Cauchy filters can also be grouped into proximity classes, using the same idea. Thus the third section focuses on Cauchy spaces. Here Cauchy space is used in the sense of Keller [4]. It turns out that a Cauchy system is contigual iff it is the set of

Cauchy filters for a totally bounded uniformity on the space. Thus if the proximity π is not an Efremovich proximity then its system of π-filters cannot be a Cauchy system.

This leads us to consider the more general case of a filter system for a convergence space. The final section introduces this investigation. A filter system is simply a set Θ of non-convergent filters. There is no obvious generalization of the construction of a corresponding extension. However, one such extension, κ_Θ, is presented. For this extension it turns out that Θ is contigual iff κ_Θ is compact. This result is due to D.C. Kent. Moreover, a filter system induces a proximity relation as before. With a mild restriction on the proximity, it turns out that each proximity class has a largest member, which is contigual. Thus the application to convergence spaces of the theory already developed for topological spaces appears promising.

1. Extensions and filter systems

This section focuses on the relation between extensions of topological spaces and their corresponding trace systems. The main result is the development of the 1-1 correspondence between principal T_1-compactifications and contigual filter systems (Cor. 1.11).

Let X be a T_0-space. An extension $\kappa = (e,Y)$ induces a system of open filters on X via the pullbacks of the neighborhood filters under the map e. This system is known as the trace system of κ, and the corresponding map will be denoted by Tr ; that is,

$$Tr(\kappa) = \{e^{-1}(\mathcal{N}_y) : y \in Y\},$$

where $\mathcal{N}_y$ is the neighborhood filter at y in Y. It turns out that any system of open filters on X which includes all the neighborhood filters is the trace system for some T_0-extension. For this reason we make the following definition.

1.1. Definition. A filter system on a space X is a family Θ of open filters which includes all the neighborhood filters.

1.2. Construction. Given a filter system Θ on a T_0-space X we define an extension κ_Θ in the usual way. (See Thron [9].)

$Y = \Theta$.

$G^\wedge = \{\mathcal{F} \in Y : G \in \mathcal{F}\}$, where $G \subset X$.

$\mathcal{T}^\wedge$ is generated by $\{G^\wedge : G \text{ is open in } X\}$.

$j(x) = \mathcal{N}_x$.

Then $\kappa_\Theta = (j,Y)$, where $\mathcal{T}^\wedge$ is the topology on Y. We will also denote κ_Θ by Ext(Θ).

1.3.Theorem (Thron). Let Θ be a filter system on a T_0-space X.

(1) κ_Θ is a T_0-extension of X with trace system Θ.

(2) κ_Θ is T_1 iff each filter of Θ is minimal in Θ.

(3) κ_Θ is T_2 iff for $\mathcal{F} \neq \mathcal{G}$ in Θ we have that $\mathcal{F} \vee \mathcal{G}$ is not proper.

Proof. See Thron [8]. ■

1.4. Definition. Because of this theorem we will define a filter system Θ to be T_1 iff each filter in Θ is minimal in Θ. We will say that Θ is T_2 iff distinct filters in Θ are disjoint.

It turns out that non-equivalent extensions can have the same trace system. However, for any filter system Θ the extension κ_Θ is in some sense the "smallest" extension with trace system Θ. For this reason we will call it a principal extension.

1.5. Definition. Two extensions $\kappa_1 = (e,Y)$ and $\kappa_2 = (f,Z)$ are equivalent iff there is a homeomorphism $h: Y \to Z$ such that $h \circ e = f$. In this case we will write $\kappa_1 \cong \kappa_2$. We call $\kappa = (e,Y)$ a principal extension of X iff $\kappa \cong \kappa_\Theta$, for some filter system Θ in X.

1.6. Proposition. If $\kappa = (e,Y)$ is a principal T_0-extension of X with trace system Θ, then $\kappa \cong \kappa_\Theta$ under the map $h(y) = e^{-1}(\mathcal{N}_y)$. Thus $\mathrm{Ext} \circ \mathrm{Tr}(\kappa) \cong \kappa$.

Proof. Since (e,Y) is a principal extension, there is a filter system Θ such that (e,Y) is equivalent to $\kappa_\Theta = (j,\Theta)$. Let $h: Y \to \Theta$ be a homeomorphism such that $h \circ e = j$. Note that for $\mathcal{F} \in \Theta$ and $\mathcal{F} = h(y)$ we have

$$\mathcal{F} = j^{-1}(\mathcal{N}_{\mathcal{F}}) = e^{-1}(\mathcal{N}_y)$$

This establishes that Θ must be the trace system of κ. ■

1.7. Theorem..Let X be a T_0-space. The map $\kappa \to \mathrm{Tr}(\kappa)$ is a 1-1 map from (equivalence classes of) principal T_0 extensions of X to filter systems on X.

Proof. Note that equivalent extensions have identical trace systems. Thus the proof follows easily from the preceding result and from Thm. 1.3. ■

This correspondence works beautifully on the T_2-compactifications of X. Every compact T_2 space has a unique compatible uniformity. For a given T_2-compactification of X, the corresponding trace system is the set of minimal Cauchy filters for the induced uniformity on X. Thus we have the following motivating theorem.

1.8. Theorem. Let X be a completely regular T_2-space.

(1) The T_2-compactifications of X are in 1-1 correspondence with the totally bounded uniformities on X.

(2) The trace system of a T_2-compactification of X consists of the minimal Cauchy filters for the associated uniformity.

Proof. See Page [6], Thm. 6.16, for a proof of the 1-1 correspondence. Bourbaki [1] constructs the completion X^ of a separated uniform space X by using minimal Cauchy filters. (See Bourbaki [1], p. 192.) It is easy to check that X^ regarded as a topological extension is simply κ_Θ, where Θ consists of the minimal Cauchy filters on X. Given a T_2-compactification $\kappa = (e,Y)$ the unique uniformity on Y is obtained by taking sets of the form

$$\cup\{A_i x A_i\colon 1 \le i \le n\},$$

where $\{A_i\colon 1 \le i \le n\}$ is a finite cover of Y which is "fat"; that is, there is a cover $\{B_i\colon 1 \le i \le n\}$ such that for each i we have $cl(B_i) \subset int(A_i)$. Restricting this uniformity to X yields the desired totally bounded uniformity on X. ■

This result has been generalized to the T_1 case by Herrlich [3], who expressed these results in terms of nearnesses. Herrlich proved that principal T_1-compactifications are in 1-1 correspondence with contigual nearnesses. The following theorem expresses the same result in terms of filter systems, which can be regarded as the duals of nearnesses.

1.9. Definition. A filter $\mathcal{F}$ is near a filter system Θ iff for any finite family $\mathcal{A}$ of subsets of X, if $\mathcal{A}$ meets every filter in Θ, then $\mathcal{A}$ meets $\mathcal{F}$. More precisely, if $\mathcal{A} \cap \mathcal{F} = \emptyset$ then there is a filter $\mathcal{G}$ in Θ with $\mathcal{A} \cap \mathcal{G} = \emptyset$. A filter system Θ is contigual iff every filter near Θ actually contains a member of Θ.

1.10. Theorem. Let Θ be a filter system on X. Then Θ is contigual iff κ_Θ is compact.

Proof. ($\Rightarrow$) Let Θ be a contigual filter system on X. Recall that $\kappa_\Theta = (j,Y)$, where $Y = \Theta$, and $j(x) = \mathcal{N}_x$. Let $\mathcal{U}$ be an ultrafilter on Y. We wish to show that $\mathcal{U}$ converges. Set

$$\mathcal{F} = \{A\colon A\hat{} \in \mathcal{U}\}.$$

It is straightforward to check that $\mathcal{F}$ is a filter on X. Notice that for $A \subset X$ we have

$$A \subset B \Rightarrow A\hat{} \subset B\hat{}$$

$$\text{and } (A \cap B)\hat{} = A\hat{} \cap B\hat{}.$$

(1) $\mathcal{F}$ is near Θ. Let $\mathcal{A}$ be a finite family of subsets of X, with $\mathcal{A} \cap \mathcal{F} = \emptyset$. For $A \in \mathcal{A}$ we have $A\hat{} \notin \mathcal{U}$ and hence $\sim(A\hat{}) \in \mathcal{U}$. Since $\mathcal{A}$ is finite,

$$\cap\{\sim(A\hat{})\colon A \in \mathcal{A}\} \neq \emptyset.$$

Let $\mathcal{G}$ be a member. Clearly $\mathcal{G}$ is a member of Θ for which $\mathcal{G} \cap \mathcal{A} = \emptyset$.

(2) Let $\mathcal{F}^*$ be a member of Θ for which $\mathcal{F} \supset \mathcal{F}^*$. We claim that $\mathcal{U} \to \mathcal{F}^*$.

Let G be an open set in $\mathcal{F}^*$. Then $G \in \mathcal{F}$ and hence $G^\wedge \in \mathcal{U}$. This is sufficient to establish that $\mathcal{U} \to \mathcal{F}^*$.

($\Leftarrow$) Now suppose κ_Θ is compact. Let $\mathcal{F}$ be a filter near Θ. We need to show that $\mathcal{F}$ contains a filter in Θ. Let

$$\mathcal{S} = \{\sim(G^\wedge)\colon \text{ G is open in X and } G \notin \mathcal{F}\}.$$

(1) $\mathcal{S}$ has the finite intersection property. Let $\mathcal{A}$ be a finite family of open sets not in $\mathcal{F}$. Since $\mathcal{A} \cap \mathcal{F} = \emptyset$, and $\mathcal{F}$ is near Θ, there must be some filter $\mathcal{G}$ in Θ for which $\mathcal{A} \cap \mathcal{G} = \emptyset$. Clearly then $\mathcal{G} \in \cap\{\sim(G^\wedge)\colon G \in \mathcal{A}\}$.

(2) Since κ_Θ is compact, we may conclude $\cap\mathcal{S} \neq \emptyset$. Let $\mathcal{F}^*$ be a member. It is easy to check that $\mathcal{F}^*$ is a member of Θ for which $\mathcal{F} \supset \mathcal{F}^*$. ■

1.11. Corollary. Let (e,Y) be a principal T_0-extension of X.

(1) Y is compact iff its trace system is contigual.

(2) The map $\kappa \to Tr(\kappa)$ defines a 1-1 correspondence between the principal T_0 compactifications of X and the contigual filter systems on X.

Proof. Let Θ denote the trace system of κ. By Prop. 1.6, $\kappa \cong \kappa_\Theta$. If κ is compact, then κ_Θ must be compact. Then by the preceding theorem, Θ is contigual. Conversely, if Θ is contigual then κ_Θ must be compact, and hence κ is compact. Thus (1) holds.

The map $\kappa \to Tr(\kappa)$ is 1-1, by Thm. 1.7, and it maps the principal T_0-extensions of X onto the filter systems of X. From (1) it follows that compactifications get mapped onto contigual filter systems. ■

Notice that the minimal Cauchy filters have been replaced by contigual filter systems. Thus contigual filter systems can be regarded as generalizations of totally bounded uniformities; in fact, when we choose a filter system we have chosen the "Cauchy" filters -- those which "ought to" converge in the extension.

Since uniformities can be grouped into proximity classes, and filter systems are in a sense a generalization of uniformities, it is natural to ask what happens when filter systems are grouped into proximity classes. This is the content of the next section.

2. Proximities and extensions

Given a filter system Θ we can define two sets to be near if they are both members of the dual of some filter in Θ. In this way, filter systems can be grouped into proximity classes. Each proximity class of T_1-filter systems has a largest member, which yields a clan-complete T_1-compactification (Thm. 2.11). The principal T_1-clan-complete compactifications are in 1-1 correspondence with the Lodato proximities on the space (Thm. 2.16).

2.1. Definition. A proximity on a set X is a relation π on the subsets of X which satisfies the following axioms. Let A, B, and C be subsets of X.

(P1) $A \not\pi \emptyset$.

(P2) $A \cap B \neq \emptyset \Rightarrow A \pi B$.

(P3) $A \pi (B \cup C)$ iff $A \pi B$ or $A \pi C$.

(P4) $A \pi B \Rightarrow B \pi A$.

A proximity relation π on X induces a closure operator as follows:

$$cl_\pi A = \{x: \{x\} \pi A\}.$$

We say π is a Lodato proximity iff

(P5) $(cl_\pi A) \pi (cl_\pi B) \Rightarrow A \pi B$.

For any proximity π, if cl_π agrees with the closure operator of a given topology $\mathcal{T}$ on X we say that π is compatible with $\mathcal{T}$, or that π is a proximity on $(X,\mathcal{T})$.

We will now define what is meant by a proximity class of filter systems. For this we need the concept of the dual of a filter, which is also called a grill.

2.2. Definition. For any filter $\mathcal{F}$ we define the dual of $\mathcal{F}$ as follows:

$$d\mathcal{F} = \{A: \sim A \notin \mathcal{F}\}.$$

Using the fact that every proper filter is contained in some ultrafilter, we can easily see that the dual of $\mathcal{F}$ is the union of all the ultrafilters which contain $\mathcal{F}$. From this it follows easily that $d\mathcal{F}$ is closed under supersets and is prime with respect to unions. (So if $A \cup B$ is in $d\mathcal{F}$, then A or B is in $d\mathcal{F}$.) Such families are called grills, and have been studied extensively by Thron [10].

2.3. Definition. Let Θ be any family of filters on X. Let $[\Theta]$ denote those filters which contain members of Θ. Suppose $[\Theta]$ contains all filters of the form

$$\dot{x} = \{A: x \in A\}.$$

Such filters will be called point filters. We define a proximity π_Θ on X as follows:

$$A \,\pi_\Theta\, B \text{ iff } \exists\, \mathcal{F} \in \Theta \text{ such that } A,B \in d\mathcal{F}.$$

We say that two filter systems are in the same proximity class iff they induce the same proximity relation on X.

2.4. Proposition. Let Θ be a family of filters on X such that $[\Theta]$ contains all the point filters. Then π_Θ is a proximity on X.

Proof. Notice that for $x \in A \cap B$ we have that $\dot{x}$ contains some filter $\mathcal{F} \in \Theta$. Then A and B are in $d\mathcal{F}$ and hence $A \,\pi_\Theta\, B$. The other required properties now follow easily from the properties of filter duals. ■

2.5. Proposition. Let Θ be a T_1 filter system on a T_1-space X. Then π_Θ is a Lodato proximity compatible with the topology on X.

Proof. We have seen that π_Θ is at least a proximity on X. Notice that for $x \in X$ and $A \subset X$ we have

$$x \in cl(A) \text{ iff } A \in d\mathcal{N}_x.$$

(1) π_Θ is compatible with the topology on X.

Let $x \in cl(A)$. Then A and $\{x\}$ are both in $d\mathcal{N}_x$, and hence $\{x\} \,\pi_\Theta\, A$. Conversely, suppose $\{x\} \,\pi_\Theta\, A$. Then there is a filter $\mathcal{F}$ in Θ such that A and $\{x\}$ are in $d\mathcal{F}$. Since $\mathcal{F}$ is an open filter, we have $\mathcal{N}_x \supset \mathcal{F}$. Using that Θ is T_1, we may conclude $\mathcal{F} = \mathcal{N}_x$. Since $A \in d\mathcal{N}_x$, we have $x \in cl(A)$.

(2) π_Θ is a Lodato proximity. In view of (1) we simply need to show that

$$cl(A) \,\pi_\Theta\, cl(B) \Rightarrow A \,\pi_\Theta\, B.$$

The key to the proof is the fact that for any open filter $\mathcal{F}$, we have

$$cl(A) \in d\mathcal{F} \Rightarrow A \in d\mathcal{F}.$$

Notice that if cl(A) is in $d\mathcal{F}$ then $\sim cl(A)$ is not in $\mathcal{F}$. Since $\mathcal{F}$ is open, we have $\sim A \notin \mathcal{F}$, and hence $A \in d\mathcal{F}$. ■

Given a Lodato proximity π there is always a largest T_1 filter system which induces π. This filter system is contigual, and consists of all the minimal π-filters.

2.6. Definition . Let π be a proximity on a set X. A filter $\mathcal{F}$ on X is a π-filter iff for A,B in $d\mathcal{F}$ we have $A \,\pi\, B$. We define

$$\Theta(\pi) = \{\mathcal{F}\colon \mathcal{F} \text{ is an open } \pi\text{-filter}\}.$$

$$\Theta^*(\pi) = \{\mathcal{F}\colon \mathcal{F} \text{ is a minimal } \pi\text{-filter}\}.$$

We will use Θ and Θ^* when the meaning is clear.

2.7. Lemma. Let π be a Lodato proximity on a T_1-space X.

(1) Every π-filter contains a minimal π-filter.

(2) Every minimal π-filter is open.

(3) Every neighborhood filter is a minimal π-filter.

Proof.

(1) Every π-filter contains a minimal π-filter. Let $\mathcal{F}$ be a π-filter. Set

$$\mathcal{A} = \{\mathcal{G}: \mathcal{G} \text{ is a } \pi\text{-filter and } \mathcal{F} \supset \mathcal{G}\}.$$

Notice that for $\mathcal{G}_2 \supset \mathcal{G}_1$ we have $d\mathcal{G}_1 \supset d\mathcal{G}_2$. Thus if $\mathcal{C}$ is a non-empty chain in $\mathcal{A}$ then $\cap\mathcal{C}$ is in $\mathcal{A}$. Then by Zorn's lemma, $\mathcal{A}$ has a minimal element $\mathcal{G}^*$. Clearly then $\mathcal{G}^*$ is a minimal π-filter.

(2) Every minimal π-filter is open. This follows easily from the fact that if $\mathcal{F}$ is a π-filter then $\text{int}(\mathcal{F})$ is a π-filter, where $\text{int}(\mathcal{F})$ is generated by the open sets in $\mathcal{F}$. The fact that $\text{int}(\mathcal{F})$ is a π-filter follows from the fact that π is Lodato. Notice that if $\sim A$ is not in $\text{int}(\mathcal{F})$ then $\sim \text{cl}(A)$ is not in $\mathcal{F}$.

(3) Every neighborhood filter is a minimal π-filter. Recall that $A \in d\mathcal{N}_x$ iff $x \in \text{cl}(A)$. Hence for A,B in $d\mathcal{N}_x$ we have $\text{cl}(A) \cap \text{cl}(B) \neq \emptyset$ and hence $\text{cl}(A)\ \pi\ \text{cl}(B)$. Since π is Lodato, this requires that $A\ \pi\ B$.

Suppose now that $\mathcal{F}$ is a π-filter with $\mathcal{N}_x \supset \mathcal{F}$. We wish to show $\mathcal{F} = \mathcal{N}_x$. Let $x \in G$, where G is open. Then $\{x\} \not\pi \sim G$. Clearly $\{x\} \in d\mathcal{N}_x \subset d\mathcal{F}$. Thus $\sim G \notin d\mathcal{F}$ and hence $G \in \mathcal{F}$. ■

2.8. Definition. We can define a partial order on filter systems as follows:

$$\Theta_1 < \Theta_2 \text{ iff every filter in } \Theta_1 \text{ contains a filter in } \Theta_2.$$

Notice that for T_1 filter systems, this relation is antisymmetric.

2.9. Theorem. Let π be a Lodato proximity on a T_1-space X. Then $\Theta(\pi)$ and $\Theta^*(\pi)$ are contigual filter systems in the proximity class of π. Moreover, $\Theta(\pi)$ is the largest filter system in the class of π, and $\Theta^*(\pi)$ is the largest T_1-system.

Proof. From the preceding lemma it follows easily that Θ^* is a filter system on X, and that $\Theta < \Theta^* \subset \Theta$. From this it follows that Θ and Θ^* must be in the same proximity class, and that Θ is contigual iff Θ^* is contigual.

(1) Θ is compatible with π. Suppose $A\ \pi_\Theta\ B$. Let $\mathcal{F}$ be a member of Θ such that $A,B \in d\mathcal{F}$. Since $\mathcal{F}$ is a π-filter, clearly $A\ \pi\ B$. Conversely, suppose $A\ \pi\ B$. Using a Zorn's lemma argument, it can be established that there are ultrafilters $\mathcal{U}$ and $\mathcal{V}$ such that $A \in \mathcal{U}$, $B \in \mathcal{V}$, and $\mathcal{U}\ \pi\ \mathcal{V}$; that is, $U\ \pi\ V$ for $U \in \mathcal{U}$ and $V \in \mathcal{V}$. (For details of the proof, see Thron[10].) From this it is easy to establish that $\mathcal{F} = \mathcal{U} \cap \mathcal{V}$ is a π-filter. Clearly $A,B \in d\mathcal{F}$ and hence $A\ \pi_\Theta\ B$.

(2) Θ is contigual. Let $\mathcal{F}$ be a filter near Θ. We wish to show that $\mathcal{F}$ is a π-filter. Let $A,B \in d\mathcal{F}$. Let $\mathcal{A} = \{\sim A, \sim B\}$. Then $\mathcal{A} \cap \mathcal{F} = \emptyset$, and so there is a filter $\mathcal{G}$ in Θ with $\mathcal{A} \cap \mathcal{G} = \emptyset$. Since $\mathcal{G}$ is a π-filter, clearly $A\ \pi\ B$.

At this point we have established that Θ and Θ^* are contigual filter systems in the proximity class of π. Since Θ^* consists of minimal π-filters, clearly Θ^* is T_1. Moreover, every filter system in the proximity class is a subset of Θ, and so Θ is the largest filter system in the proximity class. But $\Theta = [\Theta^*]$ and so we have that Θ^* is the largest T_1-system in the class. ■

From this theorem it follows that the extension associated with the filter system $\Theta^*(\pi)$ is in some sense the largest principal T_1-compactification yielding π. Such compactifications can be characterized as the clan-complete compactifications of X. They are essentially the compactifications of Gagrat and Naimpally [2], Thm. 3.13.

2.10. Definition. For any topological space X, a clan is defined to be a filter dual with the property that the closures of any two sets in the dual have non-empty intersection. We say X is clan-complete iff the dual of every clan converges. Notice that every clan-complete space is compact.

For π a Lodato proximity on X, we will denote the compactification $\mathrm{Ext}(\Theta^*(\pi))$ by $\kappa_G(\pi)$, and refer to it as the the Gagrat-Naimpally compactification.

2.11. Theorem. If π is a Lodato proximity on a T_1-space X then the Gagrat-Naimpally compactification is clan-complete.

Proof. Let (j, Y) denote the extension determined by $\Theta^*(\pi)$, and let Φ be a filter on the space $Y = \Theta^*(\pi)$ such that $d\Phi$ is a clan. We wish to show Φ converges.

Let $\mathcal{F} = \{F: F^\wedge \in \Phi\}$. We claim that $\mathcal{F}$ is a π-filter. Clearly, $\mathcal{F}$ is a filter. Now notice that for $A \subset X$ we have that $A^\wedge = (\mathrm{int}\, A)^\wedge$ and hence $A^\wedge$ is open. Thus for $A \in d\mathcal{F}$ we have that $(\sim A)^\wedge$ is open, and hence $\sim(\sim A)^\wedge$ is a closed set in $d\Phi$. Thus if $A,B \in d\mathcal{F}$ then since Φ is a clan, we have $\sim(\sim A)^\wedge \cap \sim(\sim B)^\wedge \neq \emptyset$. Let $\mathcal{G}$ be a member. Then clearly $A,B \in d\mathcal{G}$ and hence $A \,\pi\, B$.

Since $\mathcal{F}$ is a π-filter, $\mathcal{F}$ must contain a minimal π-filter $\mathcal{G}$, which is thus a member of Y. Clearly $\mathcal{G}^\wedge \subset \Phi$ and so $\Phi \to \mathcal{G}$. ■

2.12. Definition. For κ an extension of a space X we define a proximity π_κ as follows:

$$A \,\pi_\kappa\, B \text{ iff } \mathrm{cl}(eA) \cap \mathrm{cl}(eB) \neq \emptyset.$$

2.13. Proposition. If κ is a T_1-extension of X then π_κ is a Lodato proximity on X and $\mathrm{Tr}(\kappa)$ is in the proximity class of π_κ.

Proof. It is easy to check that π_κ is a Lodato proximity. The fact that κ is T_1 insures that π_κ is compatible with the topology on X.

Let $\Theta = \mathrm{Tr}(\kappa)$,. We will show $\pi_\Theta = \pi_\kappa$. Notice that for $\mathcal{F} = e^{-1}(\mathcal{N}_y)$ we have

$$A \in d\mathcal{F} \text{ iff } y \in \mathrm{cl}(eA).$$

From this it is easy to see that $\mathcal{F}$ is a π_κ-filter and hence $\pi_\kappa \supset \pi_\Theta$. Now suppose $A\ \pi_\kappa\ B$. Then $cl(eA) \cap cl(eB) \neq \emptyset$. Let y be a member, and let $\mathcal{G} = e^{-1}(\mathcal{N}_y)$. Then $A,B \in d\mathcal{G}$. Since $\mathcal{G} \in \Theta$ we have $A\ \pi_\Theta\ B$. ■

2.14. Theorem. If κ is a clan-complete principal T_1-extension of X then $\kappa \cong \kappa_G(\pi_\kappa)$.

Proof. Let Θ denote the trace system of $\kappa = (e,Y)$; and let $\Theta^* = \Theta^*(\pi_\kappa)$. Since κ is a principal extension we have from Prop. 1.6 that $\kappa \cong \kappa_\Theta$. The desired equivalence will follow from the fact that $\Theta = \Theta^*$. Recall that Θ and Θ^* are T_1-filter systems (Thm 1.3 and Thm. 2.9). In view of Definition 2.8 it is sufficient to show that $\Theta < \Theta^* < \Theta$.

(1) $\Theta < \Theta^*$. Because of Lemma 2.7 it is sufficient to show that for $y \in Y$ we have that $\mathcal{F} = e^{-1}(\mathcal{N}_y)$ is a π_κ-filter. But for $A,B \in d\mathcal{F}$ we have $y \in cl(eA) \cap cl(eB)$ and hence $A\ \pi_\kappa\ B$.

(2) $\Theta^* < \Theta$. Let $\mathcal{F}$ be any filter on X. Notice that for $A \in de(\mathcal{F})$ we have that $e^{-1}(A) \in d\mathcal{F}$. Thus if $\mathcal{F}$ is a π_κ-filter and $A,B \in de(\mathcal{F})$ then $e^{-1}(A)\ \pi_\kappa\ e^{-1}(B)$ and hence $cl(A) \cap cl(B) \neq \emptyset$. This establishes that $de(\mathcal{F})$ is a clan, and hence $e(\mathcal{F})$ must converge to some $y \in Y$. It follows easily that $\mathcal{F} \supset e^{-1}(\mathcal{N}_y) \in \Theta$. ■

2.15. Theorem. If π is a Lodato proximity on a T_1-space X and $\kappa = \kappa_G(\pi)$ then $\pi = \pi_\kappa$.

Proof. Let Θ^* denote $\Theta^*(\pi)$. From Thm. 2.9 we have that Θ^* is in the proximity class of π (Thm. 2.9). But Θ^* is the trace system of κ (Def 2.10 and Thm. 1.3) and so Θ^* is also in the proximity class of π_κ (Prop. 2.13). Thus we have $\pi_\kappa = \pi_{\Theta^*} = \pi$. ■

2.16. Theorem. Let X be a T_1-space. If we identify equivalent extensions, then the map $\kappa \rightarrow \pi_\kappa$ defines a 1-1 correspondence between clan-complete principal T_1-extensions of X and the Lodato proximities on X.

Proof. Notice that equivalent extensions yield the same proximity on X. For suppose (e_1,Y_1) and (e_2,Y_2) are equivalent extensions of X under the homeomorphism $h: Y_1 \rightarrow Y_2$. It is easy to check that for $A \subset X$ we have $h[cl(e_1A)] = cl(e_2A)$. Thus the map $\kappa \rightarrow \pi_\kappa$ is well-defined. If π is Lodato then $\kappa_G(\pi)$ is a clan-complete extension mapping onto π (Thm. 2.11 and Thm. 2.15). Finally, from Thm. 2.14, we see that the map is 1-1. ■

When π is an Efremovich proximity the family Θ^* turns out to have two interesting characterizations. It is the set of filters which are maximal round with respect to π. It is also the set of minimal Cauchy filters for the unique totally bounded uniformity in the proximity class of π. This is the content of the next theorem.

2.17. Definition. For π a proximity on X we define a relation $<_\pi$ as follows:

$$A <_\pi B \text{ iff } A \not\pi \sim B.$$

A filter $\mathcal{F}$ is round with respect to π iff for $B \in \mathcal{F}$ there is a set A in $\mathcal{F}$ with $A <_\pi B$. For $A \subset X$ we defiine

$$r_\pi(A) = \{B: A <_\pi B\}.$$

Note that $r_\pi(A)$ is a filter on X. For any filter $\mathcal{F}$ on X we set

$$r_\pi(\mathcal{F}) = \{A: \exists F \in \mathcal{F} \text{ such that } F <_\pi A\}.$$

Notice that $\mathcal{F} \supset r_\pi(\mathcal{F})$. We will call $r_\pi(\mathcal{F})$ the round hull of $\mathcal{F}$. We will denote it by $r(\mathcal{F})$ when the meaning is clear. A maximal round filter is one which is not properly contained in any proper round filter.

For purposes of what is to follow, we will define an Efremovich proximity to be a proximity π for which $r_\pi(A)$ is always a round filter. Equivalently, the relation $<_\pi$ is dense; i.e. $<_\pi \subset <_\pi \circ <_\pi$. Using this it is easy to show that every Efremovich proximity is Lodato. For example, see Thron [9], Thm. 21.9.

2.18. Definition. A uniformity $\mathcal{H}$ is in the proximity class of an Efremovich proximity π iff we have $<_\pi = <_\mathcal{H}$, where

$$A <_\mathcal{H} B \text{ iff } \exists H \in \mathcal{H} \text{ such that } H(A) \subset B.$$

2.19. Theorem. Let π be an Efremovich proximity on a completely regular T_2 space X, and let $\mathcal{H}^*$ be the totally bounded uniformity in the proximity class of π. Then

(1) $\Theta(\pi)$ is the set of Cauchy filters

(2) $\Theta^*(\pi)$ is the set of minimal Cauchy filters.

(3) $\Theta^*(\pi)$ is the set of maximal round filters.

Proof. Let $< = <_\pi$ and let r denote r_π.

(1) The π-filters are identical with the Cauchy filters.

($\Leftarrow$) Suppose $\mathcal{F}$ is Cauchy, and suppose $A \not\pi B$. We will show that A or B is not in $d\mathcal{F}$. Choose H in $\mathcal{H}^*$ so that $H(A) \cap B = \emptyset$, and pick $F \in \mathcal{F}$ such that $F x F \subset H$. If $A \in d\mathcal{F}$ then $\sim A \notin \mathcal{F}$, and so $F \cap A \neq \emptyset$. Then $F \subset H(A)$ and so $F \cap B = \emptyset$. This establishes that $B \notin d\mathcal{F}$.

($\Rightarrow$) Let $\mathcal{F}$ be a π-filter, and let $H \in \mathcal{H}^*$. Then there are finite families B_i and A_i so that

$$\cup B_i = X,\ B_i <_\pi A_i, \text{ and } A_i x A_i \subset H_i,$$

for $1 \le i \le n$. (See Thron [9], Thm. 21.20.) Since $\mathcal{F}$ is a π-filter, then for each i, either $\sim B_i$ or A_i must be in $\mathcal{F}$. But $\cap \sim B_i = \emptyset$, and so we have that some A_i must be in $\mathcal{F}$. Since $A_i x A_i \subset H$ this establishes that $\mathcal{F}$ is Cauchy with respect to $\mathcal{H}^*$.

(2) A filter $\mathcal{F}$ is a minimal π-filter iff it is minimal Cauchy. This follows easily from (1).

(3) Every minimal Cauchy filter is maximal round. Let $\mathcal{F}$ be a minimal Cauchy filter. Notice that $r(\mathcal{F})$ is also Cauchy. Thus $\mathcal{F} = r(\mathcal{F})$ and $\mathcal{F}$ is round. Moreover, if $\mathcal{F} \subset \mathcal{G}$ it is easy to show that $r(\mathcal{G}) \subset \mathcal{F}$; and hence if $\mathcal{G}$ is round then $\mathcal{F} = \mathcal{G}$.

(4) Every maximal round filter is a minimal π-filter. Let $\mathcal{F}$ be a maximal round filter. Suppose $A \not\pi B$. We wish to show A or $B \notin d\mathcal{F}$. Suppose $A \in d\mathcal{F}$. Then $\sim A \notin \mathcal{F}$ and so $\mathcal{F} \vee r(A)$ is a proper round filter containing $\mathcal{F}$. Since $\mathcal{F}$ is maximal round we have $r(A) \subset \mathcal{F}$. But $A \not\pi B$ and hence $A < \sim B$. Thus $\sim B \in \mathcal{F}$ and $B \notin d\mathcal{F}$.

This establishes that $\mathcal{F}$ is a π-filter. If $\mathcal{F}$ contains a π-filter $\mathcal{G}$ then it is easy to check that $r(\mathcal{F}) \subset \mathcal{G}$. Since $\mathcal{F}$ is round, clearly $\mathcal{F} = \mathcal{G}$. Thus $\mathcal{F}$ is a minimal π-filter.■

3. Cauchy spaces

The Cauchy filters of a uniform convergence space can be thought of as a filter system. Thus it is natural to ask what happens when we extend the correspondence between extensions, filter systems, and proximities to Cauchy spaces. It turns out that every contigual Cauchy system $\mathcal{C}$ consists of all the $\pi_{\mathcal{C}}$-filters. Moreover, whenever all the π-filters form a Cauchy system then π must be an Efremovich proximity. Thus every contigual Cauchy system consists of the Cauchy filters for some totally bounded uniformity. This would seem to indicate that if we try to generalize the T_1 case to Cauchy systems we instead find ourselves back in the known correspondence between T_2 compactifications and totally bounded uniformities.

3.1. Definition. (Keller) We call a collection $\mathcal{C}$ of filters on a set X a Cauchy system iff the following axioms are satisfied.

(C1) $x \in X \Rightarrow \dot{x} \in \mathcal{C}$.

(C2) $\mathcal{F} \in \mathcal{C}$ and $\mathcal{F} \subset \mathcal{G} \Rightarrow \mathcal{G} \in \mathcal{C}$.

(C3) $\mathcal{F}, \mathcal{G} \in \mathcal{C}$ and $\mathcal{F} \vee \mathcal{G}$ proper $\Rightarrow \mathcal{F} \wedge \mathcal{G} \in \mathcal{C}$.

Notice that we no longer assume that X is a topological space. However, it still makes sense to call $\mathcal{C}$ contigual, since that definition had nothing to do with the topology on X. The construction $\pi_{\mathcal{C}}$ also still makes sense in this set-theoretic context. Note that $\pi_{\mathcal{C}}$ is a proximity, although it may no longer be a Lodato proximity.

3.2. Theorem. If $\mathcal{C}$ is contigual, then $\mathcal{C}$ consists of all the $\pi_{\mathcal{C}}$-filters; that is, $\mathcal{C} = \Theta(\pi_{\mathcal{C}})$.

Proof. Clearly every filter in $\mathcal{C}$ is a $\pi_{\mathcal{C}}$-filter. This follows simply from the definition of $\pi_{\mathcal{C}}$. Now we need to establish that every $\pi_{\mathcal{C}}$-filter is near $\mathcal{C}$. Since $\mathcal{C}$ is contigual, this will guarantee that every $\pi_{\mathcal{C}}$-filter is actually in $\mathcal{C}$.

(1) If $\mathcal{U}$ and $\mathcal{V}$ are ultrafilters, with $\mathcal{U}\,\pi_{\mathcal{C}}\,\mathcal{V}$ then $\mathcal{U}\wedge\mathcal{V} \in \mathcal{C}$. We will show that $\mathcal{U}\wedge\mathcal{V}$ is near $\mathcal{C}$. Let $\mathcal{A}$ be a finite family of subsets of X with $\mathcal{A} \cap (\mathcal{U}\wedge\mathcal{V}) = \emptyset$. Divide $\mathcal{A}$ into $\mathcal{A}_1$ and $\mathcal{A}_2$, where $\mathcal{A}_1 \cap \mathcal{U} = \emptyset$ and $\mathcal{A}_2 \cap \mathcal{V} = \emptyset$. Let

$$U = \cap\{\sim A:\ A \in \mathcal{A}_1\} \text{ and } V = \{\cap \sim A:\ A \in \mathcal{A}_2\}.$$

Clearly $U \in \mathcal{U}$ and $V \in \mathcal{V}$. Thus $U\,\pi_{\mathcal{C}}\,V$, and so there is a filter $\mathcal{F}$ in $\mathcal{C}$ with $U,V \in d\mathcal{F}$. Then $\mathcal{A} \cap \mathcal{F} = \emptyset$, so that $\mathcal{U}\wedge\mathcal{V}$ is near $\mathcal{C}$.

(2) If $\mathcal{F}$ is a $\pi_{\mathcal{C}}$-filter and $\mathcal{F} \subset \mathcal{U}_1,...,\mathcal{U}_n$, where each $\mathcal{U}_i$ is an ultrafilter, then $\mathcal{U}_1\wedge...\wedge\mathcal{U}_n$ is in $\mathcal{C}$. The proof uses induction on n. Suppose $\mathcal{F} \subset \mathcal{U}_1,...,\mathcal{U}_{n+1}$, and $\mathcal{U}_1\wedge...\wedge\mathcal{U}_n$ is in $\mathcal{C}$. Since $\mathcal{F}$ is a $\pi_{\mathcal{C}}$-filter, we have $\mathcal{U}_n\,\pi_{\mathcal{C}}\,\mathcal{U}_{n+1}$ and hence $\mathcal{U}_n\wedge\mathcal{U}_{n+1} \in \mathcal{C}$. Then by (C3) we have $\mathcal{U}_1\wedge..\wedge\mathcal{U}_{n+1} \in \mathcal{C}$.

(3) If $\mathcal{F}$ is a $\pi_{\mathcal{C}}$-filter then $\mathcal{F}$ is near $\mathcal{C}$. Suppose $\mathcal{A} \cap \mathcal{F} = \emptyset$ and $\mathcal{A} = \{A_1,...,A_n\}$. Then there are ultrafilters $\mathcal{U}_1,...,\mathcal{U}_n$ such that $A_i \in \mathcal{U}_i$ and $\mathcal{F} \subset \mathcal{U}_i$. Let $\mathcal{G} = \cap\mathcal{U}_i$. By (2), we have that $\mathcal{G} \in \mathcal{C}$. Clearly $\mathcal{G} \cap \mathcal{A} = \emptyset$. ∎

3.3. Theorem. Let π be a proximity on X. If $\Theta(\pi)$ is a Cauchy system then π is an Efremovich proximity.

Proof. Let Θ denote $\Theta(\pi)$.

(1) π is transitive on ultrafilters. If $\mathcal{U}\,\pi\,\mathcal{V}\,\pi\,\mathcal{W}$ then $\mathcal{U}\wedge\mathcal{V}$ and $\mathcal{V}\wedge\mathcal{W}$ are π-filters, and hence in Θ. By (C3), since $\mathcal{V}$ is a common upper bound, we have $\mathcal{U}\wedge\mathcal{V}\wedge\mathcal{W} \in \Theta$. Thus $\mathcal{U}\pi\mathcal{V}$.

(2). If $r(A)\vee r(B)$ is proper then $A\,\pi\,B$. Let $\mathcal{F} = r(A)\vee r(B)$ and suppose $\mathcal{F}$ is proper. Let $\mathcal{V}$ be an ultrafilter containing $\mathcal{F}$. Since $r(A) \subset \mathcal{V}$ we have

$$A \in \pi\mathcal{V} = \{D:\ D\,\pi\,\mathcal{V} \text{ for all } V \in \mathcal{V}\}.$$

Notice that $\pi\mathcal{V}$ is a grill, and so there is an ultrafilter $\mathcal{U}$ such that $A \in \mathcal{U} \subset \pi\mathcal{V}$. Similarly, $B \in \pi\mathcal{V}$ and so there is an ultrafilter $\mathcal{W}$ such that $B \in \mathcal{W} \subset \pi\mathcal{V}$. Then we have $\mathcal{U}\,\pi\,\mathcal{V}\,\pi\,\mathcal{W}$, and hence from (1) it follows that $\mathcal{U}\,\pi\,\mathcal{W}$. Thus $A\,\pi\,B$, as desired.

(3) π is an Efremovich proximity. We will show that $r(A)$ is always round. Let $B \in r(A)$. Then $A \not\pi \sim B$, and so by (2), $r(A)\vee r(\sim B)$ is not proper. Choose sets D and E such that

$$A <_{\pi} D,\ \sim B <_{\pi} E, \text{ and } D \cap E = \emptyset.$$

Then D is a member of $r(A)$ for which $D <_{\pi} B$. ∎

3.4. Corollary. If $\mathcal{C}$ is a contigual Cauchy system on a set X then $\mathcal{C}$ is the set of Cauchy filters for a totally bounded uniformity on X.

Proof. We have seen that if C is contigual then $C = \Theta$, the set of π_C-filters on X (Thm. 3.2). Thus Θ is Cauchy, and so π_C is an Efremovich proximity (Thm. 3.3). Then Θ must be the set of Cauchy filters for the totally bounded uniformity in the proximity class of π_C (Thm. 2.19). ■

To summarize, we have seen that for an Efremovich proximity π, the set $\Theta^*(\pi)$ is the set of minimal Cauchy filters of the totally bounded uniformity $\mathcal{H}^*$ in the class of π. The set $\Theta^*(\pi)$ is also the trace system of the compact T_2 extensions associated with π. In case π is only a Lodato proximity, then $\Theta^*(\pi)$ is the largest T_1 filter system in the proximity class of π, and is the trace system for the clan-complete T_1-extension associated with π. If we try to apply these results to the proximity class π_C induced by a Cauchy system C we find that C coincides with $\Theta(\pi)$ iff π is an Efremovich proximity, and C is the set of Cauchy filters for the totally bounded uniformity in the proximity class of π. Thus whenever a set of filters is both contigual (precompact) and Cauchy (T_2) then it must be the trace system for a compact T_2-extension. In the next section we will investigate what happens when we generalize to convergence space extensions.

4. Extensions of Convergences Spaces

Many of these correspondences carry over to the convergence space context. Given a filter system Θ appropriate for a convergence space, we can construct an extension κ_Θ in such a way that κ_Θ is a compactification iff Θ is contigual. This particular result is due to D.C. Kent, in a letter dated 6/5/89. Moreover, the proximity π_Θ still has a largest member, which yields a compactification.

4.1. Definition. A convergence space is a pair (X,q), where q(x) is thought of as the set of filters which converge to x. For $\mathcal{F} \in q(x)$ we write $\mathcal{F} \to x$.

(L1) $\dot{x} \to x$.

(L2) $\mathcal{F} \to x$ and $\mathcal{G} \supset \mathcal{F} \Rightarrow \mathcal{G} \to x$.

(L3) $\mathcal{F}, \mathcal{G} \to x \Rightarrow \mathcal{F} \wedge \mathcal{G} \to x$.

In what follows, the term "space" will denote a convergence space.

4.2. Definition. A filter system on a space X is a family Θ of non-convergent filters on X.. A T_1-filter system is one for which each filter in Θ is minimal in Θ. We will further require that no filter in Θ is contained in a point filter.

4.3. Construction of κ_Θ (Kent). Let Θ be a set of non-convergent filters on X. Let Y_Θ be the filters in Θ, together with all the point filters of X. Let $j(x) = \dot{x}$. For Φ a filter on Y_Θ, define a convergence p on Y_Θ as follows:

$$\Phi \to \dot{x} \text{ iff } \exists\, \mathcal{F} \to x \text{ such that } \mathcal{F}^\wedge \subset \Phi, \text{ and}$$

$$\Phi \to \mathcal{G} \text{ iff } \mathcal{G}^\wedge \subset \Phi.$$

Here $\mathcal{F}^\wedge$ is the filter generated by $\{F^\wedge\colon F \in \mathcal{F}\}$ and $F^\wedge$ is defined as in Construction 1.2. Let $\kappa_\Theta = (j,(Y_\Theta, p))$.

4.4. Theorem. For any filter system Θ on a space X, κ_Θ is an extension of X. If X is T_0 then so is κ_Θ. If X and Θ are T_1, then so is κ_Θ.

Proof. It is easy to check that p is a convergence structure on X. Now notice that for $A \subset X$ we have

$$j(A) \subset A^\wedge \text{ and } j^{-1}(A^\wedge) = A.$$

From this it follows easily that for any filter $\mathcal{F}$ on X,

$$j(\mathcal{F}) \supset \mathcal{F}^\wedge \text{ and } j^{-1}(\mathcal{F}^\wedge) = \mathcal{F},$$

and hence

$$\mathcal{F} \to x \text{ iff } j(\mathcal{F}) \to j(x) \text{ and}$$

$$\mathcal{F} \in \Theta \Rightarrow j(\mathcal{F}) \to \mathcal{F}.$$

This establishes that j is a dense homeomorphic embedding of X into Y_Θ.

Now suppose that X is T_0; that is, different points have different sets of convergent filters. Clearly then different point filters cannot have the same set of convergent filters. Moreover, if $\mathcal{F} \in \Theta$ then $\mathcal{F}$ is non-convergent, and hence $j(\mathcal{F})$ cannot converge to any point filter. Finally, different members of Θ cannot have the same set of convergent filters. The key relation here is that $A^\wedge \subset B^\wedge \Rightarrow A \subset B$.

Finally, suppose that X is T_1; that is, if $\dot{x} \to y$ then $x = y$. We claim that Y_Θ is T_1. Suppose $\mathcal{F}^\bullet \to \mathcal{G}$, where $\mathcal{F}, \mathcal{G} \in Y_\Theta$. We wish to show that $\mathcal{F} = \mathcal{G}$. Notice that

$$\mathcal{F}^\wedge \subset \mathcal{G}^\bullet \Rightarrow \mathcal{F} \subset \mathcal{G}.$$

Case I. $\mathcal{F} = \dot{x}$. From the construction of the convergence structure on Y_Θ, we have for some filter $\mathcal{H} \to x$ that $\mathcal{H}^\wedge \subset \mathcal{G}^\bullet$. Then $\mathcal{H} \subset \mathcal{G}$ and thus $\mathcal{G} \to x$. Therefore $\mathcal{G} \notin \Theta$, and so $\mathcal{G}$ must be a point filter. Since X is T_1, we have that $\mathcal{G} = \dot{x} = \mathcal{F}$.

Case II. $\mathcal{F} \in \Theta$. Since $\mathcal{G}^\bullet \to \mathcal{F}$, we have that $\mathcal{F}^\wedge \subset \mathcal{G}^\bullet$ and hence $\mathcal{F} \subset \mathcal{G}$. Since Θ is T_1, $\mathcal{G}$ cannot be a point filter. Thus $\mathcal{G} \in \Theta$. But the members of Θ are minimal in Θ, and so we have $\mathcal{F} = \mathcal{G}$. ■

4.5. Definition. Let Θ be a filter system on a convergence space X, and let $\mathcal{F}$ be a filter on X. We say that $\mathcal{F}$ is near Θ iff for any finite family $\mathcal{A}$ of subsets of X we have that if $\mathcal{A} \cap \mathcal{F} = \emptyset$ then there is a filter $\mathcal{G}$ in Y_Θ such that $\mathcal{A} \cap \mathcal{G} = \emptyset$. We say that Θ is <u>contigual</u> iff whenever $\mathcal{F}$ is near Θ, either $\mathcal{F}$ is convergent or $\mathcal{F} \supset \mathcal{G}$ for some $\mathcal{G} \in \Theta$.

4.6. Theorem. A filter system Θ on a space X is contigual iff κ_Θ is compact.

<u>Proof</u>. ($\Rightarrow$) Let Θ be contigual, and let $\mathcal{U}$ be an ultrafilter on Y_Θ. We will show that $\mathcal{U}$ converges. Let

$$\mathcal{F} = \{A: A^\wedge \in \mathcal{U}\}.$$

As in the proof of Thm. 1.10, it is easy to check that $\mathcal{F}$ is a filter on X. We claim that $\mathcal{F}$ is near Θ. Suppose $\mathcal{A}$ is a finite family of subsets of X for which $\mathcal{A} \cap \mathcal{F} = \emptyset$. Then for A in $\mathcal{A}$ we have that $\sim(A^\wedge)$ is in $\mathcal{U}$, and thus $\cap\{\sim(A^\wedge): A \in \mathcal{A}\} \neq \emptyset$. Let $\mathcal{G}$ be a member. Then $\mathcal{G}$ is a member of Y_Θ for which $\mathcal{A} \cap \mathcal{G} = \emptyset$.

Since Θ is contigual, either $\mathcal{F}$ converges or else $\mathcal{F} \in [\Theta]$. If $\mathcal{F} \to x$, then since $\mathcal{F}^\wedge \subset \mathcal{U}$, we have that $\mathcal{U} \to \dot{x}$. If $\mathcal{F}$ contains a filter $\mathcal{G}$ in Θ then clearly $\mathcal{U} \to \mathcal{G}$.

($\Leftarrow$) Suppose now that κ_Θ is compact. Let $\mathcal{F}$ be a filter near Θ. We wish to show that $\mathcal{F}$ converges, or $\mathcal{F} \in [\Theta]$. Let

$$\mathcal{S} = \{\sim(A^\wedge): A \notin \mathcal{F}\}.$$

We claim that $\mathcal{S}$ has the f.i.p. Let $\mathcal{A}$ be a finite family of sets not in $\mathcal{F}$. Since $\mathcal{F}$ is near Θ, there is some filter $\mathcal{G}$ in Y_Θ such that $\mathcal{A} \cap \mathcal{G} = \emptyset$. Clearly $\mathcal{G} \in \cap\{\sim(A^\wedge): A \in \mathcal{A}\}$.

Now let $\mathcal{U}$ be an ultrafilter containing $\mathcal{S}$. Since κ_Θ is compact, there is a filter $\mathcal{G}$ in Y_Θ for which $\mathcal{U} \to \mathcal{G}$. Notice that for any filter $\mathcal{H}$, if $\mathcal{H}^\wedge \subset \mathcal{U}$ then $\mathcal{H} \subset \mathcal{F}$.

<u>Case I.</u> $\mathcal{G} = \dot{x}$, for some x in X. We claim that $\mathcal{F} \to x$. Choose $\mathcal{H} \to x$ such that $\mathcal{H}^\wedge \subset \mathcal{U}$. Then $\mathcal{H} \subset \mathcal{F}$, and thus $\mathcal{F} \to x$.

<u>Case II.</u> $\mathcal{G} \in \Theta$. Since $\mathcal{U} \to \mathcal{G}$ we have $\mathcal{G}^\wedge \subset \mathcal{U}$ and therefore $\mathcal{G} \subset \mathcal{F}$; i.e. $\mathcal{F} \in [\Theta]$. ■

4.7. Definition. A proximity π on a convergence space (X,q) is <u>compatible</u> with the convergence structure on X iff for all $A \subset X$ we have

$$cl_\pi(A) = cl_q(A) = \{x: \exists\, \mathcal{F} \to x \text{ such that } A \in \mathcal{F}\}.$$

We will call π a <u>convergence proximity</u> iff

(CP1) If $cl_\pi A \cap cl_\pi B \neq \emptyset$ then $A \,\pi\, B$.

(CP2) If $\mathcal{F}$ is a π-filter and $\mathcal{F} \subset \dot{x}$ then $\mathcal{F} \to x$.

Let $\mathcal{L}$ denote the set of convergent filters, and define

$$A \,\pi_\Theta\, B \text{ iff } \exists\, \mathcal{F} \in \Theta \cup \mathcal{L} \text{ such that } A,B \in d\mathcal{F}.$$

4.8. Theorem. Let π be a convergence proximity compatible with a convergence space X. Let Θ be the family of all the non-convergent π-filters, and let Θ^* be the family of minimal π-filters which are non-convergent. Then Θ and Θ^* are contigual filter systems in the proximity class of π. Moreover, Θ is the largest filter system in the proximity class, and Θ^* is the largest T_1-filter system in the class.

Proof. From the proof of Lemma 2.7 it follows that every π-filter contains a minimal π-filter. From this we can prove that Θ and Θ^* are in the same proximity class. Since $\Theta^* \subset \Theta$, clearly $\pi_{\Theta^*} \subset \pi_\Theta$. Now suppose $A\ \pi_\Theta\ B$. Choose $\mathcal{F} \in \Theta \cup \mathcal{L}$ such that $A,B \in d\mathcal{F}$. Let $\mathcal{G}$ be a minimal π-filter contained in $\mathcal{F}$. Note $A,B \in d\,\mathcal{G}$; moreover, $\mathcal{G} \in \Theta^* \cup \mathcal{L}$, so that $A\ \pi_{\Theta^*}\ B$.

We claim that every convergent filter is a π-filter. For if $\mathcal{F} \to x$ and $A,B \in d\mathcal{F}$ then $x \in cl_qA \cap cl_qB$. Since π is compatible with q, and since (CP1) holds, we have $A\ \pi\ B$.

From this it is easy to see that every filter near Θ is a π-filter. For suppose $\mathcal{F}$ is a filter near Θ, with A,B in $d\mathcal{F}$. Let $\mathcal{A} = \{\sim A, \sim B\}$. Then $\mathcal{A} \cap \mathcal{F} = \emptyset$, and hence there is a filter $\mathcal{G}$ in Y_Θ such that $\mathcal{A} \cap \mathcal{G} = \emptyset$. Since $\mathcal{G} \in \Theta \cup \mathcal{L}$, clearly $\mathcal{G}$ is a π–filter, and thus $A\ \pi\ B$.

This establishes that Θ is contigual. We claim that Θ^* is contigual as well. For suppose $\mathcal{F}$ is a filter near Θ^*. We wish to show that $\mathcal{F}$ converges or else $\mathcal{F} \in [\Theta^*]$. Clearly $\mathcal{F}$ is near Θ, and therefore it must be a π-filter. Let $\mathcal{G}$ be a minimal π-filter contained in $\mathcal{F}$. If $\mathcal{G}$ converges then so does $\mathcal{F}$. If $\mathcal{G}$ does not converge then $\mathcal{G} \in \Theta^*$ and hence $\mathcal{F} \in [\Theta^*]$.

Since every convergent filter is a π-filter, we can use the proof of Thm. 2.9 to estasblish that $\pi_\Theta = \pi$. From (CP2) it follows easily that no filter in Θ^* can be contained in a point filter. Thus Θ^* is a T_1-filter system. Finally, every filter system in the proximity class of π is a subset of $\Theta = [\Theta^*]$. Thus Θ is the largest filter system in the proximity class, and Θ^* is the largest T_1 system.

This result leads one to expect that the correspondence between extensions, filter systems, and proximities ought to carry over in some form into the realm of convergence-space extensions. We have replaced Cauchy structures on X with the family Θ of non-convergent π-filters. Does Θ have a nice characterization, similar to the axioms for a Cauchy structure? What convergence-space extensions correspond to the principal topological extensions?

One potentially fruitful approach would proceed along the lines of function classes. Recently Eva Lowen-Colebunders [5] published a beautifully-written monograph on function classes of real-valued Cauchy continuous functions. Any such function class can be regarded as the set of continuously extendible functions for some extension Y of X. This observation helps provide deep insight into the relationship between extensions, function classes, and Cauchy structures. In particular, Lowen-Colebunders was able to obtain two very nice characterizations of those function classes Φ on X which are the Cauchy-continuous functions for some Cauchy structure on X. Since systems of π-filters share several of the nice features of Cauchy spaces, one would expect that this work on Cauchy functions would help provide insight into the more general situation of π-filters.

References

[1] N. Bourbaki, *General Topology*, Part I, Addison-Wesley, Reading, Mass., 1966.

[2] M.S. Gagrat and S.A. Naimpally, *Proximity approach to extension problems*, Fund. Math. 71 (1971), 63-76.

[3] H. Herrlich, *A Concept of nearness*, General Topology and Appl. 4 (1974), 191-212.

[4] H.H. Keller, *Die Limes-Uniformisierbarkeit der Limesraume*, Math. Ann. 176 (1968), 334-341.

[5] Eva Lowen-Colebunders, *Function Classes of Cauchy Continuous Mapr*, Marcel Dekker, New York, 1989.

[6] W. Page, *Topological Uniform Structures*, John Wiley & Sons, New York, 1978.

[7] Ellen E. Reed, *Nearnesses, proximities, and T_1-compactifications*, Trans. Amer. Math. Soc. 236 (1978), 193-207.

[8] Yu M. Smirnov, *On proximity spaces*, Math. Sb. N.S. 31 (73) (1952), 543-574; English transl. Amer. Math.Soc. Transl. (2) 38 (1964), 5-35.

[9] W.J. Thron, *Topological Structures*, Holt, Rinehart and Winston, New York, 1966.

[10] ________, *Proximity structures and grills*, Math. Ann. 206 (1973), 35-62.

Idealized Piccard-Pettis Theorems

David A. Rose and T. R. Hamlett East Central University, Ada, Oklahoma

1 INTRODUCTION

H. Steinhaus (1920) proved that the sum A+B and difference A-B of two sets of real numbers A and B contain intervals if A and B have positive Lebesgue measure. A topological analogue of this theorem was found by S. Piccard (1939). She proved that for a set A of real numbers, A + A contains an interval if A is nonmeager (i.e. second category) and a Baire set (i.e. for some open set U, the symmetric difference $A \triangle U$ is meager). B. J. Pettis (1950) generalized a theorem of S. Banach (1931) by showing that in a topological group (G, ·), if a subset A contains a nonmeager Baire set, then the interior of $A^{-1}A$ contains the group identity e. Later, B. J. Pettis (1951) extended his result to generalize some results of E. J. McShane (1950). Without noting it, Pettis simultaneously proved a strong version of Piccard's theorem. He showed that in a semitopological group (G,·), (i.e. a topological group except that the multiplication is only assumed to be separately continuous in each variable) if a subset A of G contains a nonmeager Baire set then for any nonmeager subset B of G, both $B^{-1}A$ and AB^{-1} have nonvoid interior. Other generalizations of Piccard's theorem have appeared such as in the papers of Z. Kominek (1971) and J. Smital (1980). Wolfgang Sander (1975) generalized the Pettis result in two ways. He replaced the group operation of "multiplication by the inverse" with a more general type of function from X×Y into Z where X, Y, and Z are arbitrary topological spaces, and he widened the scope from the ideal of meager sets to that of a general ideal satisfying certain conditions. In this note we also use a general ideal approach and identify two properties that an ideal may possess sufficient to allow an ideal Piccard-Pettis theorem. This general version of the theorem admits application to measure theory by which we obtain a theorem similar in flavor to the aforementioned theorem of H.Steinhaus.

2 PRELIMINARIES

2.1 IDEAL SPACES

An ideal space is a triple $(X,\tau,\mathcal{I})$ where (X,τ) is a topological space and $\mathcal{I}$ is an ideal in X, i.e. $\mathcal{I}$ is a nonempty family of subsets of X satisfying heredity ($A\subseteq B\in\mathcal{I}\Rightarrow A\in\mathcal{I}$) and finite additivity ($\{A,B\}\subseteq\mathcal{I}\Rightarrow A\cup B\in\mathcal{I}$). (Such families are called ideals since they are precisely the algebraic ideals in the Boolean ring $(\mathcal{P}(X),\Delta,\cap)$ where $\mathcal{P}(X)$ is the power set of X, Δ denotes symmetric difference, and $\cap$ is ordinary intersection.) If also $\mathcal{I}$ has countable additivity ($\{A_n|n<\omega\}\subseteq\mathcal{I}\Rightarrow\cup A_n\in\mathcal{I}$), $\mathcal{I}$ is called a σ-ideal. Of course, each family $\mathcal{A}$ of subsets of X is contained in a unique smallest (σ-) ideal, $\mathcal{I}(\mathcal{A})$ ($\mathcal{I}_\sigma(\mathcal{A})$), "generated" by $\mathcal{A}$. And if $\mathcal{A} = \{A\}$, $\mathcal{I}(\mathcal{A}) = \mathcal{P}(A)$ is the "principal" ideal generated by A.

If (X,τ) is a topological space let us denote by $\tau(x)$ the family of all open sets containing x, for each $x\in X$. If also $\mathcal{I}$ is an ideal in X, for each $A\subseteq X$, let $A^*(\tau,\mathcal{I})$, written A^* if τ and $\mathcal{I}$ are understood, be defined by $A^* = \{x\in X|U\in\tau(x)\Rightarrow U\cap A\notin\mathcal{I}\}$, i.e. A^* is the set of all points of X at which A is not locally in $\mathcal{I}$. Then $Cl^*A = A\cup A^*$ defines a Kuratowski closure operator which induces a topology $\tau^*(\mathcal{I})$, written τ^* if $\mathcal{I}$ is understood, with $\tau\subseteq\tau^*$. It is useful to note that for all subsets A and B of X, $(A\cup B)^* = A^*\cup B^*$. Dually, for each $A\subseteq X$, let $A^\circ(\tau,\mathcal{I})$, written A° if τ and $\mathcal{I}$ are understood, be defined by $A^\circ = \cup\{U\in\tau|U\text{-}A\in\mathcal{I}\}$. Then $A^\circ = X\text{-}(X\text{-}A)^*$ and for all subsets A and B of X, $(A\cap B)^\circ = A^\circ\cap B^\circ$. Also, for each $A\subseteq X$, the τ^*-interior of A is given by $Int^*A = A\cap A^\circ$, and $\tau^* = \{A\subseteq X|A\subseteq A^\circ\}$.

2.2 IDEAL CONDITIONS

There are two important conditions that an ideal may satisfy. Given an ideal space $(X,\tau,\mathcal{I})$, we say that $\mathcal{I}$ is τ-codense, or simply $\mathcal{I}$ is codense if τ is understood, when each member of $\mathcal{I}$ is codense in (X,τ), i.e. each $A\in\mathcal{I}$ has empty interior. Clearly, $\mathcal{I}$ is τ-codense if and only if $\tau\cap\mathcal{I} = \{\emptyset\}$. For example, the ideal of countable subsets is codense in any space with uncountable dispersion character. A perhaps more important observation is that the ideal $\mathcal{N}(\tau)$ of nowhere dense subsets of (X,τ) is always τ-codense. We also mention that (X,τ) is a Baire space if and only if the σ-ideal $\mathcal{M}(\tau)$ of meager subsets is τ-codense. Other characterizations of codenseness of $\mathcal{I}$ can be found in D. Jankovic′ and T. R. Hamlett (1990) and T. R. Hamlett and Dragan Jankovic′ (1990). One in particular that is easily verified but useful is that $\mathcal{I}$ is τ-codense if and only if $\emptyset^\circ(\tau,\mathcal{I}) = \emptyset$. Thus, if $\mathcal{I}$ is codense and $A^\circ\neq\emptyset$ then $A\neq\emptyset$.

A second very useful condition for an ideal $\mathcal{I}$ to satisfy is now described. We say that $\mathcal{I}$ is τ-local, or simply that $\mathcal{I}$ is local when τ is understood, if $\mathcal{I}$ contains all subsets of X which are locally in $\mathcal{I}$. Here, we say that a subset A of X is locally in $\mathcal{I}$ if there is an open cover $\mathcal{U}\subseteq\tau$ for A so that for each $U\in\mathcal{U}$, $U\cap A\in\mathcal{I}$. It follows easily that principal ideals are τ-local for every topology τ. Also, easily obtained but important is the fact that each σ-ideal is τ-local when (X,τ) is hereditarily Lindelöf [D. Jankovic' and T. R. Hamlett (1990)]. Less easily, $\mathcal{N}(\tau)$ is always τ-local. Extremely important is the (non-trivial) Banach Category Theorem which asserts that $\mathcal{M}(\tau)$ is always τ-local. For our purposes the following weak version of τ-locality will be sufficient. An ideal $\mathfrak{I}$ on a topological space (X,τ) is weakly τ-local if $A^* = \emptyset$ implies $A \in \mathfrak{I}$ for each $A \in \mathcal{P}(X)$. This condition is called compactness for $\mathfrak{I}$ in [Vaidyanathaswamy (1945)] where it is shown to be weaker than τ-locality.

2.3 IDEAL TOPOLOGICAL GROUPS

By an ideal (semitopological) topological group is meant a quadruple $(X,\tau,\mathfrak{I},\cdot)$ where $(X,\tau,\mathfrak{I})$ is an ideal space and $(X,\tau,\cdot)$ is a (semitopological) topological group. We say that the ideal $\mathfrak{I}$ is "left translation invariant" or simply "left invariant" if for each $x \in X$, $x\mathfrak{I} = \{xA | A \in \mathfrak{I}\} \subseteq \mathfrak{I}$, in which case, $x\mathfrak{I} = \mathfrak{I}$ for each $x \in X$. Likewise, $\mathfrak{I}$ is "right translation invariant", or simply "right invariant", if $\mathfrak{I}x = \{Ax | A \in \mathfrak{I}\} = \mathfrak{I}$ for each $x \in X$. Of course, ideals which are preserved by homeomorphism from X to X are both left and right invariant. The following useful result is easily obtained.

PROPOSITION 1 If $(X,\tau,\mathfrak{I},\cdot)$ is a left invariant ideal semitopological group, then for each $x \in X$ and $A \subseteq X$, $xA^\circ = (xA)^\circ$ and consequently, $xA^* = (xA)^*$.

In order to find ideal versions of the theorems of Piccard and Pettis, it is necessary to generalize the notion of a Baire set. If $(X,\tau,\mathfrak{I})$ is an ideal space and A is any subset of X, let us agree that A is a Baire set relative to τ and $\mathfrak{I}$, or simply a Baire set if τ and $\mathfrak{I}$ are understood, if for some $U \in \tau$, $A \Delta U \in \mathfrak{I}$. Let $\mathcal{B}r(X,\tau,\mathfrak{I})$, or simply $\mathcal{B}r(X)$, denote the family of all Baire subsets of X. Trivially, $\mathfrak{I} \subseteq \mathcal{B}r(X,\tau,\mathfrak{I})$, and the family of "nontrivial" Baire sets is denoted $\mathcal{B}(X,\tau,\mathfrak{I})$, or simply $\mathcal{B}(X)$, with $\mathcal{B}(X,\tau,\mathfrak{I}) = \mathcal{B}r(X,\tau,\mathfrak{I}) - \mathfrak{I}$. Among the subsets of X, let the family of those containing a nontrivial Baire set be denoted $\mathcal{U}(X,\tau,\mathfrak{I})$, or simply $\mathcal{U}(X)$.

Let $(X,\cdot)$ be a group. Let us say that a family of subsets $\mathcal{F} \subseteq \mathcal{P}(X)$ is inversion invariant if $\mathcal{F}^{-1} = \{F^{-1} | F \in \mathcal{F}\} = \mathcal{F}$.

PROPOSTION 2 If $(X,\tau,\mathfrak{I},\cdot)$ is an ideal topological group and $\mathfrak{I}$ is inversion invariant, then $(A^{-1})^\circ = (A^\circ)^{-1}$ and consequently $(A^{-1})^* = (A^*)^{-1}$ for each $A \in \mathcal{P}(X)$. Further, $\mathcal{B}r(X)$, $\mathcal{B}(X)$, and $\mathcal{U}(X)$ are each inversion invariant.

3 IDEAL PICCARD-PETTIS THEOREMS

We state our theorems for left invariant ideals and note that similar results hold for right invariant ideals. We begin with the following general result.

THEOREM 1 Let $(X,\tau,\mathfrak{I},\cdot)$ be a left invariant semitopological group. Then $A^\circ(B^*)^{-1} \subseteq AB^{-1}$ for all A, $B \in \mathcal{P}(X)$.

Proof: If $A^\circ(B^*)^{-1} = \emptyset$ there is nothing to prove. If $x \in A^\circ(B^*)^{-1}$, then $x = yz^{-1}$ for some

$y \in A^{\circ}$ and $z \in B^*$. Let $U \in \tau$ with $y \in U$ and $U-A \in \mathcal{I}$. Since $y = xz \in xB^* = (xB)^*$, $U \cap xB \notin \mathcal{I}$. Thus, $U \cap xB = (U \cap A \cap xB) \cup ((U-A) \cap xB) \subseteq (A \cap xB) \cup (U-A)$ implies $A \cap xB \notin \mathcal{I}$. Apparently, $A \cap xB \neq \emptyset$ so that $x \in AB^{-1}$. □

COROLLARY 1.1 [Sander (1975)] If $(X,\tau,\mathcal{I},\cdot)$ is a left invariant ideal semitopological group, then for all A, $B \in \mathcal{P}(X)$, $A^{\circ}(B^*)^{-1}$ is a nonempty open subset of AB^{-1} if and only if $A^{\circ} \neq \emptyset$ and $B^* \neq \emptyset$.

COROLLARY 1.2 Let $(X,\tau,\mathcal{I},\cdot)$ be an ideal topological group with $\mathcal{I}$ both left invariant and inversion invariant. Then $A^{\circ}B^* \subseteq AB$ for all $A,B \in \mathcal{P}(X)$ and $A^{\circ}B^*$ is a nonempty open subset of AB if and only if $A^{\circ} \neq \emptyset$ and $B^* \neq \emptyset$.

3.1 THE WEAK PICCARD-PETTIS THEOREM AND SOME CONSEQUENCES

THEOREM 2 Let $(X,\tau,\mathcal{I},\cdot)$ be an ideal semitopological group with $\mathcal{I}$ left invariant and τ-codense. Then AB^{-1} has a nonempty interior if A, $B \in \mathcal{U}(X,\tau,\mathcal{I})$.

Proof: It suffices to show that $A^{\circ} \neq \emptyset$ and $B^* \neq \emptyset$. If $C \subseteq A$ and $U \in \tau$ with $C \Delta U \in \mathcal{I}$ and $C \notin \mathcal{I}$, then $\emptyset \neq U \subseteq U^{\circ} = C^{\circ} \subseteq A^{\circ}$. Let $D \subseteq B$ and $V \in \tau$ with $D \Delta V \in \mathcal{I}$ and $D \notin \mathcal{I}$. Clearly $V \neq \emptyset$ and since $\mathcal{I}$ is τ-codense, $V \subseteq V^* = D^* \subseteq B^*$. □

COROLLARY 2.1 Let $(X,\tau,\mathcal{I},\cdot)$ be an ideal topological group with $\mathcal{I}$ left invariant, inversion invariant, and τ-codense. Then $\text{int}(AB) \neq \emptyset$ for all $A,B \in \mathcal{U}(X,\tau,\mathcal{I})$.

COROLLARY 2.2 Let $(X,\tau,\mathcal{I},\cdot)$ be an ideal semitopological group with identity e. If $\mathcal{I}$ is left invariant and τ-codense and if $A \in \mathcal{U}(X,\tau,\mathcal{I})$, then e belongs to the interior of AA^{-1}.

COROLLARY 2.3 Under the hypotheses of Corollary 1.2, if also A is a subgroup of X, then A is clopen (i.e. both open and closed).

COROLLARY 2.4 Let $(X,\tau,\cdot)$ be a locally compact topological group with identity e and left invariant Haar measure μ. Let $H_o = \{A \subseteq X | \mu(A) = 0\}$ be the family of Haar null subsets of X. Then H_o is a left invariant σ-ideal and whenever $A,B \in \mathcal{U}(X,\tau,H_o)$, AB has nonempty interior and e belongs to the interior of AA^{-1}.

Proof: Since μ is a complete measure, H_o has heredity and is thus a σ-ideal. Also, H_o is τ-codense since $H_o \cap \tau = \{\emptyset\}$. Further, H_o is inversion invariant and $A \in \mathcal{U}(X,\tau, H_o)$ if and only if $A^{-1} \in \mathcal{U}(X,\tau,H_o)$.□

COROLLARY 2.5 Let m be Lebesgue measure on the usual real line (R,τ) and let $L_o = \{A \subseteq R | m(A) = 0\}$ be the family of Lebesgue null sets. If $A \in \mathcal{U}(R,\tau,L_o)$, 0 belongs to the interior of A-A and A+A has nonempty interior (i.e. contains an interval).

3.2 THE STRONG PICCARD-PETTIS THEOREM AND SOME CONSEQUENCES

For our next theorem observe that a weakly τ-local left invariant ideal $\mathcal{I}$ on a semitopological group $(X,\tau,\cdot)$ is τ-codense if $X \notin \mathcal{I}$. For if $\mathcal{I}$ is not τ-codense, there exists a set $U \in \tau \cap \mathcal{I}$ with $e \in U$ where e is the group identity. By the weak τ-locality of $\mathcal{I}$, $X^* \neq \emptyset$. If $x \in X^*$, $xU \cap X = xU \notin \mathcal{I}$. But this contradicts the left invariance of $\mathcal{I}$. The Sorgenfrey plane S^2

with ordinary vector addition and the ideal $\mathcal{C}$ of countable subsets is an example of a left invariant semitopological group and $\mathcal{C}$ is τ-codense since S^2 has uncountable dispersion character. However, $\mathcal{C}$ is not weakly τ-local for the minor diagonal $D = \{(x,-x)|x \in S\}$, where S is the Sorgenfrey line, is an uncountable closed discrete subspace of S^2 so that $D^* = \emptyset$ but $D \notin \mathcal{C}$.

THEOREM 3 Let $(X,\tau,\mathcal{I},\cdot)$ be an ideal semitopological group with $\mathcal{I}$ left invariant and weakly τ-local. If $A \in \mathcal{U}(X,\tau,\mathcal{I})$ and $B \in \mathcal{P}(X)-\mathcal{I}$, then AB^{-1} has nonempty interior.

Proof: With $A \in \mathcal{U}(X,\tau,\mathcal{I})$ and $B \in \mathcal{P}(X)-\mathcal{I}$, $X \notin \mathcal{I}$ so that $\mathcal{I}$ is τ-codense by the remarks above. Thus, as in the proof of Theorem 2, $A^{\circ} \neq \emptyset$. Of course, $B^* \neq \emptyset$ since $\mathcal{I}$ is weakly τ-local. The conclusion follows from Corollary 1.1. □

Theorem 3 above (with its proof) generalizes Korollar 2 (Corollary 2) of Sander (1975) and Corollary 3 of Pettis (1951), since these results deal specifically with the case when $\mathcal{I}$ is the ideal of meager sets. Since the usual space of real numbers $(\mathcal{R},\tau)$ is hereditarily Lindelöf and the ideal of Lebesgue null sets L_o is a σ-ideal, it follows that L_o is τ-local. Since also Lebesgue measure is additively translation and inversion invariant, the following special case of Theorem 3 follows.

COROLLARY 3.1 If $A \in \mathcal{U}(\mathcal{R},\tau,L_o)$ and $B \in \mathcal{P}(\mathcal{R})-L_o$, then $A+B$ contains an interval (has nonempty interior).

If f: $[0,1] \rightarrow [0,1]$ is the classical Cantor function which is constant on each component of the complement of the nowhere dense Cantor set K, then $h(x) = f(x) + x$ defines a homeomorphism so that the nowhere dense image h(K) has positive Lebesgue measure (of 1). Clearly $h(K) \notin \mathcal{U}(\mathcal{R},\tau,L_o)$. Apparently, Corollary 3.1 above is independent from the classical theorem of H. Steinhaus (1920) and it would appear that the so-called topological analogues of Piccard and Pettis are also independent from the theorem of Steinhaus.

One amusing application of Corollary 3.1 is as follows. If A is an "interval" of irrationals and B is a nonmeasurable subset of $\mathcal{R}$, then $A+B$ contains an interval.

The following question asks whether Corollary 3.1 holds more generally in the setting of topological groups and Haar measure.

QUESTION 1 Let $(X,\tau,\cdot)$ be a locally compact topological group and let H_o be the σ-ideal of Haar null sets. Is it true that for any $A \in \mathcal{U}(X,\tau,H_o)$ and $B \in \mathcal{P}(x)-H_o$, that AB^{-1} has nonempty interior? If so, then also AB and BA have nonempty interiors since Haar measure is inversion invariant.

By Theorem 3, the answer is yes if H_o is weakly τ-local and hence in the special case that (X,τ) is hereditarily Lindelöf.

In closing we remark that Theorem 3 can be applied to a wider setting than just category (with the ideal of meager sets) and measure. For example one could use the ideal of scattered sets in a T_1 semitopological group. However, we have limited our attention to these two applications since we feel that some of the symmetry observable between "measure and category" (the title and topic of an excellent book by J. Oxtoby (1980)) is due to the common properties shared by the ideals involved.

ACKNOWLEDGEMENT We would like to thank Professor Janusz Matkowski for making us aware of several important references, and the referee for supplying Theorem 1 which enabled us to simplify the proofs of Theorems 2 and 3. The referee also pointed out that stronger versions exist for some of our measure-theoretic applications. For example, in Corollary 2.4, $e \in \text{int}(AA^{-1})$ if A has positive inner Haar measure. Also, though Corollary 3.1 is independent from the Classical Steinhaus theorem, both are implied by the "strong Steinhaus theorem" which asserts that A+B, A-B, and B-A each have nonempty interior when A has positive inner Lebesgue measure and B has positive outer Lebesgue measure.

REFERENCES

Banach, S. (1931). Über metrische gruppen, Studia Math., III, p. 101-113.

Hamlett, T. R. and Jankovic', Dragan (1990). Ideals in topological spaces and the set operator ψ, to appear in the Boll. U.M.I..

Hamlett, T.R. and Rose, David (submitted 1989). Remarks on some theorems of Banach, McShane, and Pettis, submitted to the Rocky Mountain J..

Jankovic', D. and Hamlett, T. R. (1990). New topologies from old via ideals, American Math. Monthly, p. 295-310.

Kelley, John L. (1955). General Topology, D. Van Nostrand, Princeton, New Jersey.

Kominek, Zygfryd (1971). On the sum and difference of two sets in topological vector spaces, Fund. Math., LXXI, p. 165-169.

Kuratowski, K. (1966). Topology, Vol. 1, Academic Press and PWN, New York, New York.

McShane, E. J. (1950). Images of sets satisfying the condition of Baire, Annals of Math., 51, p. 380-386.

Oxtoby, J. C. (1980). Measure and Category, second edition, Springer-Verlag, New York, New York.

Pettis, B. J. (1951). Remarks on a theorem of E. J. McShane, Proc. Amer. Math. Soc., 2, p. 166-171.

Piccard, S. (1939). Sur les ensembles de distances des ensembles de points d'un espace euclidien, Mëm. Univ. Neuchâtel, 13, Secrétariat de l'université Neuchâtel.

Sander, Wolfgang (1975). Verallgemeinerungen eines satzes von S. Piccard, Manuscripta Math., 16, p. 11-25.

Smital, Jaroslav (1980). Generalization of a theorem of S. Piccard, Universitas Com. Acta Math. Univ. Comenianae, XXXVII, p. 173-181.

Steinhaus, H. (1920). Sur les distances des points des ensembles de mesure positive, Fund. Math., 1, p. 99-104.

Vaidyanathaswamy, R. (1945). The localization theory in set topology, Proc. Indian Acad. Sci., 20, p. 51-61.

Graph Convergence for Monotone Functions on the Circle

Frank Rhodes and Christopher L. Thompson University of Southampton, Southampton, England

1 INTRODUCTION

If a continuous flow on the torus has no singularities then the Poincaré map, the first return map on a circular section, is a homeomorphism. The characteristics of the flow are determined by those of the Poincaré map, in particular by the rotation number of this map. If the set of all homeomorphisms of the circle is given the topology of uniform convergence then the rotation number varies continuously with the map. Cherry (1938) studied continuous flows on the torus with two singularities, a sink and a saddle. In this case the Poincaré map is a continuous non-decreasing map of the circle with a flat section. Again the rotation number exists and depends continuously on the map (Palis, de Melo, 1982). A flow on the torus with more singularities may have a cross section map that is non-decreasing and has flat sections and discontinuities. The rotation number is still defined for such maps (Rhodes and Thompson, 1986). In the second section of this paper we give examples which show that in this case the rotation number need not depend continuously on the map.

It seems to be widely believed that for a family of strictly increasing maps whose points of discontinuity do not vary, the rotation number does depend continuously on the map. The general assumption seems to be that the desired result for strictly increasing functions can be deduced from the known result for continuous non-decreasing functions by running orbits backwards. We are not aware of any published results of this kind, though the rotation numbers of strictly increasing maps with just one discontinuity have been studied by other means (Keener, 1980). If the points of discontinuity vary from one map to another then some mode of convergence other than that of uniform convergence is needed. In this paper we introduce a suitable mode of convergence, graph convergence, which reduces to uniform convergence when the limit function is continuous, and we use it to show that for all one-parameter families of strictly increasing functions, the rotation number depends continuously on the parameter. The proof depends on using left inverses to run orbits backwards. The left inverse of a strictly increasing function is a continuous non-decreasing function whose flat sections correspond to the jumps of the strictly increasing function. Graph convergence of strictly increasing functions guarantees graph convergence, and so uniform convergence, of the left inverse functions, which ensures convergence of the rotation numbers.

An alternative approach to the problem is to associate a relation with each non-decreasing function by filling in the vertical jumps in the graph at the discontinuities of the function. It turns out that if the condition defining graph convergence is imposed at the points of continuity of the limit function rather than at every point, then the resulting mode of convergence corresponds to the Hausdorff metric on the graphs of these relations. This mode of convergence is equivalent to one studied in probability theory — uniform convergence at each point of continuity of the limit function. Details will appear elsewhere (Rhodes and Thompson, 1990).

We draw attention to some unresolved questions. 1) It is not difficult to show that graph convergence (when formulated in terms of nets) generates a convergence class in the sense of Kelley (1955) and therefore corresponds to a topology on the appropriate set of functions. Is there a simple description of the basic open sets of this topology? 2) To establish continuity of the rotation number one needs a convergence mode on the function space that guarantees pointwise convergence of each family of powers of the functions (on a non-empty set of points). Desirable convergence modes from this point of view would have the properties (i) the convergence of a family is inherited by powers of that family, and (ii) convergence of a family in the given mode implies pointwise convergence at a suitable set of points. Graph convergence has both these properties. If we work with topologies on families of continuous functions, property (i) above follows from the joint continuity of the evaluation map with respect to the compact-open topology. It seems that continuity of the evaluation map may also play a role in the inheritance of convergence modes, but we have not been able to determine quite how, nor to find an appropriate topology on the space of discontinuous functions.

The structure of the rest of this paper is as follows. In Section 2 we give two examples. In Section 3 we introduce graph convergence and left inverses and prove the properties that we need. In Section 4 we prove the continuity of the rotation number. In the final section we use left inverses to investigate limit sets of functions with flats or jumps. We show that if F has irrational rotation number and if F is strictly increasing but not continuous then the ω-limit sets of points under F and under the left inverse of F are all equal, and are a Cantor set.

2 NOTATION AND EXAMPLES

We study functions on the circle via their lifts to functions on the line. We define the sets of functions $\mathcal{M}$, $\mathcal{S}$ and $\mathcal{C}$ as follows:

$$\mathcal{M} = \{F : \mathbb{R} \longrightarrow \mathbb{R} \mid F \text{ is non-decreasing; } \forall t\in\mathbb{R},\ F(t+1) = F(t) + 1\}$$

$$\mathcal{S} = \{F \in \mathcal{M} \mid F \text{ is strictly increasing}\}$$

$$\mathcal{C} = \{F \in \mathcal{M} \mid F \text{ is continuous}\}.$$

Functions $f : S^1 \longrightarrow S^1$ which possess at least one lift F in $\mathcal{M}$ are said to be *of degree one and non-decreasing*. If F is in $\mathcal{M}$ then for each t in $\mathbb{R}$, $F^n(t)/n$ has a limit as $n \to \infty$ and this limit is independent of t. The common value is the rotation number $\rho(F)$ of F. The rotation number of a non-decreasing function f of degree one on the circle is $\hat{\rho}(f) = \exp\{2\pi i\rho(F)\}$ where F in $\mathcal{M}$ is any lift of f.

Properties of rotation numbers for families of functions F_λ in $\mathcal{M}\cap\mathcal{C}$ or in $\mathcal{S}$ are studied in later sections of this paper. It is shown that in these cases rotation numbers vary continuously with the parameter for a suitable concept of convergence of functions. In this section we give examples to show that if the functions are in $\mathcal{M}$ but are neither continuous nor strictly increasing then the rotation number need not vary continuously with the parameter. The phenomenon seems to be associated with the fact that for such a function a change in the function value at just one point of discontinuity may change the rotation number and that each point may be mapped eventually to a point of discontinuity. The situation is discussed in Rhodes and Thompson (1986).

In the first example we let the value of the function vary continuously at just one point, keeping the values at the other points fixed.

Example 2.1 For $0 \le \lambda \le 1$ let F_λ in $\mathcal{M}$ be such that

$$F_\lambda(t) = \begin{cases} \lambda, & t = 0, \\ 1, & 0 < t < 1. \end{cases}$$

Then

$$\rho(F_\lambda) = \begin{cases} 0, & \lambda = 0, \\ \frac{1}{2}, & 0 < \lambda < 1, \\ 1, & \lambda = 1. \end{cases}$$

□

In the second example $\rho(F_\lambda)$ tends to 0 through rational values. The flat sections become arbitrarily small, the points of discontinuity vary with λ and the limit function is continuous.

Example 2.2 For $0 \le \lambda < 1$ let $F_\lambda \in \mathcal{M}$ be such that

$$F_\lambda(t) = \begin{cases} t+\lambda, & 0 \le t < 1-\lambda, \\ 1+\lambda, & 1-\lambda \le t < 1. \end{cases}$$

Then for $\frac{1}{n} \le \lambda < \frac{1}{n+1}$ we have $\rho(F_\lambda) = \frac{1}{n}$.

Proof. If $\frac{1}{n} \le \lambda < \frac{1}{n+1}$ then $(n-2)\lambda < 1-\lambda \le (n-1)\lambda$. Thus

$$F_\lambda^{n-2}(0) = (n-2)\lambda < 1-\lambda$$

so that $F_\lambda^{n-1}(0) = (n-1)\lambda$ and $1-\lambda \le (n-2)\lambda < 1$. Hence $F_\lambda^{n}(0) = 1+\lambda$ and $\rho(F_\lambda) = \frac{1}{n}$. □

3 GRAPH CONVERGENCE

The concept of graph convergence of functions that will be introduced in this section proves to be particularly helpful in investigating the relationship between families of strictly increasing functions and families of left inverse continuous functions which have flats corresponding to the jump discontinuities of the original functions. It may also be a physically more realistic concept of convergence than pointwise or uniform convergence in that it makes allowance for errors in measurement in the domain as well as errors in measurement in the range of a function.

Let (F_λ) be a one parameter family of functions on $\mathbb{R}$. For uniform convergence of F_λ to F_μ one requires that when λ is near enough to μ the graph of F_λ intersects the interval $x=x_0$, $F_\mu(x_0)-\varepsilon < y < F_\mu(x_0)+\varepsilon$ for each x_0. For graph convergence of F_λ to F_μ we require that given ε, for λ near enough to μ the graph of F_λ intersects the square

$$x_0-\varepsilon < x < x_0+\varepsilon,\ F_\mu(x_0)-\varepsilon < y < F_\mu(x_0)+\varepsilon.$$

Definition 3.1. *Let* (F_λ) *be a one parameter family of functions in* $\mathcal{M}$. *We say that* F_λ *is* **graph convergent** *to* F_μ *as* λ *tends to* μ *if*

$$\forall\varepsilon>0\ \exists\delta>0\ \forall\lambda\ \left[|\lambda-\mu|<\delta \Rightarrow \forall x_0\ \exists x\colon |x-x_0| < \varepsilon \text{ and } |F_\lambda(x)-F_\mu(x_0)| < \varepsilon\right].$$

For continuous functions, graph convergence and uniform convergence are equivalent. Indeed, in the next proposition we prove slightly more.

Proposition 3.1 *Let* (F_λ) *be a one parameter family of maps in* $\mathcal{M}$ *and let* F_μ *be in* $\mathcal{M}\cap\mathcal{C}$. *If* F_λ *is graph convergent to* F_μ *as* λ *tends to* μ *then* F_λ *is uniformly convergent to* F_μ.

Proof. Let $\varepsilon>0$. Because F_μ-id is periodic, F_μ is uniformly continuous on $\mathbb{R}$ and therefore σ can be chosen so that $\sigma < \varepsilon/2$ and so that

$$|x_1-x_2| < 4\sigma \Rightarrow |F_\mu(x_1)-F_\mu(x_2)| < \varepsilon/2. \tag{1}$$

Cover $\mathbb{R}$ with adjacent non-overlapping closed intervals of length 2σ with centres c_i, $i\in\mathbb{Z}$. Pick any $x\in\mathbb{R}$. Then for some i, $|x-c_i| \le \sigma$. By graph convergence, there exists $\delta>0$ such that if $|\lambda-\mu| < \delta$ then

$$\forall t\ \exists s \text{ such that } |t-s| < \varepsilon \text{ and } |F_\lambda(s)-F_\mu(t)| < \varepsilon \tag{2}$$

Choose such a δ and let λ satisfy $|\lambda-\mu| < \delta$. On applying (2) with $t=c_{i-1}$ and then with $t=c_{i+1}$ we get

$$\left.\begin{array}{l}\exists s_{i-1} \text{ such that } |c_{i-1}-s_{i-1}| < \varepsilon \text{ and } |F_\lambda(s_{i-1})-F_\mu(c_{i-1})| < \varepsilon \\ \exists s_{i+1} \text{ such that } |c_{i+1}-s_{i+1}| < \varepsilon \text{ and } |F_\lambda(s_{i+1})-F_\mu(c_{i+1})| < \varepsilon\end{array}\right\} \tag{3}$$

Now $s_{i-1} \le c_i-\sigma \le x \le c_i+\sigma \le s_{i+1}$. Because F_λ is non-decreasing, it follows from (3) that

$$F_\mu(c_{i-1})-\sigma \le F_\lambda(s_{i-1}) \le F_\mu(x) \le F_\lambda(s_{i+1}) \le F_\mu(c_{i+1})+\sigma$$

and hence

$$F_\mu(c_{i-1})-F_\mu(x)-\sigma \le F_\lambda(x)-F_\mu(x) \le F_\mu(c_{i+1})-F_\mu(x)+\sigma.$$

But $|c_{i-1}-x| \le 3\sigma$ and $|c_{i+1}-x| \le 3\sigma$ and so by (1)

$$-(\varepsilon/2+\sigma) < F_\lambda(x)-F_\mu(x) < \varepsilon/2+\sigma.$$

Consequently $|F_\lambda(x)-F_\mu(x)| < \varepsilon$. This proves the proposition. □

We turn now to the consideration of left inverses. A strictly increasing function F which is not continuous does not have an inverse function. But F does have a left inverse function G in $\mathcal{M}$ (so that $G(F(x)) = x$ for all $x\in\mathbb{R}$), and G has some additional properties which makes it behave sufficiently like a right inverse for our purposes. The link between families of strictly increasing functions and their families of left inverse functions will be graph convergence rather than uniform convergence. We state a preliminary lemma before showing that left inverses exist and are unique in $\mathcal{M}$.

Lemma 3.1 *(i)* *If* $F\in\mathcal{M}$ *and* $y\in\mathbb{R}$ *then* $\sup\{u|F(u)<y\} = \inf\{v|F(v)\geq y\}$.

(ii) *If* $F\in\mathcal{S}$ *and* $y\in\mathbb{R}$ *then* $\sup\{u|F(u)\leq y\} = \inf\{v|F(v)\geq y\}$.

Lemma 3.2 *Each function in* $\mathcal{S}$ *has a unique left inverse in* $\mathcal{M}$.

Proof. *Existence.* Suppose that F is in $\mathcal{S}$. Then the function G defined by setting $G(y)=\inf\{t|F(t)\geq y\}$ satisfies

$$G(F(x)) = x \quad \textit{for all } x\in\mathbb{R}. \tag{4}$$

and belongs to $\mathcal{M}$.

Uniqueness. Suppose that G is in $\mathcal{M}$ and satisfies (4). Then G(y) is uniquely determined if y is in the range of F. If y is not in the range of F then by Lemma 3.1(ii) we may set $x = \sup\{u|F(u)\leq y\} = \inf\{v|F(v)\geq y\}$. Choose $u_k \uparrow x$, $v_k \downarrow x$. Then $F(u_k) \leq y \leq F(v_k)$ for all k and on applying G and using (4) we obtain $u_k \leq G(y) \leq v_k$, and so $G(y)=x$. Thus G(y) is uniquely determined for all y. □

Lemma 3.3. *Let* $F\in\mathcal{S}$ *and let* G *be the left inverse of* F. *Let* $x,y\in\mathbb{R}$.

(i) *If* $G(y) < x$ *then* $y < F(x)$; *if* $x < G(y)$ *then* $F(x) < y$.

(ii) *If* G *is constant on an interval* $[\alpha,\beta]$ *and* $\alpha \leq F(x) \leq \beta$ *then* F *is discontinuous at* x *and* $F(x+0)-F(x-0) \geq \beta-\alpha$.

Proof. (i) If $y \geq F(x)$ then $G(y) \geq G(F(x)) = x$. If $y \leq F(x)$ then $G(y) \leq G(F(x)) = x$. (ii) Let $\delta>0$. Clearly $G(\alpha) = x = G(\beta)$. Hence $x-\delta < G(\alpha)$ and $G(\beta) < x+\delta$. By (i) we have $F(x-\delta) < \alpha$ and $\beta < F(x+\delta)$. The result follows. □

The left inverse G of a function F in $\mathcal{S}$ need not be a right inverse. However $F(G(y))=y$ at enough points for our purposes.

Lemma 3.4 *Let* $F\in\mathcal{S}$ *and let* G *be the left inverse of* F. *Let* $y\in\mathbb{R}$.
Then (i) $F(G(y))=y$ *if* F *is continuous at* $G(y)$,
(ii) G *is continuous everywhere.*

Proof. (i) Let $\delta>0$. By Lemma 3.3(i) $F(G(y)-\delta) < y < F(G(y)+\delta)$. If F is continuous at $G(y)$ then it follows that $F(G(y))=y$.
(ii) Let $y_0\in\mathbb{R}$ and let $\varepsilon>0$. Let $y_1=F(G(y_0)-\varepsilon)$, $y_2=F(G(y_0)+\varepsilon)$. It follows from Lemma 3.3(i) that $y_1 < y_0 < y_2$. Hence if $y_1 < y < y_2$ then $G(y_1) \leq G(y) \leq G(y_2)$, that is $G(y_0)-\varepsilon \leq G(y) \leq G(y_0)+\varepsilon$. Thus G is continuous at y_0. □

We thus have a map $\Psi : \mathcal{S} \longrightarrow \mathcal{M}\cap\mathcal{C}$ where $\Psi(F)$ is the left inverse of F. Although Ψ is surjective, it is not injective, as the next lemma shows. Functions in $\mathcal{M}$ are said to be *equivalent* if they agree except on a countable set.

Lemma 3.5 *(i) The map* Ψ *is surjective. (ii) Let* $F_1,F_2\in\mathcal{S}$. *Then* $\Psi(F_1)=\Psi(F_2)$ *if and only if* F_1 *and* F_2 *are equivalent.*

Proof. (i) Suppose that $G\in\mathcal{M}\cap C$. By Lemma 3.1(i) we may define F by

$$F(x) = \inf\{y \mid G(y)\geq x\} = \sup\{y \mid G(y)<x\}.$$

We shall show that $G(F(x))=x$. Let $\varepsilon>0$. By the definition of F there exist y_1 and y_2 such that $y_1>F(x)-\varepsilon$ and $G(y_1)<x$ and also $y_2<F(x)+\varepsilon$ and $G(y_2)\geq x$. Hence

$$G(F(x)-\varepsilon) \leq G(y_1) < x \leq G(y_2) \leq G(F(x)+\varepsilon).$$

Since G is continuous it follows that $G(F(x))=x$. Thus $G=\Psi(F)$.

(ii) Let $G_i=\Psi(F_i)$. If G_1 and G_2 are not equal we may suppose that $G_1(y)<G_2(y)$ for some y. If t satisfies $G_1(y)<t<G_2(y)$ then $y<F_1(t)$ and $y>F_2(t)$ by Lemma 3.3(i). Consequently F_1 and F_2 differ on an interval and so are not equivalent. Conversely, suppose that $G_1=G_2=G$. Let x be such that $F_1(x)\neq F_2(x)$, say $F_1(x)<F_2(x)$. Then G is constant on the interval $[\alpha, \beta] = [F_1(x), F_2(x)]$. By Lemma 3.3(ii) F_1 and F_2 are discontinuous at x. Denoting by $\mathbb{D}(F)$ the set of discontinuities of F, we see that if x is not in the countable set $\mathbb{D}(F_1)\cap\mathbb{D}(F_2)$ then $F_1(x)=F_2(x)$. Thus F_1 and F_2 are equivalent. □

The proof of the next result is straightforward and is omitted.

Lemma 3.6. *(i) If* G_i *is the left inverse of* F_i *(i=1,2), then the left inverse of* $F_2\circ F_1$ *is* $G_1\circ G_2$.
(ii) If G *is the left inverse of* F *and* n>0 and k *are integers, then the left inverse of* F^n+k *is* G^n-k.

There now follows the main result of this section.

Proposition 3.2 *Let* (F_λ) *be a one parameter family of functions in* $\mathcal{S}$ *and for each* λ *let* G_λ *be the left inverse of* F_λ. *If* F_λ *is graph convergent to* F_μ *as* λ *tends to* μ *then* G_λ *is graph convergent to* G_μ *as* λ *tends to* μ.

Proof. Suppose that F_λ is graph convergent to F_μ. Let $\varepsilon>0$. There is a $\delta>0$ such that for all λ satisfying $|\lambda-\mu|<\delta$ we have

$$\forall x_0\ \exists x \text{ such that } |x-x_0| < \varepsilon \text{ and } |F_\lambda(x)-F_\mu(x_0)| < \varepsilon. \tag{5}$$

To prove the proposition, it suffices to show that for any such λ,

$$\forall y_0\ \exists y \text{ such that } |y-y_0| < \varepsilon \text{ and } |G_\lambda(y)-G_\mu(y_0)| < \varepsilon. \tag{6}$$

Choose any $y_0\in\mathbb{R}$ and let $x_0=G_\mu(y_0)$. We consider two cases separately.
Case (i). Suppose F_μ is continuous at x_0. By Lemma 3.4(i), $F_\mu(x_0)=F_\mu G_\mu(y_0)=y_0$. By graph convergence of F_λ to F_μ there exists $x=x_1$ such that (5) holds. Let $y_1=F_\lambda(x_1)$. Then

$$|y_1-y_0| = |F_\lambda(x_1)-F_\mu(x_0)| < \varepsilon \text{ and } |G_\lambda(y_1)-G_\mu(y_0)| = |x_1-x_0| < \varepsilon$$

and so (6) holds with $y=y_1$.

Case (ii). Suppose F_μ is discontinuous at x_0. Let B_- and B_+ be the open boxes centered at $(x_0-\varepsilon/2,\ F_\mu(x_0-\varepsilon/2))$ and at $(x_0+\varepsilon/2,\ F_\mu(x_0+\varepsilon/2))$ whose sides have length ε and are parallel to the axes. Each box contains a point of the graph of F_λ and hence of the graph of G_λ. The line $y=y_0$ lies above the centre of B_- and below the centre of B_+. If this line intersects one of these boxes then we pick a point $(x_1,\ F_\lambda(x_1))$ in that box. Let $y_1=F_\lambda(x_1)$. Then $G_\lambda(y_1)=x_1$ and

$$|y_1-y_0| = |F_\lambda(x_1)-y_0| < \varepsilon \quad \text{and} \quad |G_\lambda(y_1)-G_\mu(y_0)| = |x_1-x_0| < \varepsilon$$

so that (6) holds with $y=y_1$. If, on the other hand, the line $y=y_0$ does not intersect either box, then we let $x_1 = \sup\{x|F_\lambda(x)\le y_0\} = G_\lambda(y_0)$. Because the graph of F_λ in the interval $|x-x_0| < \varepsilon$ has points below and points above the line $y=y_0$ it follows that $|x_1-x_0| < \varepsilon$. Consequently (6) holds with y equal to y_0. This completes the proof that G_λ is graph convergent to G_μ. □

4 CONTINUITY OF THE ROTATION NUMBER

The examples discussed in Section 2 show that for one parameter families of non-decreasing functions the rotation number need not be a continuous function of the parameter. In this section we show that if all the functions are continuous, or if all the functions are strictly increasing, then the rotation number is a continuous function of the parameter. The second case is deduced from the first using the idea of left inverses. Moreover, for increasing families of functions each irrational rotation number is attained for a unique parameter value.

We start with some preliminary results. The first lemma is Lemma 2 of Rhodes and Thompson (1986) with its contrapositive. The second and third lemmas follow from the definition of rotation number.

Lemma 4.1 *Suppose that* $F\in\mathcal{M}$, $\beta\in\mathbb{R}$ *and* $z\in\mathbb{Z}$.

(a) *If* $F(t) \geq t+\beta$ *for all* $t\in\mathbb{R}$ *then* $\rho(F) \geq \beta$.
If $\rho(F) < \beta$ *then there exists* $t_0\in\mathbb{R}$ *such that* $F(t_0) < t_0+\beta$.

(b) *If there exists* $t_0\in\mathbb{R}$ *such that* $F(t_0) > t_0+z$ *then* $\rho(F) \geq z$.
If $\rho(F) < z$ *then* $F(t) < t+z$ *for all* $t\in\mathbb{R}$.

The statements continue to hold if all the inequalities are reversed.

Lemma 4.2 *If* F_1 *and* F_2 *are in* $\mathcal{M}$ *and* $F_1(x) \leq F_2(x)$ *for all* x *then* $\rho(F_1) \leq \rho(F_2)$.

Lemma 4.3 *Let* F *be in* $\mathcal{M}$ *and let* q *be a positive integer. Then* $\rho(F) > p/q$ *if and only if* $\rho(F^q) > p$.

The proof by induction of the next proposition is straightforward and is omitted.

Proposition 4.1 *Let* (F_λ) *be a one parameter family of functions from* $\mathbb{R}$ *to* $\mathbb{R}$. *Suppose that* F_μ *is uniformly continuous and* F_λ *converges uniformly to* F_μ *as* λ *tends to* μ. *Then for each* $n>0$, F_λ^n *converges uniformly to* F_μ^n *as* λ *tends to* μ.

Theorem 4.1 *Let* (F_λ) *be a one parameter family of functions in* $\mathcal{M}$ *and let* F_μ *be in* $\mathcal{M}\cap\mathcal{C}$. *If* F_λ *is uniformly convergent to* F_μ *then* $\rho(F_\lambda)$ *tends to* $\rho(F_\mu)$ *as* λ *tends to* μ.

Proof. Given $\epsilon>0$ there exist rational numbers p/q and r/s such that

$$\rho(F_\mu)-\varepsilon < p/q < \rho(F_\mu) < r/s < \rho(F_\mu)+\varepsilon.$$

Then by Lemma 4.3, $\rho(F_\mu^q) > p$. So by Lemma 4.1, $F_\mu^q(t) > t+p$ for all $t\in\mathbb{R}$. Now F_μ is uniformly continuous on $\mathbb{R}$ and by Proposition 4.1, F_λ^q converges uniformly to F_μ^q. Thus for λ near enough to μ and for all $t\in\mathbb{R}$, we have $F_\lambda^q(t) > t+p$. Hence $\rho(F_\lambda^q) \geq p$ and $\rho(F_\lambda) \geq p/q$.

Similarly, for λ near enough to μ, $\rho(F_\lambda) \le r/s$. It follows that $\rho(F_\lambda)$ tends to $\rho(F_\mu)$ as λ tends to μ. □

As a consequence of Proposition 3.1 the condition of uniform convergence in Theorem 4.1 can be replaced by graph convergence. Theorem 4.1 shows that for a one parameter family of functions in $\mathcal{M}\cap\mathcal{C}$, the rotation number is a continuous function of the parameter relative to uniform convergence, or equivalently, to graph convergence. We shall deduce that for a one parameter family of functions in $\mathcal{S}$, the rotation number is a continuous function of the parameter relative to graph convergence. First we need to study the relationship between the rotation number of a strictly increasing function and the rotation number of its left inverse function.

Proposition 4.2 *If the function F is in $\mathcal{S}$ and its left inverse function is G then $\rho(F) = -\rho(G)$.*

Proof. By the last line of the proof of Theorem 1 of Rhodes and Thompson (1986) we have

$$\left|\frac{F^n(y)-y}{n} - \rho(y)\right| \le \frac{1+2|y|}{n},$$

and by the periodicity of $F^n(y)-y$,

$$\left|\frac{F^n(y)-y}{n} - \rho(y)\right| \le \frac{3}{n}$$

for all y in $\mathbb{R}$. So for any sequence of points $\{x_n\}$,

$$\lim \frac{F^n(x_n)-x_n}{n} = \rho(F).$$

Similarly, for any sequence of points $\{y_n\}$,

$$\lim \frac{G^n(y_n)-y_n}{n} = \rho(G).$$

Now choose any x in $\mathbb{R}$ and let $y_n = F^n(x)$. Then as $n \to \infty$ we have

$$\rho(G) = \lim \frac{G^n(y_n)-y_n}{n} = \lim \frac{G\ (F^n(x))-F^n(x)}{n}$$

$$= \lim \frac{x - F^n(x)}{n} = -\rho(F). \qquad \square$$

From Theorem 4.1 and Propositions 3.1, 3.2 and 4.2 we can now deduce the continuity of the rotation number for families of strictly increasing functions.

Theorem 4.2 *Let* (F_λ) *be a one parameter family of functions in* $\mathcal{S}$. *If* F_λ *is graph convergent to* F_μ *then* $\rho(F_\lambda)$ *tends to* $\rho(F_\mu)$ *as* λ *tends to* μ.

Proof. For each λ let G_λ be the left inverse of F_λ. Then G_λ is graph convergent, and hence uniformly convergent, to G_μ. Thus $\rho(G_\lambda)$ tends to $\rho(G_\mu)$ as λ tends to μ, i.e. $-\rho(F_\lambda)$ tends to $-\rho(F_\mu)$ as λ tends to μ. □

5 LIMIT SETS

As a further application of left inverses we discuss limit sets of functions on the circle with flats or jumps. If the rotation number of a homeomorphism of the circle is irrational then the limit set of the homeomorphism is either the whole circle or a Cantor subset. If the homeomorphism is a diffeomorphism with a continuous derivative of bounded variation then the limit set must be the whole circle (Denjoy, 1932) but if there is just one point where the derivative is zero then it is possible for the limit set to be a Cantor subset (Hall, 1981). Here we show that the limit set must be a Cantor set if the map is either strictly increasing but not continuous or non-decreasing and continuous but not strictly increasing. Such functions have ω-limit sets, but their α-limit sets are not well defined. However, it is possible to work with the ω-limit sets of strictly increasing functions and their left inverses. We find it convenient to continue to work on the line rather than on the circle and to study sets on the line which are lifts of ω-limit sets on the circle.. Note first that if a function in $\mathcal{M}$ has an irrational rotation number then its ω-limit sets (in the usual sense) are empty. This is clear from the following lemma.

Lemma 5.1 *Suppose that* $F\in\mathcal{M}$ *and* $\rho(F)$ *is irrational. For a point* $a_0\in\mathbb{R}$ *define* a_k *for* $k>1$ *by* $a_k=F(a_{k-1})$. *Then* $a_0\neq a_1$. *Moreover,*

(i) if $a_0<a_1$ *then* $\bigcup_1^\infty [a_{k-1},a_k)=[a_0,\infty)$,

(ii) if $a_1<a_0$ *then* $\bigcup_1^\infty (a_k,a_{k-1}]=(-\infty,a_0]$.

Proof. The irrationality of $\rho(F)$ implies that the points a_0, a_1, $a_2\ldots$ are distinct. The monotonicity of F implies that the sequence (a_i) is strictly monotone. Suppose that $a_0<a_1$ so that (a_i) is strictly increasing. If the intervals $[a_{j-1},a_j)$ do not cover $[a_0,\infty)$ then there exists ξ such that $a_k \to \xi$. Consequently

$$\lim_{t\to\infty} F(t) = \lim_{j\to\infty} F(a_{j-1}) = \lim_{j\to\infty} a_j = \xi.$$

Since $F(a_0)=a_1>a_0$ it follows from Lemma 3 of Rhodes and Thompson (1986) that $F(t)>t$ for all t. Proposition 3(a) of that paper implies that $\rho(F)$ is rational, a contradiction. This shows that (i) is true. The proof of (ii) is similar, using Proposition 3(b). □

For functions in $\mathcal{M}$ there is a natural replacement for the usual ω-limit set.

Definition 5.1 *For* $F\in\mathcal{M}$ *and* $x\in\mathbb{R}$ *we say that* u *is an* $\tilde{\omega}$-**limit point** *of* x *under* F *if* $F^{r_i}(x)-m_i \longrightarrow u$ *for some sequences of integers* m_i *and* r_i *such that* $r_i\longrightarrow\infty$. *The* $\tilde{\omega}$-limit set $\tilde{\omega}(F,x)$ *is the set of all such* $\tilde{\omega}$*-limit points.*

Definition 5.2. *When* $F\in\mathcal{M}$ *and* $x\in\mathbb{R}$ *we say that* H *is a* **shifted power** *of* F *if* H *is of the form* F^i+m, *where* i *and* m *are integers and* $i\geq 0$. *The*

mod 1 orbit *of* x *under* F *is the set of all points of the form* $F^i(x)+m$, *where* i *and* m *are integers and* $i \geq 0$.

The shifted powers of F form a semigroup under composition, and the mod 1 orbit of x is the orbit of x under this semigroup. The set of accumulation points of the mod 1 orbit of x is $\tilde{\omega}(F,x)$. The map F of $\mathbb{R}$ projects, via $p:t \mapsto \exp 2\pi it$, to a map $f=p\circ F$ of the circle. The set $\tilde{\omega}(F,x)$ is the lift of the ω-limit set of p(x) under f. The proof of the next lemma is straightforward and is omitted.

Lemma 5.2. *Let* $F\in\mathcal{M}$, *let* $x\in\mathbb{Z}$, $k\in\mathbb{Z}$ *and let* H *be any shifted power of* F. *Then*

(i) $\tilde{\omega}(F,x+k) = \tilde{\omega}(F,x)$,

(ii) $t \in \tilde{\omega}(F,x) \Rightarrow t+k \in \tilde{\omega}(F,x)$,

(iii) $\tilde{\omega}(F,F^k x) = \tilde{\omega}(F,x)$ *if* $k\geq 0$, ,

(iv) $\tilde{\omega}(H,x) \subseteq \tilde{\omega}(F,x)$.

(v) $\rho(F)$ *and* $\rho(H)$ *are both rational or both irrational.*

Lemma 5.3 *Let* $F\in\mathcal{M}$, *let* $x_0\in\mathbb{R}$, *let* $u\in\tilde{\omega}(F,x_0)$ *and let* $\delta>0$. *Then there is a shifted power* H *of* F *and a point* a_0 *in the mod 1 orbit of* x_0 *such that*

$$|a_0-u|<\delta, \qquad |H(a_0)-u|<\delta. \tag{7}$$

Proof. Because $u\in\tilde{\omega}(F,x_0)$ there exist integers $i>0$, $j>0$, m and n such that

$$|F^i(x_0)-m-u|<\delta, \qquad |F^{i+j}(x_0)-n-u|<\delta.$$

Let $a_0 = F^i(x_0)-m$. Define $H:\mathbb{R}\to\mathbb{R}$ by $H(t) = F^j(t)-(n-m)$. Then H is a shifted power of F and (7) holds. □

Proposition 5.1 *Let the rotation number of* $F\in\mathcal{S}$ *be irrational and let* G *be the left inverse of* F. *Then* $\tilde{\omega}(F,x) = \tilde{\omega}(G,y)$ *for all* $x,y\in\mathbb{R}$.

Proof. Fix x and y in $\mathbb{R}$. We prove the inclusion $\tilde{\omega}(G,y) \subseteq \tilde{\omega}(F,x)$. Let $v\in\tilde{\omega}(G,y)$ and let $\varepsilon>0$. We shall show that there exist integers $s\geq 0$ and n such that

$$|F^s(x)-n-v| < \varepsilon. \tag{8}$$

By Lemma 5.3 there is a shifted power K of G and a point b_0 in the mod 1 orbit of y such that

$$|b_0-v|<\varepsilon, \quad |K(b_0)-v|<\varepsilon.$$

Now suppose that $b_0<b_1$. Then by Lemma 5.1(i) there exist integers $j\geq 0$ and i such that $x+i\in(b_j,b_{j+1}]$, where $b_j=K^j(b_0)$. By the irrationality of $\rho(K)$ we can choose i and j so that x+i is an interior point of this interval. Since K is a shifted power of G there are integers $k\geq 0$ and p such that

$$G^k(b_0) < x+i+p < G^k(b_1).$$

Now G^k is the left inverse of F^k and by Lemma 3.3(i) it follows that

$$b_0 < F^k(x)+i+p < b_1$$

and so (8) holds when $b_0<b_1$. A similar argument using Lemma 5.1(ii) shows that (8) holds when $b_1<b_0$. From (8) we deduce that $v\in\tilde{\omega}(F,x)$ and it follows that $\tilde{\omega}(G,y) \subseteq \tilde{\omega}(F,x)$. The reverse inclusion is proved by a similar, but simpler, argument. □

We showed elsewhere (Rhodes and Thompson, 1986) that equivalent functions have the same rotation number; by Lemma 3.5(ii) equivalent functions in $\mathcal{S}$ have the same left inverse.

Corollary 5.1

(i) *Let* $F\in\mathcal{S}$ *have irrational rotation number. Then* $\tilde{\omega}(F,x)=\tilde{\omega}(F)$ *is independent of* x.

(ii) *Let* $G\in\mathcal{M}\cap\mathcal{C}$ *have irrational rotation number. Then* $\tilde{\omega}(G,x)=\tilde{\omega}(G)$ *is independent of* x.

(iii) *Let* F_1 and F_2 *be equivalent functions in* $\mathcal{S}$, *with irrational rotation number. Then* $\tilde{\omega}(F_1)=\tilde{\omega}(F_2)$.

Proof. Part (i) is clear. For (ii) choose any $F\in\mathcal{S}$ such that G is the left inverse of F; e.g. let $F(x)=\inf\{y:G(y)\geq x\}$. Then apply part (i), recalling that $\rho(F)$ is irrational by Proposition 4.2. To prove (iii), observe that equivalent functions in $\mathcal{S}$ have the same left inverse. □

Theorem 5.1 *Let $F\in\mathcal{S}$ have irrational rotation number, let $G\in\mathcal{M}\cap\mathcal{C}$ be the left inverse of F and let $\tilde{\omega}=\tilde{\omega}(F)=\tilde{\omega}(G)$. Then (i) $\tilde{\omega}$ is closed, (ii) if $u\in\tilde{\omega}$ then $G(u)\in\tilde{\omega}$ and if $G^{-1}(0)$ is a singleton $\{c\}$ then $G(u)\in\tilde{\omega}$ implies $u+c\in\tilde{\omega}$, (iii) $\tilde{\omega}$ is perfect, and (iv) either $\tilde{\omega}$ is nowhere dense or else $\tilde{\omega}$ is equal to $\mathbb{R}$.*

Proof. Let $x,y\in\mathbb{R}$, so that $\tilde{\omega}=\tilde{\omega}(F,x)=\tilde{\omega}(G,y)$.

(i) Let $u\in\mathbb{R}$. Then u is not in $\tilde{\omega}$ if and only if

$$\exists\varepsilon>0\ \exists N\ \forall r>N\ \forall m\in\mathbb{Z}\quad |F^r(x)-m-u|\geq\varepsilon.$$

This shows that the complement of $\tilde{\omega}$ is open. Hence $\tilde{\omega}$ is closed.

(ii) The first assertion is true because G is continuous. Suppose that $G^{-1}(0)=\{c\}$, a singleton. If $G(u)\in\tilde{\omega}$ then $G(u)=\lim(G^{r_i}(y)-m_i)$ and therefore $\lim G(t_i)=0$ where $t_i=G^{r_i-1}(y)-m_i-u$. The sequence t_i is therefore bounded; it must converge to c for otherwise a subsequence converges to some $d\neq c$ and then $G(d)=0$, a contradiction. Thus $u+c\in\tilde{\omega}$.

(iii) We may suppose that $y\in\tilde{\omega}=\tilde{\omega}(G,y)$. Let $u\in\tilde{\omega}$. Then there exist sequences m_i, r_i of integers with $r_i\to\infty$ such that

$$G^{r_i}(y)-m_i\to u.$$

Let $G^{r_i}(y)-m_i=t_i$. Then $t_i\in\tilde{\omega}$ by (ii) and because $\tilde{\omega}$ is invariant under translation by $\mathbb{Z}$. It follows that u is an accumulation point of $\tilde{\omega}$. Thus $\tilde{\omega}$ is perfect.

(iv) Suppose that $\tilde{\omega}$ is dense in some interval $[\alpha,\beta]$ with $\alpha<\beta$. Let $\beta-\alpha=2\delta$ and $\beta+\alpha=2\gamma$. Then $\gamma\in\tilde{\omega}$ and by Lemma 5.3 there is a shifted power K of G and a point a_0 such that

$$|a_0-\gamma|<\delta,\qquad |K(a_0)-\gamma|<\delta.$$

Let $a_j=K^j(a_0)$. Suppose that $a_0<a_1$. By Lemma 5.1(i), $\bigcup_1^\infty [a_{j-1},a_j)=$

$[a_0,\infty)$. Now K maps $[a_{j-1},a_j)$ onto $[a_j,a_{j+1})$, and $\tilde{\omega}$ is dense in $[a_0,a_1]$. Since $\tilde{\omega}$ is invariant under applications of K it follows that $\tilde{\omega}$ is dense in $[a_0,\infty)$. Since $\tilde{\omega}$ is invariant under translation by $\mathbb{Z}$, it follows that $\tilde{\omega}$ is everywhere dense. Since $\tilde{\omega}$ is closed, it follows that $\tilde{\omega}=\mathbb{R}$. When $a_1<a_0$ the proof is similar, using Lemma 5.1 (ii). □

Corollary 5.2 *If $F\in\mathcal{S}\backslash\mathcal{C}$ or if $F\in\mathcal{M}\cap\mathcal{C}\backslash\mathcal{S}$ and the rotation number of F is irrational then $\tilde{\omega}(F)\cap[0,1]$ is a Cantor set.*

Proof. If $F\in\mathcal{S}\backslash\mathcal{C}$ then the range of F omits an interval. Consequently $\tilde{\omega}(F)$ cannot equal $\mathbb{R}$. Hence $\tilde{\omega}(F)$ is nowhere dense, and hence $\tilde{\omega}(F)\cap[0,1]$ is a Cantor set. If $G\in\mathcal{M}\cap\mathcal{C}\backslash\mathcal{S}$ then there exists $F\in\mathcal{S}\backslash\mathcal{C}$ such that G is the left inverse of F. Then $\tilde{\omega}(G)\cap[0,1] = \tilde{\omega}(F)\cap[0,1]$ is a Cantor set. □

REFERENCES

Cherry, T. M. (1938). Analytic quasi-periodic curves of discontinuous type on the torus, *Proc. London Math. Soc.* **44**, 175-215.

Denjoy, A. (1932). Sur les courbes définies par les équations différentielles à la surface du tore, *J. Math. Pures Appl.* **11**, 333-375.

Hall, G. (1981). A C^∞ Denjoy counterexample, *Ergod. Theory & Dyn. Systems* **1**, 261-272.

Keener, J. P. (1980) Chaotic behaviour in piecewise continuous difference equations, *Transactions Amer. Math. Soc.* **261**, 589-604.

Kelley, J. L. (1955). *General Topology,* Van Nostrand, Princeton, p.74.

Palis, J. and de Melo, W. (1982) *Geometric Theory of Dynamical Systems* (tr. A. K. Manning), Springer, New York.

Rhodes, F. and Thompson, C. L. (1986). Rotation numbers for monotone functions on the circle, *J. London Math. Soc.* (2) **34** (1986) 360-368.

Rhodes, F. and Thompson, C. L. (1990). Topologies and rotation numbers for families of monotone on the circle, *J. London Math. Soc.* (to appear).

Near Valuations

Niel Shell The City College of New York, The City University of New York, New York, New York

ABSTRACT. We prove a Hausdorff ring topology on a field is locally bounded if and only if it is induced by a function which we call a near valuation. Near valuations are easier to construct and to work with than the set valued generalized valuations Nakano used to induce the class of locally bounded topologies. In connection with near valuations, it is natural to use a definition of valuation due to Kowalsky and Dürbaum, which is somewhat more general than the usual one. Standard theorems of valuation theory are considered from this slightly more general point of view.

Many generalizations of the usual notion of absolute value on the complex number field have been considered (see, for example, [6], [8], [11], [15]). The earliest and most important generalizations, due to Kürschak [12] in 1913 and Krull [11] in 1932, were the notions (in the terminology of Definition 1 below) of an absolute value and a nonarchimedean valuation, respectively. Artin [1, p. 37] replaced the usual triangle inequality for absolute values by the inequality $|x + y| \leq \lambda \max(|x|, |y|)$, λ a fixed constant. Kowalsky and Dürbaum [10] used the same inequality for functions taking values in totally ordered groups, thus obtaining a common generalization for the notion of absolute value and nonarchimedean valuation. Zelinsky [20] (see also [4]) modified the definition of nonarchimedean valuation by requiring only that the value group be directed instead of totally ordered. Definition 1 below is a composite of the generalizations of Kowalsky-Dürbaum and Zelinsky.

Nakano [16] defined a set valued function, which he called a generalized valuation, that behaved like a norm (i.e., was only submultiplicative) to obtain a representation of all locally bounded topologies on a field. Nakano's values were subsets of a directed group. In Theorem 5 we show that near valuations (Definition 1) also may be used to represent all locally bounded topologies on a field. We believe the simplicity of our theorems and examples shows that near valuations are substantially more natural and easier to use than generalized valuations.

The term field here means commutative field; the symbol K is used to denote a field. The nonzero elements of a subset A of a field K will be denoted by A^*. For a subset A of a field K, $A^\times$ is defined to be the set $\{x \in A : x^{-1} \in A\}$. If A is a multiplicative semigroup with identity, then $A^\times$ is a multiplicative group; if $-1 \in A$, then $-1 \in A^\times$.

Sometimes we denote by $n \cdot 1$ (instead of just n), for $n \in \mathbf{Z}$, the image of n under the homomorphism from $\mathbf{Z}$ to a field K which carries the unit element of $\mathbf{Z}$ to the unit element of K.

A directed group is a partially ordered group which is a directed set. If $\leq_1$ and $\leq_2$ are partial orders with respect to which G is a partially ordered group, we say $(G, \leq_2)$ is an extension of $(G, \leq_1)$ if $g \leq_1 h$ implies $g \leq_2 h$. We say an isomorphism $h : G \longrightarrow H$ between partially ordered groups is an order isomorphism when $h(a) \leq h(b)$ if and only if $a \leq b$.

The set of positive elements of a subset A of an ordered field will be denoted by A^+. In particular, $\mathbf{R}^+$ and $\mathbf{Z}^+$ denote the set of positive real numbers and the set of positive integers, respectively.

The usual absolute value and topology on subfields of the complex field $\mathbf{C}$ will be denoted by $| \ |_\infty$ and $\mathcal{T}_\infty$, respectively.

A subset B of a field K with a ring topology is bounded if, for each neighborhood U of zero, there exists a neighborhood V of zero such that $VB \subset U$. A ring topology on a field (and also the field) is locally bounded if there is a bounded neighborhood of zero.

The set $\{xU : x \neq 0\}$ of all nonzero multiples of a bounded neighborhood U of zero in a field with a nondiscrete locally bounded ring topology forms a neighborhood base at zero ([9], [10]). Conversely, if all nonzero multiples of a subset U of a field K form a neighborhood base at zero for a ring topology on K, then U is a bounded neighborhood of zero in this topology ([18, p. 168]); the topology will be said to be induced by U and be denoted by $\mathcal{T}_U$.

A ring topology is called nonarchimedean if it has a neighborhood base at zero consisting of open additive subgroups.

Topologies $\mathcal{S}$ and $\mathcal{T}$ on a set are independent if $S \cap T \neq \emptyset$ whenever $S \in \mathcal{S}\backslash\{\emptyset\}$ and $T \in \mathcal{T}\backslash\{\emptyset\}$.

The adjective 'proper' will refer to a proper subset.

The author thanks Professor Gerhard Schiffels for informing the author of several errors in an earlier draft of this paper.

1. Prevaluations

DEFINITION 1: Let G be a directed abelian multiplicative group to which we adjoin a least element 0 (with multiplication extended to $G \cup \{0\}$ by defining $0 \cdot g = g \cdot 0 = 0$ for every $g \in G \cup \{0\}$).

A surjective function $v : K \longrightarrow G \cup \{0\}$ is called a *directed homomorphism* on the field K if the following conditions are satisfied:

(V1) $v(x) = 0$ if and only if $x = 0$.

(V2) $v(xy) = v(x)v(y)$ for all $x, y \in K$.

A directed homomorphism is called a *prevaluation* if also

(V3) there exists $\lambda \in G$ such that $v(x), v(y) \leq g$ implies $v(x+y) \leq \lambda g$.

A prevaluation is called a *near valuation* if

(V4) $v(-1) = 1$.

An element λ for which (V3) is satisfied is called a *dominator of* v. If G is totally ordered, v is called a *valuation.* A prevaluation v is called *nonarchimedean* if the identity of G is a dominator. The group G is called the *value group* of the prevaluation. The prevaluation is called *discrete* if its value group is infinite cyclic. A valuation v is called a *real* valuation if its value group is an ordered subgroup of $\mathbf{R}^+$.

A function $| \ | : K \longrightarrow [0, \infty)$ on a field K is called an absolute value if $|x| = 0$ if and only if $x = 0$ and, for all $x, y \in K$, $|xy| = |x||y|$ and $|x + y| \leq |x| + |y|$.

By a valued (prevalued, etc.) field we mean a pair consisting of a field together with a valuation (prevaluation, etc.).

We will use the terminology above for a function v whose domain is a ring and whose range generates G, but which otherwise meets the conditions above; we will specify explicitly that v is a valuation (near valuation, etc.) *on a ring.*

Every absolute value is a valuation. Examples 6 and 7 below are neither absolute values nor nonarchimedean valuations, but are valuations according to our definition.

Our usage of the terms absolute value and nonarchimedean valuation is common, but not universal, in the literature. (E.g., sometimes nonarchimedean valuations are called Krull valuations, and sometimes the term absolute value or valuation is used for what we call a nonarchimedean absolute value or nonarchimedean valuation, respectively.) However, our usage of the Kowalsky-Dürbaum definition of valuation is very unusual: Normally, only absolute values and nonarchimedean valuations are considered.

A nontrivial multiplicative group homomorphism v on K^* satisfies (V4) if and only if it is symmetric, i.e., $v(x) = v(-x)$ for all $x \in K$. A directed homomorphism satisfies (V3) if and only if it satisfies

(V3′) there exists $\mu \in G$ such that $v(x), v(y) \leq g$ implies $v(x - y) \leq \mu g$.

If v is a directed homomorphism to a totally ordered (or, more generally, lattice ordered) group, then (V3) is equivalent to the following:

(V3″) There exists $\lambda \in G$ such that, for all $x, y \in K$,

$$v(x + y) \leq \lambda \max[v(x), v(y)].$$

Since any prevaluation v is a multiplicative homomorphism on K^*, $v(1) = 1$ (where 1 denotes the identity of the value group as well as the identity of the field). By setting $g = 1$, $x = 1$, and $y = 0$ in (V3), we see that each dominator is greater than or equal to 1. In a totally ordered group, each element other than the identity has infinite order, which implies that $v(-1) = 1$. I.e., for maps to totally ordered groups, (V4) is superfluous. Also, if $v(-1) \leq 1$ or $v(-1) \geq 1$, then $v(-1) = 1$. Examples 4 and 5 show that (V4) is not superfluous in general.

Suppose v is a prevaluation on a ring, λ is a dominator of v, and $v(y - x) \leq v(x)$. Then $v(y) \leq \lambda v(x)$, since $y = x + (y - x)$ and $v(x), v(y - x) \leq v(x)$.

We call the sets

$$S_g^v(x) = \{y \in K : v(y - x) < g\} \qquad C_g^v(x) = \{y \in K : v(y - x) \leq g\},$$

where v is a directed homomorphism on K, *spheres* of (K, v) (or of K if v is understood). The superscripts v will be omitted when v is clear from context; $S_g(0)$ and $C_g(0)$ will be shortened to S_g and C_g, respectively.

LEMMA 1.1. *Suppose $v : K \longrightarrow G \cup \{0\}$ is a directed homomorphism, $x \in K$, and $g \in G$.*

(1) $xS_g = S_{v(x)g}$.
(2) $-S_g = S_{v(-1)g}$.
(3) $S_\delta S_\delta \subset S_\epsilon$, for $\delta \leq 1, \epsilon$.
(4) $x + S_g = S_g(x)$.
(5) v is a nonarchimedean near valuation if and only if $z \in K$ and $y \in C_h(z)$ imply $C_h(y) = C_h(z)$ (see Examples 2 and 5).

The statements obtained by substituting C's for all occurences of S's are true.

THEOREM 1. *If v is a prevaluation on a field K, then the collection of sets S_ϵ, $\epsilon \in v(K^*)$, is a neighborhood base at zero for a Hausdorff ring topology $\mathcal{T}_v$ on K. (We call $\mathcal{T}_v$ the topology induced by v.)*

If v is a valuation, then $\mathcal{T}_v$ is a field topology ([10]).

PROOF: The equality $\cap S_\epsilon = \{0\}$ follows from the fact that $v(K^*)$ has no minimal elements if $v(S_1) \neq \{0\}$. The $\{S_\epsilon\}$ form a neighborhood base at zero for a ring topology because

$$\begin{aligned}
S_\delta + S_\delta &\subset S_\epsilon, & \delta &< \frac{\epsilon}{\lambda}; \\
xS_\delta &\subset S_\epsilon, & \delta &\leq \frac{\epsilon}{v(x)}; \\
-S_\delta &\subset S_\epsilon, & \delta &\leq \frac{\epsilon}{v(-1)}; \\
S_\delta S_\delta &\subset S_\epsilon, & \delta &\leq 1, \epsilon.
\end{aligned}$$

Suppose v is a valuation. Observe that $v(x) < 1/\lambda$ implies $v(1 + x) \geq 1/\lambda$: otherwise

$$1 = v(1) = v((1 + x) - x) \leq \lambda \max(v(1 + x), v(x)) < 1.$$

Therefore,

$$\frac{S_\delta}{1 + S_\delta} \subset S_\epsilon, \qquad \delta < \min\left(\frac{1}{\lambda}, \frac{\epsilon}{\lambda}\right). \bullet$$

Since a directed group containing more than one element has no minimal elements, the trivial valuation ($v(x) = 1$ for $x \neq 0$) is the only prevaluation inducing the discrete topology.

Fuchs [4] (see also [19, p. 116]) has considered homomorphisms taking values in arbitrary partially ordered groups with a zero adjoined. Nothing of a topological nature is gained by this generalization: Suppose $v : K \longrightarrow G \cup \{0\}$, where G is a partially ordered group, satisfies (V1) and (V2); and suppose the spheres (defined as for directed homomorphisms) S_ϵ, $\epsilon \in v(K^*)$, form a neighborhood base at zero for a topology for which zero is not an isolated point (such as a nondiscrete ring topology). If $S_r \subset S_g \cap S_h$ and $x \in S_r{}^*$, then $v(x) \leq g, h$; i.e., G is directed.

It is not true that, conversely, the spheres $\{S_\epsilon\}$ of a directed homomorphism constitute a neighborhood base at zero of a ring topology. When the spheres of a directed homomorphism to a totally ordered group fail to be a base at zero, they do so in a spectacular way, as (4) of Theorem 2 indicates.

THEOREM 2. *Suppose G is totally ordered, and $v : K \longrightarrow G \cup \{0\}$ is a directed homomorphism. The following statements are equivalent:*

(1) v is a valuation.

(2) $\{S_\epsilon\}_{\epsilon \in G}$ is a neighborhood base at zero for a ring topology on K.

(3) For all $\epsilon \in G$ there exists $\delta \in G$ such that $S_\delta + S_\delta \subset S_\epsilon$.

(4) There exists $\delta \in G$ such that $S_\delta + S_\delta \neq K$.

PROOF: (1) implies (2) by Theorem 1; and obviously (2) implies (3), and (3) implies (4).

(4) implies (1): Suppose $c \notin S_\delta + S_\delta$. Then $S_\delta + S_\delta \subset S_{v(c)}$. Otherwise, there exist $a, b \in S_\delta$ such that $v(a + b) \geq v(c)$, and

$$c = \left(\frac{c}{a+b}\right) a + \left(\frac{c}{a+b}\right) b \in S_\delta + S_\delta.$$

We show that any element $v(d) \in v(K^*)$ which is greater than $v(c)/\delta$ is a dominator. For such d,

$$h := \frac{v(c)}{v(d)\delta} < 1.$$

Suppose $v(x), v(y) \leq g$ and b is an element in K such that $v(b) = h\delta/g$. Then

$$\begin{aligned} v(bx), v(by) &\leq h\delta < \delta; \\ v(bx + by) &< v(c); \\ v(x + y) &\leq \frac{v(c)}{v(b)} = v(d)g. \bullet \end{aligned}$$

EXAMPLE 1: ([8, p. 110]) Let G be infinite cyclic with generator $g < 1$. Define a function on $\mathbf{Q}$: $v(0) = 0$ and, for $x \neq 0$,

$$v(x) = g^{v_p(x) + v_q(x)},$$

where v_p and v_q are the exponential p-adic and q-adic valuations for distinct primes p and q. For each $x \in \mathbf{Q}$ and $\delta \in v(\mathbf{Q}^*)$, the terms of the sum

$$x = \frac{xp^n}{p^n + q^n} + \frac{xq^n}{p^n + q^n}$$

belong to S_δ for sufficiently large n. Thus $S_\delta + S_\delta = \mathbf{Q}$.

Statements similar to the following appear in [10, p. 140], [7, p. 528], and [8, p. 101]: "Addition is continuous in the topology induced by a directed homomorphism if and only if (V3) is satisfied." The statements suggest that, even when (V3) is not satisfied, a topology is determined by requiring that $\{S_\epsilon(x)\}_{\epsilon \in v(K^*)}$ be a neighborhood base at each $x \in K$. In fact, this is not true. If there were such a topology for this example, then $y + S_\delta = S_\delta(y) \subset S_1$ for some $y \in K^*$ and some $\delta \in v(K^*)$. But

$$y + y\left(\frac{p^{2n}}{q^n}\right) = y\left(\frac{q^n + p^{2n}}{q^n}\right) \in S_\delta(y) \backslash S_1$$

for n sufficiently large.

Nagata [15] considers directed homomorphisms for which $v(K)$ is a subset of a totally ordered field. He calls such a function a valuation if it has a dominator which is in the rational subfield of the ordered field (but not necessarily in $v(K^*)$). By letting g above be an element of an ordered field with $1 - g < r$ for each positive rational number and letting $\lambda = 2$, we see that Nagata's valuations may not induce ring topologies by taking spheres with radii in $v(K^*)$ as a base at zero. (In this example, of course, we may induce the discrete topology by taking all spheres of positive rational radius as a base at zero.)

DEFINITION 2: Let $v_i : K \longrightarrow G_i \cup \{0\}$, $i \in I$, be a family of directed homomorphisms on K, and let the cartesian product G of the G_i have the product order (i.e., an element is less than or equal to a second element if all components of the first element are less than or equal to the corresponding components of the second; see [5, pp. 22, 26]). The directed homomorphism $v(x) = \{v_i(x)\}_{i \in I}$ (and $v(0) = 0$) will be called the *product* of the v_i.

EXAMPLE 2: Let v and w be independent nonarchimedean absolute values on a field K. Then the function $\max[v(x), w(x)]$ is a norm which is not a prevaluation, and the product of v and w is a near valuation which is not a norm. Both functions induce the supremum of the topologies $\mathcal{T}_v$ and $\mathcal{T}_w$.

Suppose $v(x) = w(y) = 1$ and $v(y), w(x) < 1$. For spheres with respect to the product near valuation, $x \in S_1$ and $x + y \in S_1(x) \backslash S_1$, so $S_1(x) \neq S_1$, even though the product near valuation is nonarchimedean.

THEOREM 3. *A subset A of a prevalued field (K, v) is bounded if and only if $A \subset C_g$ for some $g \in v(K^*)$. Consequently a topology induced by a prevaluation is locally bounded.*

PROOF: If $A \subset C_g$, then, for each $\epsilon \in v(K^*)$, $S_{\epsilon/g}A \subset S_\epsilon$. Conversely, if A is bounded, $x \neq 0$, and $xA \subset C_1$, then $A \subset C_{1/v(x)}$. •

2. Preorders

DEFINITION 3: Let K be a field. A subset A of K is called a *preorder* if the following conditions are satisfied:

(O1) $0, 1 \in A$.
(O2) $AA \subset A$.
(O3) $K^* = A^*(A^*)^{-1}$.
(O4) There exists an element $z \in A^*$ such that $z(A + A) \subset A$.

An element z for which (O4) is satisfied is called an *addiator*. An element $d \in K^*$ will be called a *dominator* of a subset B of K if $B + B \subset dB$. A preorder is called a *near order* if it contains -1. If a preorder A also satisfies

(O5) for each $x \in K^*$, $x \in A$ or $x^{-1} \in A$,

then (A is a near order and) A is called a *valuation* near order (Vollfastordnung in [10]).

Different authors use slightly different axioms in the definition of near order and slightly different translations of the original term *Fastordnung*.

A near order of K is an order of K (i.e., a ring with identity whose quotient is K) if and only if 1 is an addiator. A valuation order as defined here is exactly a valuation ring.

$\{0\}$ is not a preorder of the field K. However, it induces the discrete topology.

Clearly, an element $z \in K^*$ is an addiator of a preorder A of K if and only if $z \in A$ and $1/z$ is a dominator of A. If A is a subset of a field K satisfying (O1)-(O3) and if there is a dominator of A, then A is a preorder.

The following theorem extends observations made in [10]:

THEOREM 4. *(1) If v is a prevaluation on the field K, then $A_v := C_1^v$ is a preorder. A_v is a near order if and only if v is a near valuation. (We call A_v the preorder of (or associated with) v.)*

(2) An element $d \in K^*$ is a dominator of A_v if and only if $v(d)$ is a dominator of v.

(3) A_v is a valuation near order if and only if v is a valuation.

(4) A_v is an order if and only if v is nonarchimedean and symmetric.

(5) A_v is a valuation ring if and only if v is a nonarchimedean valuation.

(6) The topology induced by a nontrivial prevaluation coincides with the topology induced by the associated preorder.

PROOF: (1) Clearly A_v satisfies conditions (O1) and (O2); (O3) is satisfied because because A_v is neighborhood of zero. We have $xA_v = xC_1 = C_{v(x)}$.

Now, suppose λ is a dominator of v and d is such that $v(d) = \lambda$ and $a, b \in A_v$. Then

$$a + b \in C_\lambda = dA_v.$$

Thus d is a dominator of A_v, and the proof that A_v is a preorder is complete.

(2) We just showed that if $v(d)$ is a dominator of v, then d is a dominator of A_v. The converse follows from an argument similar to that at the end of the proof of Theorem 2.

(3) The statements we wish to show are equivalent are both equivalent to this statement: $v(x/y) \leq 1$ or $v(y/x) \leq 1$ for all $x, y \in K^*$; (4) is the special case $d = 1$, $\lambda = 1$ of (2); (5) follows from (3) and (4); and (6) is clear. •

The trivial valuation on a field K induces the discrete topology, while K, the associated preorder, induces the trivial topology.

DEFINITION 4: Let A and B be two proper preorders of a field K. We say A and B are *equivalent* if there exist elements $a, b \in K^*$ with $aA \subset B$ and $bB \subset A$.

Two prevaluations on a field K are *equivalent* (*topologically equivalent*) if they have the same (equivalent) associated preorders.

Clearly these relations are equivalence relations. Two preorders (prevaluations) are equivalent (topologically equivalent) if and only if they induce the same topology.

COROLLARY 4.1. *If a prevaluation v is a valuation (is nonarchimedean), then any prevaluation equivalent to v is a valuation (is nonarchimedean).*

Suppose A is a proper preorder on the field K. The relation $xA^\times \leq yA^\times$ if $xy^{-1} \in A$ defines a partial order with respect to which $G = K^*/A^\times$ is a directed group (cf. [10, p. 141]): Since $A^\times$ is a multiplicative group and A is a multiplicative semigroup, the proposed relation is well-defined. Clearly the relation makes G a partially ordered group. Since the sets xA, $x \in K^*$, form a neighborhood base at zero for $\mathcal{T}_A$, there exists, for each pair $x, y \in K^*$, $w \in K^*$ such that $wA \subset (xA) \cap (yA)$. Then $wA^\times \leq xA^\times, yA^\times$.

Although we shall have no need of it, we mention a variation of the construction of the group G above. One sees that $(xA)(yA) = xyA$ on the collection H of all sets

$$\{E \subset K : E = xA \text{ for some } x \in K^*\}$$

is a well-defined operation with respect to which H is a group; and, if H is ordered by inclusion, $xA^\times \mapsto xA$ is an order isomorphism from G to H.

The canonical homomorphism, call it v, from K^* to G, extended by defining $v(0) = 0$, is easily seen to be a prevaluation for which $v(d)$, where d

is any dominator of A, is a dominator. We call this the prevaluation induced by A and denote it by v_A. Furthermore, the preorder of v_A is A. One consequence of the last statement is that a proper preorder and its induced prevaluation induce the same topology. Another consequence can be seen by considering the following diagram

$$v \longrightarrow A_v \longrightarrow v_{A_v} \longrightarrow A_v.$$

The diagram says that a prevaluation and the prevaluation induced by its preorder are equivalent.

The above discussion, together with Theorem 4 and Corollary 4.1, imply the following theorem:

THEOREM 5. *The correspondence which sends a proper preorder A on a field K into the equivalence class of all prevaluations equivalent to the prevaluation induced by A is a bijection between the set of all proper preorders on K and all equivalence classes of nontrivial prevaluations on K. The inverse correspondence is obtained by assigning to each equivalence class the common preorder of all its members.*

The restriction of the above correspondence establishes a 1–1 correspondence between valuation near orders (valuation rings) and equivalence classes of valuations (nonarchimedean valuations).

THEOREM 6. *([10, Theorems 5 and 6]) $\mathcal{T} \longmapsto \{A : A$ is a preorder of K and a $\mathcal{T}$-bounded $\mathcal{T}$-neighborhood of zero$\}$ is a 1–1 correspondence between the set of all nondiscrete Hausdorff locally bounded ring topologies on a field K and the set of all classes of equivalent proper preorders of K, and*

$$\{B : A \text{ and } B \text{ are equivalent preorders of } K\} \longmapsto \mathcal{T}_A$$

is the inverse correspondence.

Theorems 5 and 6 together establish a 1–1 correspondence between proper Hausdorff locally bounded topologies and topological equivalence classes of nontrivial prevaluations.

If A is a preorder, then $A \cup (-A)$ is an equivalent near order. Therefore, each prevaluation is topologically equivalent to a near valuation. Thus, for most purposes, it would seem that near orders and near valuations are the appropriate entities to use to study locally bounded fields (just as one normally uses valuation rings, rather than arbitrary bounded neighborhoods, to study nonarchimedean valuations). This is our motivation for using "near valuation" rather than "prevaluation" for the title.

EXAMPLE 3: Let K be any subfield of $\mathbf{R}$, and let $A = (-1/3, 1/3) \cup \{\pm 1/2, \pm 1\}$. Then A is a near order inducing the usual topology. One readily checks that the mapping $\{\pm x\} \mapsto |x|_\infty$ is an isomorphism of the value group of v_A onto the group K^+ of positive elements of K. If K^+ is provided with the partial order

$$a <_A b \quad \Leftrightarrow \quad a < \frac{1}{3} b \text{ or } a = \frac{1}{2} b,$$

then the isomorphism is order preserving. The elements of K^+ greater than 6 are the dominators of v_A.

We observe that no sphere of positive radius is either open or closed. We also observe that, although v_A is not a valuation, it is topologically equivalent to the usual absolute value.

THEOREM 7. *If v is a prevaluation on K with dominator λ and $g \in v(K^*)$, then*

$$S_{g/\lambda} \subset S_g^\circ \subset \overline{S_g} \subset C_{\lambda g};$$
$$C_{g/\lambda} \subset C_g^\circ \subset \overline{C_g} \subset C_{\lambda g}.$$

The closure (in the induced topology) of a preorder is an equivalent preorder.

EXAMPLE 4: Let K be any subfield of $\mathbf{R}$, and let $B = (-1/2, 1/2) \cup \{1\}$. Then B is not symmetric, and $B^\times = \{1\}$. The induced prevaluation v is the identity map from the field K onto K as a directed group (with zero adjoined and) with the order

$$a <_B b \quad \Leftrightarrow \quad |a| < \frac{1}{2}|b|.$$

EXAMPLE 5: Let w be a nonarchimedean valuation such that $w(\mathbf{Z}^* \cdot 1) = \{1\}$, and let

$$A = \cup_{n=0}^{\infty} (n \cdot 1 + S_1^w).$$

Then $v = v_A$ is nonarchimedean but not symmetric. For $k \in \mathbf{Z}^+$,

$$C_1^v(k \cdot 1) = k \cdot 1 + C_1^v = \cup_{n=k}^{\infty} (n \cdot 1 + S_1^w),$$

a proper subset of C_1^v.

3. Valuations

The Kowalsky-Dürbaum definition of valuation allows a presentation of valuation theory in a more unified way (cf. comments in [10, p. 140]; see also [9, Section 8]) than some standard approaches. The definition of equivalence of valuations in Section 2 is an example of this unification. Frequently, two absolute values are defined to be equivalent if they are powers of one another, and two nonarchimedean valuations are defined equivalent if they have the same associated valuation ring. Separate definitions are not necessary.

We consider here some standard theorems from our slightly more general point of view. We also give some alternate proofs, using topological methods, of standard theorems.

DEFINITION 5: A prevaluation v is called *weakly nonarchimedean* if $\mathbf{Z} \cdot 1$ is a $\mathcal{T}_v$-bounded set; v is called *archimedean* if $\mathbf{Z} \cdot 1$ is $\mathcal{T}_v$-unbounded.

THEOREM 8. *(See, e.g., [2, Chapter 1 and p. 127].) Let $|\ |$ be an absolute value on a field K.*

(1) The following statements are equivalent:
- *(a) $|\ |$ is nonarchimedean.*
- *(b) $|\ |$ is weakly nonarchimedean.*
- *(c) $|n \cdot 1| \leq 1$ for each $n \in \mathbf{Z}$.*
- *(d) $\mathcal{T}_{|\ |}$ is nonarchimedean.*

(2) The following statements are equivalent:
- *(a) $|n \cdot 1| > 1$ for some $n \in \mathbf{Z}$.*
- *(b) $|\ |$ is archimedean.*
- *(c) $|n \cdot 1| > 1$ for each integer $n \cdot 1 \neq 0, \pm 1$.*
- *(d) $\mathcal{T}_{|\ |}$ is additively generated.*

EXAMPLE 6: ([8], [15]) We use the multiplicative group of positive elements of the ordered field $\mathbf{Q}(t)$, where $0 < t < r$ for all nonnegative rational numbers r, as the range of a valuation on $\mathbf{Q}$ defined as follows: Let ($v(0) = 0$ and, for $x \neq 0$)

$$v(x) = |x|_p(1+t)^{v_q(x)},$$

where p and q are distinct primes, $|\ |_p$ is the p-adic absolute value such that $|p|_p = 1/p$, and $v_q(x)$ is the exponential q-adic valuation. Intuitively, v is an infinitesmal adjustment of the p-adic absolute value. Clearly,

$$\frac{1}{p}|x|_p < v(x) < p|x|_p.$$

Therefore,

$$v(x+y) < p|x+y|_p \leq p \max(|x|_p, |y|_p) < p^2 \max(v(x), v(y)),$$

which shows that v satisfies (V3). Another immediate consequence of the first displayed inequalities is that $\mathcal{T}_v = \mathcal{T}_p$.

All the displayed strict inequalities can be sharpened; however, v does not satisfy the ultrametric inequality, since, for example,

$$v(p + (q-p)) = v(q) > 1 = \max(v(p), v(q-p)).$$

Among the conditions in Theorem 1, v satisfies only (1)(b), (1)(d), and (2)(a).

We observe that v is not continuous when $v(\mathbf{Q}^*)$ is given its order topology. If a_n is a solution in $\mathbf{Z}$ of

$$1 + p^n x \equiv 0 \quad (q^n),$$

then $x_n = 1 + p^n a_n \longrightarrow 1$, but $v(x_n)$ is a sequence without a limit.

LEMMA 9.1. *If the topology of a valued field (K, v) is additively generated, then the value group of v is the convex subgroup generated by each dominator of v.*

PROOF: If λ is a dominator of v and the convex group $[\lambda]_{cg}$ generated by λ is a proper subset of the value group, then $B := \{x : v(x) \leq \lambda^n$, for some $n \in \mathbf{Z}\}$ is a proper open subgroup of K. •

Lemma 9.1 becomes false if the words "dominator of v" are replaced by "element other than one" (i.e., a valuation inducing an additively generated topology need not be rank 1):

EXAMPLE 7: Let F be any subfield of $\mathbf{R}$ for which there is $z \in \mathbf{C}$ such that $|z|_\infty = 1$ and z is transcendental over F. Let $K = F(z)$.

Let t be an element transcendental over $\mathbf{R}$, and let $\mathbf{R}(t)$ be made an ordered field by requiring that t be infinitely small. We define

$$v(x) = |x|_\infty (1+t)^{v_z(x)},$$

where v_z is the exponential z-adic valuation on K (and $v(0) = 0$). Then v is a rank two valuation inducing the usual topology on K.

LEMMA 9.2. *Let (K, v) be a nontrivially valued field. An element $x \in K^*$ is not neutral (i.e., x is nilpotent or inversely nilpotent) if and only if the convex group generated by $v(x)$ equals $v(K^*)$.*

PROOF: It suffices to prove the theorem for elements x such that $v(x) > 1$. In this case the convex group generated by $v(x)$ equals $v(K^*)$ if and only if, for each $v(a) \in v(K^*)$, there exists $n \in \mathbf{Z}^+$ such that $v(a) \leq v(x)^n$. The following statements are evidently equivalent:

$$v(a) \leq v(x)^n;$$
$$v(ax^{-n}) \leq 1;$$
$$x^{-n} \in a^{-1} A_v.$$

Our assumption that $1 < v(x)$ implies that, if the displayed equivalent statements hold for some n, then they hold for all integers greater than or equal to that n. Since the sets $a^{-1}A_v$, $a \in K^*$, form a neighborhood base at zero, the proof is complete. •

The following corollary follows immediately from Lemma 9.2.

COROLLARY 9.1. *Each valuation inducing the usual topology on a subfield of $\mathbf{R}$ has rank one.*

PROOF: In a subfield of $\mathbf{R}$, with a valuation inducing the usual topology, all elements other than ± 1 are nilpotent or inversely nilpotent. Thus, the value group is the convex subgroup generated by each element other than the identity. This implies the value group has rank one.

THEOREM 9. *Let (K, v) be a valued field.*

(1) The following statements are equivalent:

(a) v is weakly nonarchimedean.

(b) $\mathcal{T}_v$ is nonarchimedean.

(2) The following statements are equivalent:

(a) v is archimedean.

(b) $\mathcal{T}_v$ is additively generated.

(c) $(K, \mathcal{T}_v)$ is topologically isomorphic to a subfield of $(\mathbf{C}, \mathcal{T}_\infty)$.

PROOF: (1) Minimal fields are additively generated or nonarchimedean ([3, p. 37], [14, Theorem 6]), and valued fields are minimal ([9, p. 172], [14, Lemma 5]). Therefore, the equivalence of (a) and (b) in (1) follows from the equivalence of (a) and (b) in (2). We now establish (2).

(a) implies (b): If $\mathcal{T}_v$ is not additively generated, there is a bounded open additive group B in $\mathcal{T}_v$. Then, for $x \in B^*$, $x^{-1}B$ is bounded and $\mathbf{Z} \cdot 1 \subset x^{-1}B$.

(b) implies (c): If $\mathcal{T}_v$ is additively generated, then, by Lemma 9.1, the value group is the convex subgroup generated by some element $\lambda > 1$. By Lemma 9.2, $(K, \mathcal{T}_v)$ has a nonzero nilpotent element. The set M of nilpotent elements is bounded and open since $M \subset A_v$ and, for all $y \in M$, $yA_v \subset M$. Also a product of a nilpotent element and an element which is neither nilpotent nor has nilpotent inverse is again nilpotent. Now, [7, Theorem 1] states that the above cited properties of M imply $\mathcal{T}_v$ is induced by an absolute value. By Ostrowski's theorem, (c) holds.

(c) implies (a) obviously. •

THEOREM 10. *(cf. [9, pp. 171–172]) Every proper valuation near order inducing a nonarchimedean topology is contained in a valuation ring inducing the same topology.*

Every topology induced by a weakly nonarchimedean valuation is induced by a nonarchimedean valuation.

PROOF: Let A be a proper valuation near order inducing a nonarchimedean topology. There is an open additive group D contained in A. Since A is bounded, D is bounded. Now $A \subset x^{-1}D$ for some x; and $x^{-1}D$ is a bounded open additive group. Since A is a multiplicative semigroup, the ring B generated by A coincides with the additive group generated by A. Thus $B \subset x^{-1}D \neq K$. Certainly B is a valuation ring, since $A \subset B$; $A \subset B$ also implies $\mathcal{T}_B \leq \mathcal{T}_A$, hence, $\mathcal{T}_B = \mathcal{T}_A$. •

Theorem 11 follows from Theorems 9 and 10:

THEOREM 11. *A Hausdorff ring topology $\mathcal{T}$ on a field K is induced by a valuation if and only if $(K, \mathcal{T})$ is topologically isomorphic to a subfield of $(\mathbf{C}, \mathcal{T}_\infty)$ or $\mathcal{T}$ is induced by a nonarchimedean valuation.*

COROLLARY 11.1. *A nontrivially valued field K is topologically isomorphic to $\mathbf{R}$, with the usual topology, if and only if K is complete and, for each near order A inducing the topology of K, $A^\times$ is finite.*

PROOF: If $K = \mathbf{R}$, then $A^\times \subset [-1, 1]$. Conversely, the conditions imply K is archimedean and $K \neq \mathbf{C}$. •

For each preorder A of K we define $M(A) := A \backslash A^\times$. Certainly the sets $M(A)$, $A^\times$, and $[M(A)^*]^{-1}$ are disjoint.

One readily sees that, for a prevaluation v,

$$\begin{aligned} M(A_v) &= \{x \in K : v(x) < 1\}; \\ A_v^\times &= \{x : v(x) = 1\}; \\ [M(A_v)^*]^{-1} &= \{x \in K : v(x) > 1\}. \end{aligned}$$

Thus, v is a valuation if and only if

$$K = M(A_v) \cup A_v^\times \cup [M(A_v)^*]^{-1} \qquad \text{(disjoint)}.$$

Since every preorder A is the preorder of some prevaluation, a product of elements in $M(A)$ is again in $M(A)$.

For each ***absolute value*** v one sees also that $M(A_v)$ is the set of nilpotent elements, $A_v^\times$ is the set of neutral elements, and $[M(A_v)^*]^{-1}$ is the set of inversely nilpotent elements.

THEOREM 12. *Let $|\ |_1$ and $|\ |_2$ be nontrivial absolute values on a field K with valuation near orders A_1 and A_2, and induced topologies $\mathcal{T}_1$ and $\mathcal{T}_2$, respectively. The following statements are equivalent:*

(1) $|\ |_1$ and $|\ |_2$ are equivalent.

(2) $A_1 \subset A_2$.

(3) $\mathcal{T}_1 = \mathcal{T}_2$.

(4) $\mathcal{T}_1 \subset \mathcal{T}_2$.

(5) $\mathcal{T}_1$ and $\mathcal{T}_2$ are not independent.

(6) $M(A_1) = M(A_2)$.

(7) $M(A_1) \subset M(A_2)$.

(8) There exists an order isomorphism σ from the multiplicative group $|K^*|_1$ onto $|K^*|_2$ such that $|\ |_2 = \sigma \circ |\ |_1$.

(9) There exists a positive real number r such that $|\ |_2 = |\ |_1^r$.

PROOF: (1) implies (2) obviously.

(2) implies (3): If $A_1 \subset A_2$, then $\mathcal{T}_2 \leq \mathcal{T}_1$ since all nonzero multiples of a near order form a base at zero for the topology it induces. But a topology induced by a nontrivial valuation is minimal, so $\mathcal{T}_2 = \mathcal{T}_1$.

(3), (4), and (5) are equivalent: Obviously, (3) implies (4). Since comparable topologies are not independent ([17, 1.3(b)]), (4) implies (5); and, since distinct topologies induced by valuations are independent ([17, 2.3]), (5) implies (3).

(3) implies (6): $M(A_1) = M(A_2)$, since, by our observations above, both of these sets coincide with the set of elements nilpotent with respect to the common induced topology.

(6) implies (7) obviously.

(7) implies (9): Use standard methods (see, e.g., [2, p. 17]).

(8) is equivalent to (9) easily.

(9) implies (1) obviously. •

The following example shows that the statement $A_1^\times \subset A_2^\times$ cannot be added to the list of equivalences in Theorem 12: $A_\infty^\times$, the set of neutral elements of $\mathbf{Q}$ with the usual absolute value, equals $\{\pm 1\}$. Therefore, $A_\infty^\times$ is contained in $A_p^\times$, the set of neutral element of $\mathbf{Q}$ with a p-adic absolute value.

However, if we consider only nontrivial nonarchimedean absolute values, then $A_i^\times$ is a nonempty $\mathcal{T}_i$-open subset with nonempty open complement. Thus, if we restrict our attention to nonarchimedean absolute values, $A_1^\times \subset A_2^\times$ can be appended to the list of equivalences in Theorem 12, since the containment $A_1^\times \subset A_2^\times$ implies $\mathcal{T}_1$ and $\mathcal{T}_2$ are not independent.

For arbitrary valuations, the relationship between the statements in Theorem 12 is as follows:

THEOREM 13. *Let v_1 and v_2 be nontrivial valuations on a field K with valuation near orders A_1 and A_2 and induced topologies $\mathcal{T}_1$ and $\mathcal{T}_2$, respectively. The following statements are equivalent:*

(1) v_1 and v_2 are equivalent.

(6) $M(A_1) = M(A_2)$.

(8) There exists an order isomorphism σ from $v_1(K^*)$ onto $v_2(K^*)$ such that $v_2 = \sigma \circ v_1$.

Statements (2) and (7′) below are equivalent and are implied by the statements above:

(2) $A_1 \subset A_2$.

(7′) $M(A_2) \subset M(A_1)$.

The statements below are equivalent to each other, and are implied by all of the statements above:

(3) $\mathcal{T}_1 = \mathcal{T}_2$.

(4) $\mathcal{T}_1 \subset \mathcal{T}_2$.

(5) $\mathcal{T}_1$ and $\mathcal{T}_2$ are not independent.

PROOF: (1) implies (6) obviously; and (6) implies (1), according to the displayed decomposition before Theorem 12.

(8) implies (1): $v_1(x) \le 1$ if and only if $v_2(x) = \sigma(v_1(x)) \le \sigma(1) = 1$.

(1) and (6) imply (8): Let i equal 1 or 2. Let G_i be the value group of v_i. Then $v_i : K^* \longrightarrow G_i$ is a homomorphism with kernel $U = A_1^\times = A_2^\times$. Let $\hat{v}_i$ be the associated isomorphism from K^*/U onto G_i, and let $\sigma = \hat{v}_2 \circ \hat{v}_1^{-1}$.

(2) is equivalent to (7′), according to the displayed decomposition before Theorem 12.

(3), (4), and (5) are proved equivalent as in Theorem 12.

(1) implies (2) obviously.

(2) implies (3) is proved as in Theorem 12. •

Standard examples of nonarchimedean valuations show that neither implication between the groups of statements in Theorem 13 reverses. One can easily find examples showing that only (1) and (8) are equivalent for arbitrary near valuations.

A prevaluation on a field is determined by its values on any subring whose quotient is the field.

THEOREM 14. *Let v be a prevaluation on the commutative integral domain R. Then there is exactly one prevaluation w on the quotient field K whose restriction to R is v.*

Moreover, v and w have the same set of dominators and the same value group. Thus, if v is nonarchimedean, then so is w. If v is a valuation on the ring R, then w is a valuation.

PROOF: Certainly

$$w(a/b) = w(a)w(b)^{-1} = v(a)v(b)^{-1},$$

for $a \in R$ and $b \in R^*$. We verify that w satisfies (V3): If λ is a dominator of v and

$$w\left(\frac{a}{b}\right), w\left(\frac{c}{d}\right) \le g,$$

then multiplying by $w(bd)$ (and later dividing) on both sides of this inequality yields

$$v(ad), v(bc) \le gv(bd);$$
$$v(ad + bc) \le \lambda gv(bd);$$
$$w\left(\frac{a}{b} + \frac{c}{d}\right) = w\left(\frac{ad+bc}{bd}\right) \le \lambda g. \bullet$$

Not every discrete prevaluation is a valuation. For example, we may modify a discrete valuation by reordering the value group. Let $\{c^k : k = 0 \text{ or } k \ge n\}$, where c^{-1} is the value of a uniformizer and $n > 1$, be the positive cone. However, each discrete prevaluation is obtained, as above, by modifying the ordering of the value group of a discrete valuation:

LEMMA 15.1. *A subsemigroup P with identity is the positive cone of a directed infinite cyclic group G if and only if there is a generator c of G and a nonnegative integer n such that*

$$\{c^k : k \geq n\} \subset P \subset \{c^k : k \geq 0\}.$$

Each order making G a directed group has a unique extension to a total order.

PROOF: Let P be a positive cone making G a directed group, and let c be the generator of G such that P consists of only nonnegative powers of c. If, for some $r > 1$, $P \subset \{c^{kr} : k \geq 0\}$, then no element would be bigger than both 1 and c. Since an additive subsemigroup of $\mathbf{Z}^+$ not contained in a proper ideal of $\mathbf{Z}$ is a cofinite subset of $\mathbf{Z}^+$, P satisfies the desired containments. •

THEOREM 15. *If $v : K \longrightarrow (G, \leq) \cup \{0\}$ is a discrete prevaluation, then ιv, where ι is the identity map from $(G, \leq)$ to G with the total order extending $\leq$ (and $\iota(0) = 0$), is a nonarchimedean valuation topologically equivalent to v.*

Thus, the equivalence classes of discrete prevaluations on K are in 1–1 correspondence with the set of pairs $(v, \leq)$, where v is a member of a maximal collection of pairwise inequivalent discrete absolute values on K and $\leq$ is an order which makes $\mathbf{Z}$ a directed group (under addition).

PROOF: ιv is a valuation with dominator λc^n, where λ is a dominator of v and n is as in Lemma 15.1. Viewing ιv as a real valuation, we may apply Artin's theorem to obtain that a power of ιv is a (necessarily nonarchimedean) absolute value. Then ιv is also nonarchimedean. •

REFERENCES

1. E. Artin, *Theory of algebraic numbers*, Göttinger, 1959.
2. G. Bachman, *Introduction to p-adic numbers*, Academic Press, New York, 1964, **MR** 30 #90.
3. E. Correl, *On Topologies for Fields*, Ph.D. Dissertation, Purdue University, Lafayette, Indiana, 1958.
4. L. Fuchs, *The generalization of the valuation theory*, Duke Math. J., **18** (1951), 19–26, **MR** 12 669.
5. L. Fuchs, *Partially ordered algebraic systems*, Pergamon Press, Oxford, 1963, **MR** 30 #2090.
6. P. Jaffard, *Corps demi-valués*, C. R. Acad. Sci. Paris, **231** (1950), 1401–1403, **MR** 12 475.
7. I. Kaplansky, *Topological methods in valuation theory*, Duke Math. J., **14** (1947), 527–541, **MR** 9 172.

8. B. Klotzek and H.J. Weinert, *Bemerkungen zur Theorie der bewerteten Körper*, Math. Nachr., **39** (1969), 97–115, **MR** 39 #1438.

9. H.J. Kowalsky, *Beiträge zur topologischen algebra*, Math. Nachr., **9** (1954), 143–185, **MR** 15 774.

10. H.J. Kowalsky and H. Dürbaum, *Arithmetische Kennzeichnung von Korpertopologien*, J. Reine Angew. Math., **191** (1953), 135–152, **MR** 15 98.

11. W. Krull, *Algemeine Bewertungenstheorie*, J. Reine Angew. Math., **167** (1932), 160–196.

12. J. Kürschak, *Über Limesbildung und allgemeine Körpertheorie*, J. Reine Angew. Math., **142** (1913), 211–253.

13. K. Mahler, *Über Pseudobewertungen I*, Acta Math., **66** (1936), 79–119.

14. A.F. Mutylin, *Completely simple commutative topological rings*, Mat. Zametki, **5** (1969), 161–171 = Math. Notes, **5** (1969), 100–105, **MR** 39 #4232.

15. M. Nagata, *A generalization of the notion of valuation*, Amer. J. Math., **101** (1979), 245-257, **MR** 80i:12017.

16. T. Nakano, *On the locally bounded fields*, Comment. Math. Univ. St. Paul., **9** (1961), 77–85, **MR** 24 #A2579.

17. H. Weber, *Zu einem Problem von H.J. Kowalsky*, Abh. Braunschweig Wiss. Gesellsch., **24** (1978), 127–134, **MR** 82m:12019.

18. H. Weber, *Topologische Charakterisierung globaler Körper und algebraischer Funktionenkörper in einer Variablen*, Math. Z., **169** (1979), 167–177, **MR** 80k:12037.

19. W. Wiesław, *Topological Fields*, Acta Univ. Wratislav. Mat. Fiz. Astronom. No. 43, 1982, **MR** 85f:12012.

20. D. Zelinsky, *Topological characterization of fields with valuations*, Bull. Amer. Math. Soc., **54** (1948), 175–183, **MR** 10 426.

The Dual of a Refinement Algebra

Rae Michael Shortt Wesleyan University, Middletown, Connecticut
K. P. S. Bhaskara Rao Indian Statistical Institute, Bangalore, India

§ 0. Introduction

The notion of cardinal algebra provides a common generalization for a number of important mathematical structures: non–negative real numbers under addition, σ–distributive σ–lattices under the operation of supremum, sets of non–negative measurable functions and countably additive measures on a measurable space under point–wise summation, and sets of Borel–isomorphism types under topological (or Borel) sum. As developed by Tarski and set forth in his masterful book [11], the distinctive feature of this theory is its essential use of an infinitary operation corresponding in most examples to summation or countable supremum. A number of applications have been assayed in the areas of algebra, descriptive set theory and measure theory, although the theory has in recent years settled into a modest quiescence. A sample set of references might include [1] – [8] and of course [11].

In [9] and [10] were offered the beginnings of a theory of countably additive functionals (cardinal algebra homomorphisms into $[0,\infty]$) on a cardinal algebra. If A is a cardinal algebra, the aggregate A^* of all such functionals forms a cardinal algebra, but may not be large enough to separate points of A. See [9]. Because of this and for reasons of mathematical instinct, it seems reasonable to consider the set $A_* \supseteq A^*$ of all finitely additive functionals on A. (The elements of A^* and A_* are naturally termed "measures" and "charges," respectively.) Unfortunately, A_* is not always a cardinal algebra (again [9]). It is therefore also reasonable to widen one's view from cardinal algebras to more general objects. We treat of "refinement algebras," which were considered by Tarski [11], and "weak cardinal algebras," which were not. The theory of refinement algebras makes no use of infinitary operations.

The principal result of this paper is proposition 2.1, to the effect that A_* is a weak cardinal algebra whenever A is, or even if A is merely a refinement algebra. The question of whether Hom (A,B) is a refinement algebra has not yet been answered.

§ 1..Definitions

A refinement algebra is, according to Tarski's definition, a set A together with a binary operation + satisfying the following five axioms:

Axiom 1: The operation + is commutative.

Axiom 2: The operation + is associative.

Axiom 3: There is an element $0 \in A$ such that $a + 0 = a$ for each $a \in A$ and such that $a + b = 0$ implies $a = 0$ for any a, b in A.

Axiom 4 (Finite refinement): If $a_1\ a_2\ b_1\ b_2$ are elements of A such that $a_1 + a_2 = b_1 + b_2$, then there are elements c_{ij} in A such that

$$a_1 = c_{11} + c_{12} \qquad b_1 = c_{11} + c_{21}$$

$$a_2 = c_{21} + c_{22} \qquad b_2 = c_{12} + c_{22}\,.$$

Axiom 5: If $a_1\ a_2\ b\ c$ are elements of A such that $(a_1 + a_2) + c = b + c$, then there are elements $b_1\ b_2\ c_1\ c_2$ in A such that

$$b = b_1 + b_2 \qquad c = c_1 + c_2$$

$$a_1 + c_1 = b_1 + c_1 \qquad a_2 + c_2 = b_2 + c_2\,.$$

Tarski includes some discussion of refinement algebras in his treatise on cardinal algebras

[11], and we shall make use of basic results concerning the arithmetic of these structures, such as are to be found in Tarski's book. In particular, it is worth noting, with Tarski, that many of the theorems he proves in the context of cardinal algebras work equally well when applied to the more general notion of refinement algebra. These he enumerates [11; 11.28].

One convenient source for examples of refinement algebras arises from the following circle of ideas. Let $(A, +)$ be a commutative semi-group with cancellation. Define a relation $\leq$ on A by setting $a \leq b$ in case $a + c = b$ for some $c \in A$.

<u>1.1 Lemma</u>: In the context outlined, if $\leq$ partially orders A, and $(A, \leq)$ forms a lattice, then $(A, +)$ is a refinement algebra.

<u>Indication</u>: See 13.20 and 13.24 in [11].

A <u>cardinal algebra</u> is defined as a set A together with a binary operation $+$ and an operation of countable sum

$$(a_0, a_1, a_2, \dots) \longrightarrow \sum_0^\infty a_i = a_0 + a_1 + a_2 + \dots$$

such that the following conditions are satisfied.

<u>Postulate 1</u>: If a_0 a_1 ... are elements of A, then

$$\sum_0^\infty a_i = a_0 + \sum_0^\infty a_{i+1}.$$

<u>Postulate 2</u>: If a_i and b_i are sequences in A, then

$$\sum_0^\infty (a_i + b_i) = \sum_0^\infty a_i + \sum_0^\infty b_i .$$

Postulate 3: There is an element $0 \in A$ such that $a + 0 = 0 + a = a$ for each $a \in A$.

Postulate 4 (Refinement): If $a, b, c_0, c_1, \ldots$ are elements of A, and

$$a + b = \sum_0^\infty c_i ,$$

then there are sequences a_i and b_i in A such that

$$a = \sum_0^\infty a_i \qquad b = \sum_0^\infty b_i$$

and $a_n + b_n = c_n$ for $n = 0, 1, \ldots$.

Postulate 5 (Remainder): If a_i and b_i are sequences in A, and $a_n = b_n + a_{n+1}$ for $n = 0, 1, \ldots$, then there is some $c \in A$ such that

$$a_n = c + \sum_{i=0}^\infty b_{n+i}$$

for $n = 0, 1, \ldots$.

Once again, Tarski's book [11] on the subject is the central reference point for the theory of such structures. A notion he does not discuss is that of a weak cardinal algebra, which we define as a structure $(A, +, \Sigma)$ satisfying Postulates 1, 2, 3, 5 as above, together with Axiom 4 for refinement algebras. Thus, we replace the general refinement property with its weaker counterpart, finite refinement. Clearly, every cardinal algebra is a weak cardinal algebra. As with refinement algebras, many of the basic theorems concerning the arithmetic of cardinal algebras apply just as well in the context of weak cardinal algebras. We include a list of results from Tarski's book [11] which are operative for weak cardinal algebras.

Chapter 1: All results. (None uses any form of refinement: see note of p.16.)

Chapter 2: 2.1 (n and p finite), 2.2 (n finite), 2.3 – 2.17, 2.18 (n finite), 2.19 – 2.37, 2.38 (m finite), 2.39, 2.40 (m finite).

Chapter 3: 3.1 – 3.9, 3.10 (n finite), 3.11 (n finite), 3.12 (n finite), 3.13 (n finite), 3.14 (n and p finite), 3.15 (n finite), 3.16 – 3.23, 3.25, 3.26, 3.27 (m finite), 3.28 – 3.30, 3.31 (n finite), 3.32 (n finite), 3.33 (n finite), 3.34 – 3.36.

It is perhaps worth adding a remark to the effect that the results of [11] numbered 2.38, 2.40 and 3.27 fail for $m = \infty$ and that 2.18 and 3.14 fail for $n = \infty$. Whether other of the results of chapters 1–3 hold in the context of weak cardinal algebras we have not settled. It is to be noted that in a weak cardinal algebra we have

$$a + b = a + b + 0 + 0 + \dots .$$

This follows from [11; 1.8] and shows that the operation $+$ may be derived from the infinite adder Σ. This said, we record the following implication.

1.2 Lemma: Suppose that $(A, +, \Sigma)$ is a weak cardinal algebra. Then $(A, +)$ is a refinement algebra.

Indication: Commutativity for $+$ is [11; 1.13], whilst associativity is given by [11; 1.14]. From [11; 1.33] it follows that $a + b = 0$ implies $a = b = 0$. Thus Axioms 1–3 are satisfied. Axiom 4 is hypothesized. Axiom 5 follows from [11; 2.18] with $n < \infty$.

There are many examples of cardinal algebras throughout mathematics. We list a few.

1.3 Example: $A = [0, \infty]$ under ordinary real addition.

1.4 Example: $A = \{0, 1, 2, ..., \infty\}$ under ordinary addition.

1.5 Example: A is a countably complete and countably distributive lattice with Σa_n defined as the supremum of the elements a_n and $a + b$ the supremum $a \vee b$. See [11; 15.10].

1.6 Example: A is the collection of all Borel–isomorphism types of separable metric spaces: if t_1 t_2 ... are the types of spaces X_1 X_2 ..., then $t_1 + t_2 + ...$ is the type of the topological sum of the X_n. See [11; pp. 234–35] and the use of this example in [8].

1.7 Example: Let $(X, \mathscr{B})$ be a measurable space and let A be the collection of all $\mathscr{B}$-measurable functions f: $X \longrightarrow [0, \infty]$. Under point–wise addition of functions, A becomes a cardinal algebra. This was proved by Chuaqui [2].

1.8 Example: Let $(X, \mathscr{B})$ be a measurable space and let A be the set of all measures on $(X, \mathscr{B})$. A is a cardinal algebra under set–wise addition of measures. This follows from results in [9].

In a refinement algebra $(A, +)$ we make the following notational conventions. For $n = 1, 2, \ldots,$ set

$$0a = 0$$

$$na = a + a + \ldots + a \ \ (n \text{ times}) .$$

If A is also a weak cardinal algebra, put

$$\omega a = a + a + \ldots .$$

Write $a \leq b$ in case $a = b + c$ for some $c \in A$. The relation $\leq$ partially orders any refinement algebra A [11; 11.28, 1.31]. A set $I \subseteq A$ is a <u>semi–ideal</u> if it is not void and satisfies

i) $a \in I$ and $b \leq a$ together imply $b \in I$;

ii) $a, b \in I$ implies $a + b \in I$.

If A is a weak cardinal algebra and additionally

iii) $a_n \in I$ implies $a_0 + a_1 + \ldots \in I$,

then I is an <u>ideal</u> of A.

A function μ: A —> $[0, \infty]$ is a charge if

i) $\mu(0) = 0$

ii) $\mu(a + b) = \mu(a) + \mu(b)$ all $a, b \in A$.

If A is a weak cardinal algebra and also

iii) $\mu(a_0 + a_1 + \ldots) = \mu(a_0) + \mu(a_1) + \cdots$,

then μ is a measure. Let A_* be the set of all charges on A. If A is a weak cardinal algebra, let A^* be the set of all measures on A. Let 0 represent the charge whose only value is zero.

1.9 Example: Let A be a cardinal algebra. Under point–wise addition of measures, A^* becomes a cardinal algebra. This is [9; Proposition 2.1]. However, A_* is not, in general, a cardinal algebra (refinement fails). See [9; Example 2.4].

1.10 Lemma: Let I be a semi–ideal of a refinement algebra A and let μ: I —> $[0, \infty]$ be a charge on I. Define μ_0: A —> $[0, \infty]$ by setting

$\mu_0(a) = \sup \{\mu(u): u \leq a, u \in I\}$.

Then μ_0 is a charge. It is the smallest charge on A agreeing with μ on I.

Proof: Certainly $\mu_0(0) = 0$. Given a_1 and a_2 in A and $u \leq a_1 + a_2$ with $u \in I$, we may write [11; 2.4] $u = u_1 + u_2$ with $u_1 \leq a_1$ and $u_2 \leq a_2$. Then

$$\mu(u) = \mu(u_1) + \mu(u_2) \leq \mu_0(a_1) + \mu_0(a_2).$$

Taking the supremum over such elements u yields $\mu_0(a_1 + a_2) \leq \mu_0(a_1) + \mu_0(a_2)$. Now suppose that $u_1 \leq a_1$ and $u_2 \leq a_2$ with $u_1, u_2 \in I$. Then $u_1 + u_2 \in I$, and $u_1 + u_2 \leq a_1 + a_2$. So

$$\mu_0(a_1 + a_2) \geq \mu(u_1 + u_2) = \mu(u_1) + \mu(u_2).$$

Taking suprema over the various choices of u_1 and u_2 yields $\mu_0(a_1 + a_2) \geq \mu_0(a_1) + \mu_0(a_2)$. We have proved that μ_0 is a charge. The final statement of the lemma is easily verified.

Q.E.D.

1.11 Lemma: Let ρ be a charge on a refinement algebra A and suppose that $I \subseteq A$ is a semi-ideal. Define ρ_0 on A by

$$\rho_0(a) = \inf\{\rho(v) : a = u + v, \ u \in I\}.$$

Then ρ_0 is a charge on A. It is the largest charge on A which vanishes on I and is majorised by ρ.

Proof: Of course, $\rho_0(0) = 0$. Given a_1 and a_2 in A and writing $a_1 + a_2 = u + v$ with $u \in I$, we apply refinement to find elements w_{ij} such that

$$\begin{aligned} a_1 &= w_{11} + w_{12} \qquad & u &= w_{11} + w_{21} \\ a_2 &= w_{21} + w_{22} \qquad & v &= w_{12} + w_{22}. \end{aligned}$$

Since w_{11} and w_{21} belong to I, we see that

$$\rho_0(a_1) + \rho_0(a_2) \leq \rho(w_{12}) + \rho(w_{22}) = \rho(v) .$$

Taking the infimum over possible decompositions $u + v$ yields

$$\rho_0(a_1) + \rho_0(a_2) \leq \rho_0(a_1 + a_2) .$$

Next, suppose that $a_1 = u_1 + v_1$ and $a_2 = u_2 + v_2$, where $u_1, u_2 \in I$. Then $u_1 + u_2 \in I$, and

$$\rho_0(a_1 + a_2) \leq \rho(v_1 + v_2) = \rho(v_1) + \rho(v_2) .$$

Taking infima over possible decompositions of a_1 and a_2 yields

$$\rho_0(a_1 + a_2) \leq \rho_0(a_1) + \rho_0(a_2) ,$$

so that ρ_0 is a charge.

Suppose now that ρ' is a charge on A such that $\rho' \leq \rho$ and $\rho'(u) = 0$ for each $u \in I$. Given $a = u + v$ with $u \in I$, we have

$$\rho'(a) = \rho'(u) + \rho'(v) = \rho'(v) \leq \rho(v) .$$

Taking the infimum over decompositions of a gives $\rho'(a) \leq \rho_0(a)$.

Q.E.D.

Let μ be a charge on a refinement algebra A. Say that μ is semi–finite at $a \in A$ if whenever $b \leq a$ is such that $\mu(b) = \infty$, then for each N, there is some $c \leq b$ with $\mu(c) > N$. Say that μ is semi–finite if μ is semi–finite at each $a \in A$. Given a charge μ on A, let I be the collection of all $a \in A$ such that μ is semi–finite at a.

1.12 Lemma: The set I is a semi–ideal of A.

Proof: If $a \in I$, and $b \leq a$, then $b \in I$: verification is routine. If a_1 and a_2 are in I, and $b \leq a_1 + a_2$ is such that $\mu(b) = \infty$, use refinement to write $b = b_1 + b_2$ with $b_1 \leq a_1$, and $b_2 \leq a_2$. Without loss of generality, we may suppose that $\mu(b_1) = \infty$. Since $a_1 \in I$, there is, for each N, some $c \leq b_1$ with $\mu(c) > N$. Of course, $c \leq b$. Thus $a_1 + a_2 \in I$.

Q.E.D.

1.13 Lemma: Let μ be a charge on a refinement algebra A. Define

$$\mu_s(a) = \sup\{\mu(b) : b \leq a, \ \mu(b) < \infty\}$$

$$\mu_\infty(a) = \begin{cases} 0 & \mu \text{ semi-finite at } a \\ \infty & \text{otherwise.} \end{cases}$$

Then μ_s and μ_∞ are charges on A, μ_s is semi–finite, and $\mu = \mu_s + \mu_\infty$.

Note: Call μ_s the semi–finite part of μ.

Proof: Let I be the set of all $a \in A$ such that $\mu(a) < \infty$. Then I is a semi–ideal.

Let ν be the restriction of μ to I. Lemma 1.10, applied to ν, shows that μ_S is a charge on A.

Let I' be the set of all $a \in A$ at which μ is semi–finite. By lemma 1.12, I' is a semi–ideal. It follows that μ_∞ is a charge.

It is simple to check that $\mu = \mu_S + \mu_\infty$.

Q.E.D.

Let α and β be charges on a refinement algebra A. We write $\alpha \lesssim \beta$ in case $\alpha(a) \leq \beta(a)$ for each $a \in A$. It is useful to reformulate this partial order on A_* as follows.

<u>1.14</u> <u>Lemma</u>: Let α and β be charges on a refinement algebra A with $\alpha \lesssim \beta$. Then

1) there is a charge γ on A such that $\beta = \alpha + \gamma$;

2) there is a least and a greatest such charge as given respectively by the formulae

$$\gamma(a) = \sup\{\beta(b) - \alpha(b) \colon b \lesssim a, \ \alpha(b) < \infty\}$$

$$\gamma(a) = \begin{cases} \beta(a) - \alpha(a) & \text{if } \alpha(a) < \infty \\ \infty & \text{if } \alpha(b) = \infty. \end{cases}$$

<u>Indication</u>: The proof given in [9; lemma 1.9] applies to refinement algebras.

From this lemma it will follow that Tarski's algebraic ordering of A_* is identical to the ordinary point–wise ordering of functions.

1.15 Lemma: Let α and β be charges on a refinement algebra A. Then their infimum and supremum exist in $(A_*, \leq)$ and are given by the formulae

$$(\alpha \wedge \beta)(a) = \inf \{\alpha(b) + \beta(c) : a = b + c\}$$

$$(\alpha \vee \beta)(a) = \sup \{\alpha(b) + \beta(c) : a = b + c\} .$$

Indication: See [9; lemma 1.12].

1.16 Lemma: Let α, β, γ be charges on a refinement algebra A. Then

$$(\alpha \wedge \beta) + \gamma = (\alpha + \gamma) \wedge (\beta + \gamma)$$

$$(\alpha \vee \beta) + \gamma = (\alpha + \gamma) \vee (\beta + \gamma) .$$

Indication: See [9; lemma 1.13].

1.17 Lemma: Suppose that δ, ν_1, ν_2 are charges on a refinement algebra A. Suppose also that $a \in A$ is such that $(\delta \wedge \nu_1)(a) = (\delta \wedge \nu_2)(a) = 0$. Then $(\delta \wedge (\nu_1 + \nu_2))(a) = 0$.

Proof: Given $\epsilon > 0$, choose decompositions $a = b_1 + c_1$ and $a = b_2 + c_2$ such that $\delta(b_1), \nu_1(c_1), \delta(b_2)$ and $\nu_2(c_2)$ are all less than $\frac{\epsilon}{4}$. Using refinement, choose elements $a_{11}, a_{12}, a_{21}, a_{22}$ such that

$$b_1 = a_{11} + a_{12} \qquad\qquad b_2 = a_{11} + a_{21}$$

$$c_1 = a_{21} + a_{22} \qquad\qquad c_2 = a_{12} + a_{22}.$$

Then $a = (a_{11} + a_{12} + a_{21}) + a_{22}$ with

$$\delta(a_{11} + a_{12} + a_{21}) \leq \delta(b_1) + \delta(b_2) < \frac{\epsilon}{2}$$

$$(\nu_1 + \nu_2)(a_{22}) \leq \nu_1(c_1) + \nu_1(c_2) < \frac{\epsilon}{2}.$$

Since $\epsilon > 0$ was arbitrary, the lemma follows.

Q.E.D.

§ 2. The dual of a refinement algebra

In [9], it was shown that if A is a cardinal algebra, then so also is A^*, but not necessarily A_*. In this section, we prove that A_* is a weak cardinal algebra, even if A is assumed to be merely a refinement algebra. We also investigate properties of $(A_*, \leq)$ as a lattice.

2.1 Proposition: Let A be a refinement algebra and let A_* be the family of all charges on A under the operation of point–wise sum. Then A_* is a weak cardinal algebra.

Demonstration: Postulates 1–3 are more or less obvious. To prove Postulate 5, suppose that μ_n and ν_n are sequences in A_* such that $\mu_n = \mu_{n+1} + \nu_n$ for $n = 0, 1, \ldots$. Define μ on A by setting $\mu(a) = \lim \mu_n(a)$. Since $\mu_n(a)$ is a decreasing sequence, the limit exists and serves to define a charge on A. We have, for each n,

$$\mu_n = \nu_n + \mu_{n+1} = \nu_n + \nu_{n+1} + \mu_{n+2}$$

$$= \nu_n + \nu_{n+1} + \cdots + \nu_N + \mu_{N+1}$$

for each $N \geq m$. Taking $N \longrightarrow \infty$ yields

$$\mu_n = \lim_N \mu_{N+1} + \lim_N (\nu_n + \ldots + \nu_N)$$

$$= \mu + \nu_n + \nu_{n+1} + \cdots$$

as desired. We now set about the more onerous task of proving that the Refinement Postulate is satisfied. The argument is divided into three cases, depending on whether the charges involved are semi–finite.

Suppose that μ, ν and γ_0, γ_1 are charges on A such that $\mu + \nu = \gamma_0 + \gamma_1$. We must construct charges μ_0, μ_1 and ν_0, ν_1 such that

$$\mu = \mu_0 + \mu_1 \qquad \mu_0 + \nu_0 = \gamma_0$$

$$\nu = \nu_0 + \nu_1 \qquad \mu_1 + \nu_1 = \gamma_1$$

Case 1: The charge μ is semi–finite. Define $\mu_0 = \mu \wedge \gamma_0$ and $\bar{\nu}_0$ and $\underline{\nu}_0$ as, respectively, the greatest and least solutions of the equations

$$\mu_0 + \bar{\nu}_0 = \gamma_0 \qquad \mu_0 + \underline{\nu}_0 = \gamma_0 .$$

Claim 1: $\underline{\nu}_0 \leq \nu$.

Proof of claim: We have, for each $a \in A$, that

$$\underline{\nu}_0(a) = \sup \{\gamma_0(b) - \mu_0(b)\colon\ b \leq a,\ \mu_0(b) < \infty\} .$$

Choose $b \leq a$ with $\mu_0(b) < \infty$ and note that

$$\mu_0(b) + \nu_0(b) = (\mu \wedge \gamma_0)(b) + \nu(b)$$

$$= [(\mu + \nu) \wedge (\gamma_0 + \nu)](b)$$

$$\geq \gamma_0 (b) .$$

Then $\nu (a) \geq \gamma_0 (b) - \mu_0 (b)$. Taking the supremum over such elements b yields $\nu (a) \geq \nu_0 (a)$, as desired.

Now put $\nu_0 = \bar{\nu}_0 \wedge \nu$. Then

$$\gamma_0 = \mu_0 + \bar{\nu}_0 \geq \mu_0 + \nu_0 = (\mu_0 + \bar{\nu}_0) \wedge (\mu_0 + \nu)$$

$$\geq \gamma_0 \wedge (\mu_0 + \underline{\nu}_0) \qquad \text{(claim 1)}$$

$$= \gamma_0 \wedge \gamma_0 = \gamma_0 ,$$

so that $\mu_0 + \nu_0 = \gamma_0$.

Next, define ν' and μ' as the greatest solutions of

$$\nu_0 + \nu' = \nu \text{ and } \mu_0 + \mu' = \mu .$$

Claim 2: $\mu' + \nu' \geq \gamma_1$.

Proof of claim: Given $a \in A$, we need consider only the case where $\mu' (a) + \nu' (a) < \infty$. Then it must be that $\mu_0 (a) + \nu_0 (a) < \infty$ and

$$\mu' (a) = \mu (a) - \mu_0 (a) \qquad \nu' (a) = \nu (a) - \nu_0 (a) ,$$

so that

$$\mu'(a) + \nu'(a) = (\mu(a) + \nu(a)) - (\mu_0(a) + \nu_0(a))$$

$$= \gamma_0(a) + \gamma_1(a) - \gamma_0(a)$$

$$= \gamma_1(a),$$

as asserted.

Now define $\mu_1 = \mu' \wedge \gamma_1$ and $\bar{\nu}_1$ and $\underline{\nu}_1$ as the greatest and least solutions of

$$\mu_1 + \bar{\nu}_1 = \gamma_1 \qquad \mu_1 + \underline{\nu}_1 = \gamma_1 .$$

Claim 3: $\underline{\nu}_1 \leq \nu'$.

Proof of claim: The formula for $\underline{\nu}_1$ evaluated at $a \in A$ is

$$\underline{\nu}_1(a) = \sup\{\gamma_1(b) - \mu_1(b) \colon b \leq a, \mu_1(b) < \infty\} .$$

Choose $b \leq a$ with $\mu_1(b) < \infty$ and note that

$$\mu_1(b) + \nu'(b) = (\mu' \wedge \gamma_1)(b) + \nu'(b)$$

$$= [(\mu' + \nu') \wedge (\gamma_1 + \nu')](b)$$

$$\geq \gamma_1(b) \qquad \text{(claim 2).}$$

Thus $\nu'(a) \geq \gamma_1(b) - \mu_1(b)$. Taking a supremum over b gives $\nu'(a) \geq \underline{\nu}_1(a)$.

Put $\nu_1 = \overline{\nu}_1 \wedge \nu'$ and observe that

$$\gamma_1 = \mu_1 + \overline{\nu}_1 \geq \mu_1 + \nu_1 = (\mu_1 + \overline{\nu}_1) \wedge (\mu_1 + \nu')$$

$$\geq \gamma_1 \wedge (\mu_1 + \underline{\nu}_1) \qquad \text{(claim 3)}$$

$$= \gamma_1 ,$$

so that $\mu_1 + \nu_1 = \gamma_1$.

From the inequality $\nu_1 \leq \nu'$ and the equations $\nu = \nu_0 + \nu'$ and $\mu = \mu_0 + \mu'$, it follows that

$$\nu_0 + \nu_1 \leq \nu \qquad \text{and} \qquad \mu_0 + \mu_1 \leq \mu .$$

It remains to show that these last two actually hold with equality. Let δ be the least solution of $\mu_0 + \mu_1 + \delta = \mu$.

<u>Claim</u> <u>4</u>: $\delta \leq \mu'$.

<u>Proof</u> <u>of</u> <u>claim</u>: Suppose instead that $\delta(a) > \mu'(a)$ for some $a \in A$. By the definition of δ, there must be some $b \leq a$ with $\mu_0(b) + \mu_1(b) < \infty$ and

$$\mu(b) - \mu_0(b) - \mu_1(b) > \mu'(a) \geq \mu'(b) .$$

If $\mu(b) < \infty$, this yields

$$\mu'(b) = \mu(b) - \mu_0(b) > \mu'(b) ,$$

a contradiction. So $\mu(b) = \infty$. Since μ is semi–finite, there is, for each $N > 0$, some $c \leq b$ with

$$\infty > \mu(c) > N + \mu_0(b) .$$

Then

$$\mu'(a) \geq \mu'(c) = \mu(c) - \mu_0(c)$$

$$\geq \mu(c) - \mu_0(b) > N .$$

Since N is arbitrary, this forces $\mu'(a) = \infty$, a contradiction. The claim is proved.

We observe that

$$\mu_0 + (\delta \wedge \nu_0) \leq \mu_0 + \nu_0 = \gamma_0$$

$$\mu_0 + (\delta \wedge \nu_0) \leq \mu_0 + \mu' = \mu \qquad \text{(claim 4)},$$

so that

$$(*) \qquad \mu_0 + (\delta \wedge \nu_0) \leq \mu \wedge \gamma_0 = \mu_0 \leq \mu_0 + (\delta \wedge \nu_0)$$

with equality. Likewise,

$$\mu_1 + (\delta \wedge \nu_1) \leq \mu_1 + \nu_1 = \gamma_1$$

$$\mu_1 + (\delta \wedge \nu_1) = (\mu_1 + \delta) \wedge (\mu_1 + \nu_1) .$$

Noting that $\mu_1 + \delta$ satisfies the equation $\mu_0 + (\mu_1 + \delta) = \mu$, we see that

$$\mu_1 + (\delta \wedge \nu_1) \leq \mu_1 + \delta \leq \mu' .$$

Then

$$(**) \quad \mu_1 + (\delta \wedge \nu_1) \leq \mu' \wedge \gamma_1 = \mu_1 \leq \mu_1 + (\delta \wedge \nu_1)$$

with equality. From $(*)$ and $(**)$ it follows that

$$(\delta \wedge \nu_0)(a) = 0 \quad \text{if} \quad \mu_0(a) < \infty$$

$$(\delta \wedge \nu_1)(a) = 0 \quad \text{if} \quad \mu_1(a) < \infty .$$

Now suppose that for some $a \in A$, we have $\mu(a) > \mu_0(a) + \mu_1(a)$. Since μ is semi–finite, there is some $a' \leq a$ with $\mu(a') < \infty$ and

$$\mu(a') > \mu_0(a) + \mu_1(a) = \mu_0(a') + \mu_1(a') .$$

Of course, this last quantity is finite, so that (by lemma 1.17), $(\delta \wedge (\nu_0 + \nu_1))(a') = 0$. Given $\epsilon > 0$, write $a' = b + c$ with $\delta(b)$ and $(\nu_0 + \nu_1)(c)$ less than ϵ. We see that

$$\mu + (\nu_0 + \nu_1) = \delta + \mu_0 + \mu_1 + \nu_0 + \nu_1$$

$$= \delta + \gamma_0 + \gamma_1$$

$$= \delta + \mu + \nu$$

$$\geq \delta + \mu + (\nu_0 + \nu_1)$$

$$\geq \mu + (\nu_0 + \nu_1)\,,$$

so that $\delta + \mu + (\nu_0 + \nu_1) = \mu + (\nu_0 + \nu_1)$. Then $\delta(c) = 0$ and $\delta(a') = \delta(b) < \epsilon$. Now ϵ was arbitrary, so we have $\delta(a') = 0$ and $\mu(a') = \mu_0(a') + \mu_1(a')$, a contradiction. We have proved that $\mu = \mu_0 + \mu_1$.

Claim 5: $\bar{\nu}_0 + \bar{\nu}_1 \geq \nu$.

Proof of claim: Suppose rather that $\nu(a) > \bar{\nu}_0(a) + \bar{\nu}_1(a)$ for some $a \in A$. Then $\mu_0(a) + \mu_1(a) < \infty$ and

$$\bar{\nu}_0(a) = \gamma_0(a) - \mu_0(a)$$

$$\bar{\nu}_1(a) = \nu_1(a) - \mu_1(a)\,.$$

Adding these, we derive

$$\gamma_0(a) - \mu_0(a) + \gamma_1(a) - \mu_1(a) < \nu(a)$$

$$\gamma_0(a) + \gamma_1(a) < \mu_0(a) + \mu_1(a) + \nu(a)$$

$$\mu(a) + \nu(a) < \mu(a) + \nu(a),$$

a contradiction. The claim is established.

Claim 6: $\nu_0 + \nu_1 = \nu$.

Proof of claim: We calculate, using claim 5 and lemma 1.16:

$$\nu_0 + \nu_1 = (\bar{\nu}_0 \wedge \nu) + (\bar{\nu}_1 \wedge \nu')$$

$$= (\bar{\nu}_0 + \bar{\nu}_1) \wedge (\bar{\nu}_0 + \nu') \wedge (\nu + \bar{\nu}_1) \wedge (\nu + \nu')$$

$$\geq \nu \wedge (\nu_0 + \nu') \wedge \nu \wedge \nu$$

$$= \nu.$$

The reverse inequality has already been shown.

Case 2: The charges μ and ν assume only the values 0 and ∞. Define semi–ideals

$$I_1 = \{a: \mu(a) = 0\} \qquad I_2 = \{a: \nu(a) = 0\}$$

$$I = I_1 + I_2 = \{a + b: a \in I_1, b \in I_2\}.$$

We also define functions M_n and N_n on I $(n = 0, 1)$ by setting $a = b + c$, $b \in I_1$, $c \in I_2$, and putting

$$M_n(a) = M_n(b + c) = \gamma_n(c)$$

$$N_n(a) = N_n(b + c) = \gamma_n(b) .$$

Claim 1: The functions M_n and N_n are well–defined on I.

Proof of claim: Suppose for $a \in A$ that $a = b + c = b' + c'$ for $b, b' \in I_1$ and $c, c' \in I_2$. There are elements d_{ij} such that

$$b = d_{11} + d_{12} \qquad b' = d_{11} + d_{21}$$

$$c = d_{21} + d_{22} \qquad c' = d_{12} + d_{22} .$$

Then $\mu(d_{12}) + \gamma(d_{12}) = \mu(d_{21}) + \nu(d_{21}) = 0$, so that $\gamma_n(d_{12}) + \gamma_n(d_{21}) = 0$ for $n = 0, 1$. We have

$$\gamma_n(c) = \gamma_n(d_{21}) + \gamma_n(d_{22})$$

$$= \gamma_n(d_{22}) = \gamma_n(d_{12}) + \gamma_n(d_{22})$$

$$= \gamma_n(c') .$$

Likewise $\gamma_n(b) = \gamma_n(b')$, proving the claim.

Clearly, M_n and N_n are charges defined on I such that for $a \in I$

$$M_n(a) + N_n(a) = \gamma_n(a).$$

Claim 2: For each $a \in I$, we have $M_0(a) + M_1(a) = \mu(a)$ and $N_0(a) + N_1(a) = \nu(a)$.

Proof of claim: We prove the first equation, the second following analogously. Suppose first that $\mu(a) = 0$. Then $M_n(a) = \gamma_n(0) = 0$ for $n = 0, 1$ by definition. If rather $\mu(a) = \infty$, write $a = b + c$ with $b \in I_1$ and $c \in I_2$. Since $\mu(b) = 0$, we have $\mu(c) = \infty$, and also $\nu(c) = 0$. Therefore $\mu(c) + \nu(c) = \gamma_0(c) + \gamma_1(c) = \infty$. But $M_0(a) + M_1(a) = \gamma_0(c) + \gamma_1(c) = \infty$, as desired.

The refinement problem has been solved on I. We wish to extend our solutions to all of A. To this end, we define $\varphi_n, \psi_n, \tau_n$ $(n = 0, 1)$ on A by

$$\varphi_n(a) = \sup\{M_n(u)\colon\ u \le a,\ u \in I\}$$

$$\psi_n(a) = \sup\{N_n(u)\colon\ u \le a,\ u \in I\}$$

$$\tau_n(a) = \inf\{\gamma_n(v)\colon\ a = u + v,\ u \in I\}.$$

By lemmata 1.10 and 1.11, these functions are charges on A. We are now in a position to define μ_n and ν_n on A as

$$\mu_n = \varphi_n + \frac{1}{2}\tau_n \qquad\qquad \nu_n = \psi_n + \frac{1}{2}\tau_n \quad (n = 0, 1).$$

Claim 3: $\mu_n + \nu_n = \gamma_n$ for $n = 0, 1$.

Proof of claim: Given $a \in A$, suppose first that $\mu_n(a) = \infty$. Then either $\varphi_n(a) = \infty$ or $\tau_n(a) = \infty$. In the former case, we know that for any N, there is some $u \leq a$, $u \in I$ such that $M_n(u) > N$. Then $\gamma_n(a) \geq \gamma_n(u) = M_n(u) + N_n(u) > N$. Since N is arbitrary, $\gamma_n(a) = \infty$ as required. If, on the other hand, $\tau_n(a) = \infty$, then certainly $\gamma_n(a) = \infty$. One deals with the case $\psi_n(a) = \infty$ similarly.

So we may suppose $\varphi_n(a) + \psi_n(a) < \infty$. Let $\epsilon > 0$ be given. Then we write $a = u_1 + v_1 = u_2 + v_2 = u_3 + v_3$ with $u_1, u_2, u_3 \in I$ and

$$(***)\quad \begin{cases} \varphi_n(a) \geq M_n(u_1) \geq \varphi_n(a) - \epsilon \\ \psi_n(a) \geq N_n(u_2) \geq \psi_n(a) - \epsilon \\ \tau_n(a) \leq \gamma_n(v_3) \leq \tau_n(a) + \epsilon . \end{cases}$$

Use refinement to find elements d_{ijk} such that

$$u_1 = d_{111} + d_{112} + d_{121} + d_{122}$$

$$v_1 = d_{211} + d_{212} + d_{221} + d_{222}$$

$$u_2 = d_{111} + d_{112} + d_{211} + d_{212}$$

$$v_2 = d_{121} + d_{122} + d_{221} + d_{222}$$

$$u_3 = d_{111} + d_{121} + d_{211} + d_{221}$$

$$v_2 = d_{112} + d_{122} + d_{212} + d_{222} \,.$$

Thus

$$a = (d_{111} + d_{112} + d_{121} + d_{122} + d_{211} + d_{212} + d_{221}) + d_{222} \,,$$

where the first grouping belongs to I. Thus we have written $a = u + v$ in such a way that u belongs to I and inequalities $({*}{*}{*})$ hold with u and v in place of u_1, u_2 and v_3. Then

$$\begin{aligned}
\mu_n(a) + \nu_n(a) &= \varphi_n(a) + \psi_n(a) + \tau_n(a) \\
&\geq M_n(u) + N_n(u) + \gamma_n(v) - \epsilon \\
&= \gamma_n(u) + \gamma_n(v) - \epsilon = \gamma_n(a) - \epsilon \,;
\end{aligned}$$

$$\begin{aligned}
\mu_n(a) + \nu_n(a) &= \varphi_n(a) + \psi_n(a) + \tau_n(a) \\
&\leq M_n(u) + \epsilon + N_n(u) + \epsilon + \gamma_n(v) \\
&= \gamma_n(u) + \gamma_n(v) + 2\epsilon \\
&= \gamma_n(a) + 2\epsilon \,.
\end{aligned}$$

Letting $\epsilon \longrightarrow 0$ establishes the claim.

Claim 4: $\mu = \mu_0 + \mu_1$ and $\nu = \nu_0 + \nu_1$.

Proof of claim: Suppose first that $a \in I$. Then

$$\mu_0 (a) + \mu_1 (a) = M_0 (a) + M_1 (a) = \mu (a)$$

$$\nu_0 (a) + \nu_1 (a) = N_0 (a) + N_1 (a) = \nu (a) .$$

Suppose next that $a \notin I$. Then $\mu (a) = \nu (a) = \infty$. For the sake of a contradiction, suppose now that $\mu_0 (a) + \mu_1 (a) < \infty$. As in the proof of claim 3, we may, for a given $\epsilon > 0$, write $a = u_n + v_n$ with $u \in I$ and

$$\varphi_n (a) \geq M_n (u_n) \geq \varphi_n (a) - \epsilon$$

$$\psi_n (a) \geq N_n (u_n) \geq \psi_n (a) - \epsilon$$

$$\tau_n (a) \leq \gamma_n (v_n) \leq \tau_n (a) + \epsilon \qquad (n = 0, 1) .$$

Apply refinement to $u_0 + v_0 = u_1 + v_1$ to find e_{ij} such that

$$u_0 = e_{11} + e_{12} \qquad u_1 = e_{11} + e_{21}$$

$$v_0 = e_{21} + e_{22} \qquad v_1 = e_{12} + e_{22}$$

and put $u = e_{11} + e_{12} + e_{21}$, $v = e_{22}$. Then $u \in I$, and

$$\mu_0(a) + \mu_1(a) = \varphi_0(a) + \varphi_1(a) + \frac{1}{2}(\tau_0(a) + \tau_1(a))$$

$$\geq M_0(u) + M_1(u) + \frac{1}{2}(\gamma_0(v) - \epsilon + \gamma_1(v) - \epsilon)$$

$$= \mu(u) + \frac{1}{2}(\mu + \nu)(v) - \epsilon .$$

Either $\mu(u) = \infty$ or $\mu(v) = \infty$: each yields a contradiction.

A similar argument shows that $\nu_0(a) + \nu_1(a) = \nu(a)$. The claim is proved.

Case 3: There are no restrictions on μ and ν. Using lemma 1.13, we write $\mu = \mu_s + \mu_\infty$ and $\nu = \nu_s + \nu_\infty$, where μ_s and ν_s are the semi–finite parts of μ and ν and μ_∞ and ν_∞ take only the values 0 and ∞. We apply case 1 to the equation

$$(\mu_s + \nu_s) + (\mu_\infty + \nu_\infty) = \gamma_0 + \gamma_1 ,$$

noting that $\mu_s + \nu_s$ is semi–finite. We obtain charges $\alpha_0, \alpha_1, \beta_0, \beta_1$ such that

$$(\mu_s + \nu_s) = \alpha_0 + \alpha_1 \qquad (\mu_\infty + \nu_\infty) = \beta_0 + \beta_1$$

$$\gamma_0 = \alpha_0 + \beta_0 \qquad \gamma_1 = \alpha_1 + \beta_1 .$$

Now apply case 1 and case 2 to the equations of the first line to obtain charges ζ_{ij} and ξ_{ij} such that

$$\mu_s = \zeta_{11} + \zeta_{12} \qquad \alpha_0 = \zeta_{11} + \zeta_{21}$$

$$\nu_s = \zeta_{21} + \zeta_{22} \qquad \alpha_1 = \zeta_{12} + \zeta_{22}$$

$$\mu_\infty = \xi_{11} + \zeta_{12} \qquad \beta_0 = \xi_{11} + \xi_{21}$$

$$\nu_\infty = \xi_{21} + \xi_{22} \qquad \beta_1 = \xi_{12} + \xi_{22} \,.$$

We define

$$\mu_0 = \zeta_{11} + \xi_{11} \qquad \mu_1 = \zeta_{12} + \xi_{12}$$

$$\nu_0 = \zeta_{21} + \xi_{21} \qquad \nu_1 = \zeta_{22} + \xi_{22} \,.$$

Then

$$\mu_0 + \nu_0 = \gamma_0 \qquad \mu_0 + \mu_1 = \mu$$

$$\mu_1 + \nu_1 = \gamma_1 \qquad \nu_0 + \nu_1 = \nu \,.$$

Q.E.D.

It follows from the proposition that the class of refinement algebras is closed under the operation of dual–taking $A \longrightarrow A_*$. The same is of course true for the class of weak cardinal algebras.

§ 5. References

[1] Chuaqui, R., Cardinal algebras and measures invariant under equivalence relations, Transactions Amer. Math. Soc. 142 (1969) 61–79

[2] Chuaqui, R., Cardinal algebras of functions and integration, Fund. Math. 71 (1971) 77–84

[3] Clarke, A. B., On the representation of cardinal algebras by directed sums, Transactions Amer. Math. Soc. 91 (1959) 161–192

[4] Deliyannis, P. C., Group representations and cardinal algebras, Canad. Journal Math. 22 (1970) 759–772

[5] Fillmore, P. A., An archimedean property of cardinal algebras, Michigan Math. Journal 11 (1964) 365–367

[6] Fillmore, P. A., The dimension theory of certain cardinal algebras, Transactions Amer. Math. Soc. 117 (1965) 21–36

[7] Jónsson, B., The contributions of Alfred Tarski to general algebra, Journal of Symbolic Logic 51 (1986) 883–889

[8] Shortt, R. M., Measurable spaces with c.c.c., Dissertationes Math. CCLXXVII (1989) 1–39

[9] Shortt, R. M., Duality for cardinal algebras, Forum Mathematicum (to appear)

[10] Shortt, R. M., A theory of integration for cardinal algebras , Real Analysis Exchange (to appear)

[11] Tarski, A., Cardinal Algebras, Oxford University Press, New York 1949

Pseudocompactness on Groups

F. Javier Trigos-Arrieta Wesleyan University, Middletown, Connecticut

ABSTRACT

If G is a locally compact Abelian group, let G^ denote its character group. We denote by G^+, the underlying group G equipped with the weakest topology which makes every $\chi \in$ G^ continuous. Extending and improving a result first proved (by other methods) by Glicksberg in 1962, we show: if F is a subset of G, then F is pseudocompact as a subspace of G if and only if F is pseudocompact as a subspace of G^+.

AMS Clasification Numbers: 54A05, 20K45, 22B99

Key words and phrases: locally compact Abelian group; compact space; pseudocompact space.

§0. ***Introduction and Notation.***

0.1 Let **G** be a locally compact Abelian group and **G^** its character group. Denote by $\mathbf{G}^+$ the underlying group **G** equipped with the weakest topology that makes every $\chi \in$ **G^** continuous. In [4] Irving Glicksberg showed that a subset of **G** is compact as a subspace of **G** if and only if it is compact as well as a subspace of $\mathbf{G}^+$ (Theorem 1.2). His proof rests mainly on a result of Grothendieck "in which" as Glicksberg himself states "compactness in a given topology implies compactness in a stronger one" [5; Théorème 5]. In this work, we give a new proof of Glicksberg's Theorem and we go beyond this to prove a similar result involving pseudocompactness.

0.2 NOTATION. All groups considered are Abelian and all spaces are completely regular and Hausdorff. **N**, **Z**, **Q**, **R** and **T** denote the natural numbers, integers, rationals, reals and the unit circle respectively. If **A** is a set, |**A**| will denote its cardinality. For example, |**N**| = ω. **i** will denote $\sqrt{-1}$.

Taking x to be real or in **T**, we denote by $\mathbf{N}_\varepsilon(x)$, where $\varepsilon > 0$, the usual open ε-ball around x. If **G** is a locally compact Abelian group, we denote by **G^** its character group. Open intervals in **R** are denoted by]a,b[. Finally, if **X** is a space and **F** is one of its subsets, then $\mathrm{Cl}_\mathbf{X}\mathbf{F}$ will denote the closure of **F** in **X**.

If **G** is discrete, then we write $\mathbf{G}^\#$ instead of $\mathbf{G}^+$. The following result is achieved in [2]: the pseudocompact subsets of $\mathbf{G}^\#$ are finite.

0.3 REMARK. Prof. W. W. Comfort, Prof. D. Dikranjan and Prof. D. Remus have pointed out that other proofs of Glicksberg's result are available: [3; 3.4.3], [8] and [9; 3].

§1. ***Definitions.***

1.1 Let **G** be a (T_0) locally compact Abelian group and **G^** its group of continuous characters. We define a homomorphism **e** from **G** into the product of circles $\mathbf{T}^{\mathbf{G}\wedge}$ as follows: The h-th coordinate of **e**(g) is

$$\mathbf{e}(g)_h = h(g),$$

where h ∈ **G^**.

Because $G^\wedge$ separates points [6; 23.26], **e** is an embedding (i.e. a one-one, continuous homomorphism). We denote by G^+ the image of G under **e** as a subspace of $T^{G^\wedge}$. Then G^+ is a totally bounded, completely regular, topological group. We identify the elements of the groups G and G^+.

1.2 THEOREM. *A locally compact Abelian group* G *is topologically isomorphic to* G^+ *if and only if* G *is compact.*

PROOF: Sufficiency is because **e** is a one-one continuous homomorphism, hence closed by compactness of G [13; 7.9]. To prove necessity, we see first that G is totally bounded, hence it is dense in a compact group, namely its Weil completion $\overline{G}$ [1; 1.13]. Being locally compact, G has to be closed in $\overline{G}$ [6; 5.11]. It follows that $G = \overline{G}$. □

For further reference, see [7; 33.16].

1.3 DEFINITION. Let G be a locally compact Abelian group. We say that G *respects (pseudo)compactness* if for any subspace F of G the following holds: the subspace F of G is (pseudo)compact if and only if the subspace F of G^+ is (pseudo)compact.

For example: if G is discrete or compact, then it respects both compactness and pseudocompactness ([2] and 1.2 above). Also note that if $F \subseteq G$ is compact then it is homeomorphic to $F \subseteq G^+$.

1.4 In the language of the preceding definition the following will be proved as Corollary 3.9:

THEOREM. (Glicksberg 1962) *Every locally compact Abelian group respects compactness.*

§2. *Compactness and Pseudocompactness on* R

2.1 If $x \in \mathbf{R}$ we denote by $[x]$ the integer z such that $z \leq x < z+1$. We define (x) by setting $(x) = x - [x]$. (x) is called *the decimal part of* x. The following result is proved in [10]:

PROPOSITION. *Let* $\langle x_n \rangle_{n\in \mathbf{N}}$ *be a sequence of real numbers such that* $x_1 > 0$ *and* $x_{n+1} \geq (n+1)x_n$ *for all* $n \in \mathbf{N}$. *If* $\langle u_n \rangle_{n\in \mathbf{N}} \subseteq [0,1[$, *there is an* $\alpha \in \mathbf{R}$ *such that*

$$\lim_{n\to\infty} (x_n\alpha) - u_n = 0,$$

where $(x_n\alpha)$ *is the decimal part of* $x_n\alpha$.

2.2 Note that if $\mathbf{G} = \mathbf{R}$ then (in the notation of 1.1) we have $\mathbf{G}^+ = \mathbf{R}^+ = \mathbf{e}[\mathbf{R}]$.

LEMMA. *Let* $\langle x_n\rangle_{n\in\mathbf{N}}$ *be as in 2.1 with the additional requirement* $x_1 \geq 1$. *Let* $\mathfrak{I}$ *be equal to* $\{\mathbf{J} \subseteq]0,\infty[: \mathbf{J} = \{y_n : n \in \mathbf{N}\},\ y_{n+1} \geq (n+1)y_n\}$. *If* $\mathbf{K} = \{x_n : n \in \mathbf{N}\}$ *and the element 0 is not in the closure of any element* $\mathbf{J}$ *of* $\mathfrak{I}$ *as subspaces of* $\mathbf{R}^+$, *then* $\mathbf{K}$ *is closed in* $\mathbf{R}^+$.

PROOF: Note that $x_n \geq n!$ for all $n \in \mathbf{N}$, because $x_1 \geq 1$. We proceed by contradiction, supposing that p is an accumulation point of $\mathbf{K}$ in $\mathbf{R}^+$. We have two cases.

I) $p > 0$. Put $x_0 = 0$. Let $N_0 \in \omega$ be such that $x_{N_0} \leq p < x_{N_0+1}$. Define $y_n = x_{N_0+n} - p$. Let $\mathbf{J} = \{y_n : n \in \mathbf{N}\}$. Plainly, 0 is an accumulation point of $\mathbf{J}$. We will reach a contradiction if we show that $\mathbf{J} \in \mathfrak{I}$. Let $n \in \mathbf{N}$. Then

$$\begin{aligned} y_{n+1} &= x_{N_0+n+1} - p \\ &\geq (N_0+n+1)x_{N_0+n} - p \\ &= (n+1)x_{N_0+n} + N_0 x_{N_0+n} - p \\ &\geq (n+1)x_{N_0+n} + N_0(N_0+n)! - p \\ &\geq (n+1)x_{N_0+n} - (n+1)p \\ &= (n+1)(x_{N_0+n} - p) \\ &= (n+1)y_n. \end{aligned}$$

Because n was arbitrary, it follows that $\mathbf{J} \in \mathfrak{I}$.

II) $p < 0$. Choose $N_0 \in \mathbf{N}$ such that $N_0 \geq |p|$. Define $y_n = x_{N_0+n} - p$. Again 0 is an accumulation point of $\mathbf{J} = \{y_n : n \in \mathbf{N}\}$ in $\mathbf{R}^+$. But $\mathbf{J} \in \mathfrak{I}$ because, if $n \in \mathbf{N}$ then

$$\begin{aligned} y_{n+1} &= x_{N_0+n+1} - p \\ &\geq (N_0+n+1)x_{N_0+n} - p \\ &\geq (n+1)x_{N_0+n} + N_0 x_{N_0+n} - p \\ &\geq (n+1)x_{N_0+n} + N_0(N_0+n)! - p \\ &\geq (n+1)x_{N_0+n} - np - p \\ &= (n+1)y_n. \end{aligned}$$

□

2.3 LEMMA. *Let* $\mathbf{J} = \{y_n : n \in \mathbf{N}\}$ *be as in 2.2. Then* $0 \notin \mathrm{Cl}_{\mathbf{R}^+}\mathbf{J}$.

PROOF: Define $u_n = 1/2$ for all $n \in \mathbf{N}$. Use 2.1 to find $\alpha \in \mathbf{R}$ such that $(\alpha y_n) - u_n \to 0$. Define $\phi: \mathbf{R} \to \mathbf{T}$ by

$$\phi(t) = \mathbf{exp}(2\pi i\alpha t)$$

Then $\phi \in \mathbf{R}^\wedge$ and all but finitely many $\phi(y_n)$ lie in $\mathbf{N}_1(-1)$. Choose $\delta > 0$ such that

$(N_\delta(1)\backslash\{1\}) \cap \phi[J] = \varnothing$, so that **J** and $\phi^{\leftarrow}[N_\delta(1)] \backslash\{0\}$ are disjoint. Note that, since $\phi^{\leftarrow}[N_\delta(1)]$ is an open neighborhood of 0 in $\mathbf{R}^+$, the point 0 cannot be an accumulation point of **J**. □

2.4 We note that $\mathbf{R}^+$ is Lindelöf and separable [13; 16.4a, 16.6a]. Hence $\mathbf{R}^+$ is normal [13; 16.8].

LEMMA. *Let* **F** *be an unbounded subset of* **R**. *Then* **F** *as a subspace of* $\mathbf{R}^+$ *is not pseudocompact.*

PROOF: We treat the case $\mathbf{F} \subseteq [0,\infty[$. Choose $\mathbf{K} = \{x_n : n \in \mathbf{N}\} \subseteq \mathbf{F}$ as in 2.2. By 2.2 and 2.3 **K** is closed in $\mathbf{R}^+$. Let $\langle u_n \rangle_{n\in \mathbf{N}}$ be an enumeration of $\mathbf{Q} \cap]3/8,5/8[$. 2.1 assures the existence of $\alpha \in \mathbf{R}$ such that $(x_n\alpha) - u_n \to 0$. Write $\mathbf{D} = \{(x_n\alpha): n \in \mathbf{N}\}$. We claim that $\mathbf{D} \cap]3/8,5/8[$ is dense in $]3/8,5/8[$.

Let $x \in]3/8,5/8[$ and let $\varepsilon > 0$ be given. Define $\eta = (1/2)\min\{\varepsilon, x-3/8, 5/8-x\}$. There exists $N \in \mathbf{N}$ such that $n \geq N$ implies $(x_n\alpha) \in]u_n - \eta , u_n + \eta[$. Now $|\mathbf{Q} \cap]x - \eta , x + \eta[\,| = \omega$. Hence there exists $n \geq N$ such that $q = u_n \in \mathbf{Q} \cap]x - \eta , x + \eta[$. Therefore, because $|(x_n\alpha) - u_n| < \eta$ we have that $(x_n\alpha) \in]x - \varepsilon , x + \varepsilon[$. The claim is proved.

Define $\phi: \mathbf{R} \to \mathbf{T}$ by $\phi(t) = \mathbf{exp}(2\pi \mathbf{i}\alpha t)$. Then $\phi \in \mathbf{R}^\wedge$ and because $\phi[\mathbf{K}]$ is countable and dense in some neighborhood of -1 in **T** we have that $\phi[\mathbf{K}]$ is not pseudocompact. Choose $\mathbf{f} \in C(\phi[\mathbf{K}]) \setminus C^*(\phi[\mathbf{K}])$. Then $\mathbf{f} \circ \phi \in C(\mathbf{K})\setminus C^*(\mathbf{K})$. Because **K** is closed in $\mathbf{R}^+$ and this space is normal, there is a $\mathbf{g} \in C(\mathbf{R}^+)$ such that **g** extends $\mathbf{f} \circ \phi$. But then $\mathbf{g}_{|\mathbf{F}}$ is not bounded. Hence $\mathbf{F} \subseteq \mathbf{R}^+$ is not pseudocompact. □

2.5 LEMMA. *Let* **F** *be a bounded nonclosed subset of* **R**. *Then* **F** *is not pseudocompact as a subspace of* $\mathbf{R}^+$.

PROOF: We treat the case $\mathbf{F} \subseteq [0,\infty[$. Suppose that M is an upper bound for **F** and define $\phi \in \mathbf{R}^\wedge$ by $\phi(t) = \mathbf{exp}(2\pi \mathbf{i}t/(2M))$. Then $\phi[\mathbf{F}]$ and **F** are homeomorphic ($\phi_{|[0,2M]}$ is a closed map). But $\phi[\mathbf{F}]$ is not pseudocompact. Hence **F** cannot be pseudocompact as a subspace of $\mathbf{R}^+$. □

2.6 2.4 and 2.5 imply

THEOREM. **R** *respects pseudocompactness. Furthermore, if* **F** *is pseudocompact in* $\mathbf{R}^+$

then **F** *is compact in* **R**.

2.7 COROLLARY. **R** *respects compactness.*

§3. ***Compactness in General***

3.1 We need to develop some results on operations on groups, for example products, subgroups and quotients. Our preliminary lemmas are straightforward.

3.2 LEMMA. *Suppose that* **G** *and* **H** *are locally compact Abelian groups. Then* $\mathbf{id}:(\mathbf{G}\times\mathbf{H})^+ \to \mathbf{G}^+ \times \mathbf{H}^+$ *is a topological isomorphism.*

PROOF: Recall that the topology on $(\mathbf{G}\times\mathbf{H})^+$ is the weakest topology such that every $\chi \in (\mathbf{G}\times\mathbf{H})^\wedge$ is continuous. The equation $(\mathbf{G}\times\mathbf{H})^\wedge = \mathbf{G}^\wedge \times \mathbf{H}^\wedge$ holds [6; 23.18] and each factor involved contains the trivial homomorphism **1**. So the result follows. □

3.3 LEMMA. *Let* **G** *and* **H** *be locally compact Abelian groups with* **H** *a subgroup of* **G**. *Then* **H** *as a subspace of* $\mathbf{G}^+$ *is closed.*

PROOF: **H** is closed in **G** [6; 5.11]. Suppose $a \notin \mathbf{H}$. By [6; 24.12] there is $\chi \in \mathbf{G}^\wedge$ such that $\chi[\mathbf{H}] = \{1\}$ and $\chi(a) \neq 1$. Since χ is continuous on $\mathbf{G}^+$ our lemma follows. □

3.4 LEMMA. *Let* **G** *and* **H** *be as in 3.3. Then* $\mathbf{id}_{\mathbf{H}} = \mathbf{id}_{\mathbf{G}|\mathbf{H}} : \mathbf{H} \to \mathbf{H}^+$ *is a topological isomorphism from* **H** *as a subspace of* $\mathbf{G}^+$ *onto* $\mathbf{H}^+$.

PROOF: Denote by $\mathbf{H}_+$ the subspace **H** in $\mathbf{G}^+$. We want to show that $\mathbf{id}:\mathbf{H}_+ \to \mathbf{H}^+$ is a topological isomorphism. Let $\chi \in \mathbf{H}^\wedge$ and $\varepsilon > 0$ be given. Since **H** is closed in **G**, by [6; 24.12] there is $\chi' \in \mathbf{G}^\wedge$ such that χ' extends χ. Hence $\chi'^{\leftarrow}[N_\varepsilon(1)] \cap \mathbf{H}_+ = \chi^{\leftarrow}[N_\varepsilon(1)]$. So **id** is continuous. Now, if $\chi \in \mathbf{G}^\wedge$ then $\chi_{|\mathbf{H}} \in \mathbf{H}^\wedge$ which implies that **id** is an open map. This is because $\mathbf{H}^+$ has the weakest topology such that every $\phi \in \mathbf{H}^\wedge$ is continuous. □

3.5 By 3.3, the topological group $\mathbf{G}^+/\mathbf{H}$ is completely regular.

LEMMA. *Let* **G** *and* **H** *be as in 3.3. Then* $\mathbf{id}\colon \mathbf{G}^+/\mathbf{H} \to (\mathbf{G}/\mathbf{H})^+$ *is continuous.*

PROOF: It is enough to show that every $\chi \in (\mathbf{G}/\mathbf{H})^\wedge$ is continuous as a function on $\mathbf{G}^+/\mathbf{H}$. Fix $\chi \in (\mathbf{G}/\mathbf{H})^\wedge$ and let $\varepsilon > 0$ be given. If $\phi\colon \mathbf{G} \to \mathbf{G}/\mathbf{H}$ and $\phi^+\colon \mathbf{G}^+ \to \mathbf{G}^+/\mathbf{H}$ denote the natural maps, then $\chi \circ \phi \in \mathbf{G}^\wedge$, so it is continuous on $\mathbf{G}^+$. Hence $\phi^+[(\chi \circ \phi)^{\leftarrow}[N_\varepsilon(1)]]$ is open in $\mathbf{G}^+/\mathbf{H}$. But $\phi^+[(\chi \circ \phi)^{\leftarrow}[N_\varepsilon(1)]] = \chi^{\leftarrow}[N_\varepsilon(1)]$. Therefore χ is continuous on $\mathbf{G}^+/\mathbf{H}$. Because it was arbitrary, it follows that **id:** $G^+/H \to (G/H)^+$ is continuous. □

3.6 THEOREM. *Let* G_1, G_2 *be locally compact Abelian groups, each respecting compactness. Then* $G_1 \times G_2$ *also respects compactness.*

PROOF: Let $F \subseteq G_1 \times G_2$ be compact in $(G_1 \times G_2)^+$. Then F is closed in $G_1 \times G_2$. Consider the projections π_i: $(G_1 \times G_2)^+$ $(= G_1{}^+ \times G_2{}^+)$ $\to G_i{}^+$ $(i = 1, 2)$. Then $\pi_i[F]$ is compact in $G_i{}^+$, hence compact. Now F is a closed subspace of the compact space $\pi_1[F] \times \pi_2[F] \subseteq G_1 \times G_2$. Hence F itself is compact. □

3.7 COROLLARY. *Suppose* $G_1, G_2, \ldots, G_n$ *are locally compact Abelian groups,* $n \in \mathbf{N}$. *If each* G_i $(i = 1, \ldots, n)$ *respects compactness, then* $G_1 \times \cdots \times G_n$ *also respects compacteness.*

3.8 COROLLARY. *Every compactly generated locally compact Abelian group respects compactness.*

PROOF: Let G be as hypothesized. By [6; 9.8] there exist $n, m < \omega$ and a compact Abelian group K such that G is topologically isomorphic to $\mathbf{R}^n \times \mathbf{Z}^m \times K$. Now, $\mathbf{R}$, $\mathbf{Z}$ and K respect compactness by 2.7, [2; 1.3] and 1.2 respectively, so the result follows from 3.7. □

3.9 COROLLARY. *Every locally compact Abelian group respects compactness.*

This is Theorem 1.4.

PROOF: Let G be a locally compact Abelian group and let F be a compact subset of G^+. By [6; 5.14] there is a compactly generated, open and closed, locally compact subgroup H of G. We have that F is closed in G. By Lemma 3.5, $\phi^+[F]$ is compact in $(G/H)^+ = (G/H)^\#$, hence finite [2; 1.3]. This means that F hits only finitely many translates of H.

Let $x \in G$ be such that $xH \cap F \neq \emptyset$. We claim that $xH \cap F$ is compact. If not then $x^{-1}F \cap H$ is not compact either. By 3.8 $x^{-1}F \cap H$ is not compact in H^+ and by 3.4 it is not compact as a subspace of H_+. Hence $x^{-1}F \cap H$ is not compact in G^+. But $x^{-1}F$ is compact in G^+ and H is closed in G^+ (3.3), so the intersection $x^{-1}F \cap H$ is compact in G^+. This contradiction proves our claim.

Being the finite union of compact sets, F itself has to be compact. □

§4. ***Pseudocompactness.***

Our main goal in this section is generalize 1.4 to the property of pseudocompactness.

4.1 W. W. Comfort has pointed out:

THEOREM *Let* **G** *be a locally compact Abelian group. Suppose* **F** *is a closed subspace of* $\mathbf{G}^+$ *which is pseudocompact. Then* **F** *is compact.*

PROOF: By [6; 5.14] there is a σ-compact, open and closed, locally compact subgroup **H** of **G**. As in 3.9, **F** hits only finitely many translates of **H**. Note that $\mathbf{H}^+$ is σ-compact. Therefore, if $x\mathbf{H} \cap \mathbf{F} \neq \emptyset$ then $x\mathbf{H} \cap \mathbf{F}$ is σ-compact as a subspace of $\mathbf{G}^+$. Furthermore, $x\mathbf{H} \cap \mathbf{F}$ is closed in $\mathbf{G}^+$ by 3.3, whence open and closed in the subspace **F** of $\mathbf{G}^+$. It follows that $x\mathbf{H} \cap \mathbf{F}$ is pseudocompact in $\mathbf{G}^+$. But a σ-compact, pseudocompact space is always compact. Hence $x\mathbf{H} \cap \mathbf{F}$ is compact in $\mathbf{G}^+$. Being the finite union of compact subsets of $\mathbf{G}^+$, **F** itself is compact as a subspace of $\mathbf{G}^+$. By 1.4 **F** is compact. □

4.2 REMARK. It may be instructive -as the referee points out- that 3.9 and 4.1 can be proved simultaneously as follows: First we notice as in 3.9 that if **F** is compact or closed and pseudocompact in $\mathbf{G}^+$ and **H** is an open and closed subgroup of **G** then **F** is contained in a finite union of translates of **H** in **G**. Now choose **H** to be compactly generated ([6; 5.14]) and consider the subgroup **H'** of **G** generated by the union of the translates of **H** containing **F**. It follows readily that **H'** is compactly generated, whence $\mathbf{H}'^+$ is σ-compact. Because **F** is closed in $\mathbf{H}'^+$, it is σ-compact as well. Now a σ-compact, pseudocompact space is compact. Thus **F** is compact in $\mathbf{H}'^+$. Since **H'** is compactly generated it respects compactness by 3.8. Therefore **F** is compact in **H'**, hence in **G**.

The author is grateful to the referee for letting him know this argument.

4.3 COROLLARY. *If* **F** *is pseudocompact in* $\mathbf{G}^+$, *then* $\mathrm{Cl}_{\mathbf{G}}\mathbf{F} = \mathrm{Cl}_{\mathbf{G}^+}\mathbf{F}$ *and this space is compact.*

PROOF: Since every open subset of $\mathbf{G}^+$ is open in **G**, we have $\mathrm{Cl}_{\mathbf{G}}\mathbf{F} \subseteq \mathrm{Cl}_{\mathbf{G}^+}\mathbf{F}$. Clearly $\mathrm{Cl}_{\mathbf{G}^+}\mathbf{F}$ is pseudocompact in $\mathbf{G}^+$, so it is compact (4.1). Being a closed subset of $\mathrm{Cl}_{\mathbf{G}^+}\mathbf{F}$, we have that $\mathrm{Cl}_{\mathbf{G}}\mathbf{F}$ is compact. Hence it is closed in $\mathbf{G}^+$ and $\mathbf{F} \subseteq \mathrm{Cl}_{\mathbf{G}}\mathbf{F}$. Therefore $\mathrm{Cl}_{\mathbf{G}^+}\mathbf{F} \subseteq \mathrm{Cl}_{\mathbf{G}}\mathbf{F}$. □

4.4 COROLLARY. *Let* **G** *be a locally compact Abelian group. Then* **G** *respects pseudocompactness. Furthermore, if* **F** *is pseudocompact in either* **G** *or* $\mathbf{G}^+$, *then*

$$\mathbf{e}: \mathbf{F}\ (\subseteq \mathbf{G}) \rightarrow \mathbf{F}\ (\subseteq \mathbf{G}^+)$$

is a homeomorphism.

PROOF: Suppose that **F** is pseudocompact in $\mathbf{G}^+$. Then by 4.3 the space $\mathrm{Cl}_{\mathbf{G}}\mathbf{F} = \mathrm{Cl}_{\mathbf{G}^+}\mathbf{F}$ is compact. Hence $\mathbf{e}_{|\mathrm{Cl}_{\mathbf{G}}\mathbf{F}}$ is a homeomorphism. Therefore **F** is pseudocompact and the last statement follows from 4.3. □

4.5 As an easy corollary of 4.4, we have that the pseudocompact subsets of locally compact Abelian groups have compact closure.

§**5.** ***Final Remarks and Acknowledgements.***

5.1 Of course one may ask which other topological properties are preserved by the map **e**. If **X** is a Tychonoff space, we say that a subset **B** of **X** *is (functionally) bounded in* **X** if every continuous function on **X** is bounded on **B**. With the help of [12; 1.3] is possible to show that 1.4 above is true if we replace compactenss by boundedness or the Lindelöf property [11]. Furthermore, each bounded subset of a locally compact Abelian group is homeomorphic to its image under **e** and has compact closure. This latter condition can fail for the Lindelöf property (consider **G** = **R**). We also note that the set **K** constructed in §2. is countable and discrete in both **R** and $\mathbf{R}^+$ and is bounded in neither. Hence we may ask what are the subsets of locally compact Abelian groups to which the restriction of **e** is a homeomorphism.

In [11] we study some other properties of $\mathbf{G}^+$.

5.2 ACKNOWLEDGEMENT. I want to thank Prof. W. W. Comfort for several helpful conversations related to the results of this paper and for a careful reading of an early version of this manuscript. I also want to thank Prof. L. C. Robertson for taking his time answering some questions about the subject.

List of References

1. W. W. Comfort. Topological Groups. In: Handbook of General Topology, edited by

K. Kunen and J. Vaughan, pp. 1143-1263. North-Holland Publ. Co. Amsterdam. 1984.

2. W. W. Comfort and F. J. Trigos-Arrieta. Remarks on a Theorem of Glicksberg. This volume. (See also: Abstracts Amer. Math. Soc. 9 (1988), 420-421 (= Abstract #88T-22-195)).

3. D. N. Dikranjan, I. R. Prodanov and L. N. Stoyanov. Topological Groups. Pure and applied Mathematics series, Vol. 130, Marcel Dekker, New York. 1990.

4. I. Glicksberg. Uniform Boundedness for Groups. Canadian Journal of Math. 14 (1962), 269-276.

5. A. Grothendieck. Critères de Compacité dans les Espaces Fonctionelles Generaux. Amer. J. of Math. 74 (1952), 168-186.

6. E. Hewitt and K. A. Ross. Abstract Harmonic Analysis. Volume I. Grundlehren der math. Wissenschaften, Vol. 115. Springer Verlag, Berlin-Göttingen-Heidelberg. 1963.

7. E. Hewitt and K. A. Ross. Abstract Harmonic Analysis. Volume II. Grundlehren der math. Wissenschaften, Vol. 152. Springer Verlag, Berlin-New York-Heidelberg. 1970.

8. R. Hughes. Compactness in locally compact groups. Bull. Amer. Math. Soc. 79 (1973), 122-123.

9. W. Moran. On almost periodic compactifications of locally compact groups. J. London Math. Soc. (2), 3 (1971), 507-512.

10. F. J. Trigos-Arrieta. Convergence Modulo 1 and an Application to The Bohr Compactification of Z. Submitted for publication. 1989.

11. F. J. Trigos-Arrieta. Continuity, Boundedness, Connectedness and the Lindelöf Property for Topological Groups. To appear in the Journal of Pure and Applied Algebra (Proceedings of the Summer, 1989 Curaçao Conference on Locales and Topological Groups), 1991.

12. E. van Douwen. The maximal totally bounded group topology on G and the biggest minimal G-space, for Abelian groups G. Topology and its Applications. 34 (1990), 69-91.

13. S. Willard. General Topology. Addison Wesley, 1970.

Strong Versions of Normality

Scott W. Williams and Haoxuan Zhou* State University of New York, Buffalo, New York

A space X is called *monotonically normal* (or MN for brevity) provided that for each pair (H,K) of disjoint closed sets, there is an open set $m(H,K)$ satisfying two conditions:

MN 1. $H \subseteq m(H,K) \subseteq cl(m(H,K)) \subseteq X/K$.

MN 2. If $H \subseteq K_1$ and $K \subseteq K_1$, then $m(H,K) \subseteq m(H_1,K_1)$.

Monotone normality is often said to be the "strongest form of normality" since it is a hereditary property, it implies collectionwise normal, and is possessed both by stratifiable (hence metrizable) and by orderable spaces, and it is preserved by closed continuous images **[3]**.

Here, we will use an equivalent definition **[1]** of MN: For each pair (x,G) consisting of a point and its neighborhood G, there is an open set $\mu(x,G)$ satisfying two conditions:

MN 3. $x \in \mu(x,G) \subseteq G$.

MN 4. If $\mu(x,G) \cap \mu(y,H) \neq \emptyset$, then either $x \in H$ or $y \in G$.

We will call this μ a *monotone operator* (witnessing that X is MN). Note that it suffices to define $\mu(x,G)$ when all G belong to some fixed base for the topology of X.

In a sense, our paper is a continuation of **[8]**, where we strengthened monotone normality: Let us agree to call an MN space EN, for *extremely normal*, provided its monotone operator satisfies the following strengthening of condition MN4:

**Current Affiliation*: Sichuan University, Chengdu, People's Republic of China

EN 5. $\forall x,y \in X$, if $x \neq y$ and $\mu(x,G) \cap \mu(y,H) \neq \emptyset$, then

either $\mu(x,G) \subseteq H$ or $\mu(y,H) \subseteq G$.

In [8], we proved that every compact EN space is the continuous image of a compact orderable space. Here are the main results obtained in this paper:

(1.2 & 1.6) The following are equivalent. The Souslin Hypothesis, each CCC MN space is separable, each first countable CCC EN space is metrizable.

(1.4 & 2.2) An EN space is hereditarily paracompact. A locally compact scattered hereditarily paracompact space is EN. (These improve results in [4] and [7].)

(2.4) A compact space is metrizable iff it is first countable and EN.

All spaces are assumed Hausdorff and infinite. The closure and interior operators are denoted by $cl(\cdot)$ and $int(\cdot)$. $\mathbb{Z}$ is the set of integers. $|X|$ denotes the cardinality of X.

SECTION 1. MONOTONICALLY NORMAL.

1.1. Lemma. The following are true:

1. Suppose that X is an MN space, $Y \subseteq X$, $\mathcal{R}$ is an open cover of Y. Then $cl(\cup\{\mu(y,R): y \in Y, R \in \mathcal{R}\}) = cl(Y) \cup \cup\mathcal{R}$.

2. Suppose that X is an EN space, $A,Y \subseteq X$, $\mathcal{R}$ is a family of open sets. If for each $a \in A$, there is $R_a \in \mathcal{R}$ such that $\mu(a,R_a) \cap Y \neq \emptyset$, then $cl(\cup\{\mu(a,R_a): a \in A, R \in \mathcal{R}\}) \subseteq cl(Y) \cup \cup\mathcal{R}$.

Proof. (1) is proved in [8]. For (2), suppose x is a limit point of $\cup\{\mu(a,R_a): a \in A, R \in \mathcal{R}\}$, and G is a neighborhood of x. Then $\mu(x,G)$ meets some $\mu(a,R_a)$. So either $x \in \cup\mathcal{R}$ or $\mu(a,R_a) \subseteq G$, and hence, $G \cap Y \neq \emptyset$. □

1.2 Proposition. The Souslin Hypothesis is equivalent to the statement: Each CCC MN space is separable.

Proof. Of course, a Souslin line is a non-separable CCC MN space. So we prove the converse: Assume there are no Souslin trees. Suppose X is a CCC MN space. We show that X is separable. Build a tree $\mathcal{T}$ of subsets of X ordered by reverse inclusion and satisfying:

1. X is the largest element of $\mathcal{T}$.
2. $\forall T \in \mathcal{T}$, either $|T|=1$ or T has at least two successors.
3. $\forall T \in \mathcal{T}\setminus\{X\}$, $\exists$ open G_T, contained in the set of predecessors of T, and $\exists x_T \in G_T$, such that $T=\mu(x_T,G_T)$.
4. If $\mathcal{B}$ is a branch of $\mathcal{T}$, then $\mathrm{int}(\cap\mathcal{B})=\emptyset$.
5. $\forall T \in \mathcal{T}$, $\{S \in \mathcal{T}: S$ and T have the same set of predecessors$\}$ is a family maximal with respect to being open, pairwise–disjoint closures, and the closure of each member contained in the set of predecessors of T.

Since X is CCC, $\mathcal{T}$ has no uncountable anti–chains, and by (5), no branches of uncountable length. Since $\mathcal{T}$ is not a Souslin tree, $\mathcal{T}$ is countable. Let $D=\{x_T: T \in \mathcal{T}\}$. We claim D is dense in X.

For an ordinal α let $\mathcal{T}_\alpha$ be the set of elements of $\mathcal{T}$ whose set of predecessors have order type α. Suppose $x \in G = X\setminus \mathrm{cl}(D)$. From (2) and (5), $x \in \mathrm{cl}\{S: S \in \mathcal{T}_2\}$. From (3) and 1.1, $x \in \cup\mathcal{T}_1$. Suppose $\mathcal{C}$ is a chain in $\mathcal{T}$ maximal with respect to $x \in \cap\mathcal{C}$. Since $x \notin \mathrm{int}(\cap\mathcal{C})$, (5) shows $\exists \alpha<\lambda\ \exists T \in \mathcal{T}_\alpha\setminus\mathcal{C}$, $T\cap\mu(x,G)\neq\emptyset$. But $x\notin T$, therefore, MN2 shows $x_T \in G$ – a contradiction. □

The reader might be tempted to believe the class of EN spaces is well–known, for there is (an equivalence [4] to) the notion of a *proto–metrizable* space: there is a family $\mathcal{P}$ of pairs (B_1,B_2) of open sets of X satisfying two conditions:

PM 6. $\forall$ pairs (x,G) consisting of a point x and its neighborhood G there is a $(B_1,B_2)\in\mathscr{P}$ satisfying $x\in B_1 \subseteq B_2 \subseteq G$.

PM 7. If $(B_1,G),(B_3,H)\in\mathscr{P}$ and $B_1 \cap B_3 \neq \emptyset$, then either $B_1 \subseteq H$ or $B_3 \subseteq G$.

Indeed, any Hausdorff space with just finitely many non–isolated points is clearly an EN space. However, protometrizable spaces are metrizable if they are compact, or if they are separable [4].

A space is said to be LOB (for *linearly ordered base*) provided each of its points possesses a linearly ordered, by inclusion, neighborhood base [2].

1.3. Lemma. A space is proto–metrizable iff it is an LOB EN space.

Proof. The "only if" is obvious if we define $\mu(x,G)$ to be the B_1 of a pair $(B_1,B_2)\in\mathscr{P}$ satisfying $x\in B_1 \subseteq B_2 \subseteq G$. On the other hand, EN spaces are "almost" proto–metrizable – just consider all pairs $(\mu(x,G),G)$ where $x\in G$. The "almost" refers to noticing that $(\mu(x,G),G)$ and $(\mu(x,H),H)$ need not satisfy condition PM 7. However, if each $\mu(x,G)$ is taken in the linearly ordered neighborhood base at x, then we may assume either $\mu(x,G) \subseteq \mu(x,H)$ or $\mu(x,H) \subseteq \mu(x,G)$. Thus, PM 7 is satisfied. □

We imitate the proof [4] of "Every protometrizable space is paracompact" to strengthen it.

1.4 Proposition. An EN space is (hereditarily) paracompact.

Proof. Well–order $X = \{x_\alpha:\alpha\in\kappa\}$. Suppose $\mathscr{R}$ is an open cover of X. Let $p_0 = x_0$, and $\mathscr{S}_0=\{\mu(x,R):x\in X,\ p_0\in\mu(x,R),\ R\in\mathscr{R}\}$. If $\mathscr{S}_0$ covers X, stop; otherwise, find the first β such that $p_1=x_\beta \in X\setminus\cup\mathscr{S}_0$. Continue inductively defining points p_α and families $\mathscr{S}_\alpha$, so that p_α is the first x_ν not in $\cup\cup_{\beta<\alpha}\mathscr{S}_\beta$ and $\mathscr{S}_\alpha=\{\mu(x,R):\exists x\in X,\ p_\alpha\in\mu(x,R),\ R\in\mathscr{R}\}$. It must stop eventually, say at λ. Notice that $\{p_\alpha:\alpha\in\lambda\}$ is closed discrete {Fix $y \in X$. Suppose β is the first element of λ, such that $y \in \mu(x,R) \in \mathcal{S}_\alpha$, by definition $G = (\mu(x,R) \setminus \{p_\alpha\}) \cup \{y\}$ is a neighborhood of y missing $\{x_\alpha : \alpha > \beta\}$. On the other hand, if $p_\alpha \in G$ and if $\alpha < \beta$, then $p_\beta \in \mu(x,R) \in \mathcal{S}_\alpha$ – a contradiction}. Therefore, $y \notin cl(G)$. From 1.1, for $\mathcal{S}' \subseteq \cup_{\alpha \in \lambda} \mathcal{S}_\alpha$,

$$cl(\cup \mathcal{S}') \subseteq \{R \in \mathcal{R}: \exists x \in X,\ \mu(x,R) \in \mathcal{S}'\}.$$

Thus, $\mathcal{R}$ has a cushioned pair–refinement (in the sense of [5]) of $\mathcal{R}$, and so X is paracompact. □

1.5 Lemma [8]. Let X be an EN space and $\mathcal{C} = \{C_n : n \in \omega\}$ is a chain of open subsets of X, and $\{x_n : n \in \omega\} \subseteq X$ are such that $C_{n+1} \subseteq \mu(x_n, C_n)$. For $x \in \cap \mathcal{C}$ one of the following three conditions hold:

1). There is $n \in \omega$ such that $\forall m > n\ x_m = x$.

2). $\mathcal{C}$ is a neighborhood base at x.

3). $x \in int(\cap \mathcal{C})$. □

With regards to the next result, recall that any CCC MN space is hereditarily Lindelof.

1.6. Proposition. The Souslin Hypothesis is equivalent to the statement: Each first countable CCC EN space is metrizable.

Proof. Observe that the branch space of any tree is protometrizable. Hence, any Souslin tree yields a hereditarily Lindelof first countable non–metrizable EN space.

Assume the Souslin Hypothesis. Suppose X is a first countable CCC EN space. Fix a countable neighborhood base $\{N_{x,n} : n \in \omega\}$ at each $x \in X$. It suffices to prove that X has a countable base; to do this, we build a tree by induction. Let

$\mathcal{B}_0$ be a countable open covering of X, and let $\phi:\mathcal{B}_0 \to X$ be a choice function: i.e., each $\phi(B)\in B$. $\mathcal{B}_0$ is to be the zeroth level of the tree. Now fix $B\in\mathcal{B}_0$. Find $m(B) = \min\{n\in\omega: n\geq k,\ N_{\phi(B),n} \subseteq B\}$, and find a countable family $\mathcal{C}(B)$ with $\cup\mathcal{C}(B)=B\setminus\{\phi(B)\}$, and each $\forall C\in\mathcal{C}(B)$, $C=\mu(x,G)$ for some $x\in B$ and some open $G\subseteq B(A)$. Let $\mathcal{B}_1=\{C\in\mathcal{C}(B):B\in\mathcal{B}_0\} \cup \{N_{\phi(B),m(B)+1}:B\in\mathcal{B}_0\}$. $\mathcal{B}_1$ is level one of the tree, and the successors to $B\in\mathcal{B}_0$ is the family $\mathcal{C}(B)\cup N_{\phi(B),m(B)+1}$

The tree we build is a formal object, and our proof is much simpler if X is actually zero–dimensional – then we can have differing elements at the same level pairwise–disjoint; however, in the general case, two different elements of the tree may be identical as subsets of X. To proceed from level α (a family $\mathcal{B}_\alpha$) to level $\alpha+1$, we imitate the process of going from level 0 to level 1, just making certain that if $B\in\mathcal{B}_\alpha$, then $\phi(N_{\phi(B),m(B)+1}) = \phi(B)$. Now we discuss the limit ordinal case. Suppose λ is a limit ordinal, and $\forall\alpha\in\lambda$ we have the tree $\mathcal{T}=\{\mathcal{B}_\alpha:\alpha<\lambda\}$. Define

$$\mathcal{J}=\{\cap\mathcal{M}: \mathcal{M} \text{ is a maximal, by inclusion, chain in } \mathcal{T}, \text{ if } |\cap\mathcal{M}|>1\}.$$

Notice that our choice of points $\phi(B)$ for $B\in\mathcal{T}$ implies, using 1.5, each $\cap\mathcal{M}\in\mathcal{J}$ is open. So find a countable refinement $\mathcal{B}_\lambda$ of $\mathcal{J}$.

Since X is hereditarily Lindelof, $\mathcal{T}$ has no uncountable chains or antichains. Since Souslin's Hypothesis holds, $\mathcal{T}$ is countable. Notice that this same argument shows that there is a λ for which $\mathcal{J}$ is empty. Suppose $\mathcal{M}$ is a maximal, by inclusion, chain in $\{\mathcal{B}_\alpha:\alpha<\lambda\}$. If $|\cap\mathcal{M}| = \{x\}$ and $x \neq \phi(B)$ on a tail of $\mathcal{M}$, then either x is isolated, or by 1.5, $\mathcal{M}$ is a neighborhood base at x. If $|\cap\mathcal{M}| = \{x\}$ and $x = \phi(B)$ on a tail of $\mathcal{M}$, then our choice of $N_{\phi(B),m(B)+1}$) shows $\mathcal{M}$ is a neighborhood base at x. Therefore, $\{\mathcal{B}_\alpha:\alpha<\lambda\}$ is a countable base for the topology of X.

□

SECTION 2. COMPACT EXTREMELY NORMAL SPACES.

The following is true for MN as well.

2.1. Lemma. The one–point compactification of a locally compact EN space is EN.

Proof. Suppose Y is a locally compact EN non–compact space. From 1.4, Y is the pairwise–disjoint union of a family $\mathscr{S}$ of open σ–compact subspaces. $\forall\ S \in \mathscr{S}$, we may write $S = \cup\{S(n){:}n \in \mathbb{Z}\}$, where each $S(n)$ is an open relatively compact, $cl(S(n)) \subseteq S(n+1)$. We may assume that if $x \in S$, then each $\mu(x,G) \subseteq S(m) \setminus cl(S(m-2)$, where $m = \min\{n \in \omega{:} x \in S(n)\}$. Let $X = Y \cup \{\infty\}$ be the one–point compactification of X. Then ∞ has a neighborhood base consisting of sets of the form $X \setminus \cup\{S(n_X){:}S \in \mathscr{F}\}$, where $\mathscr{F}$ is a finite subset of $\mathscr{S}$. We extend μ to X by defining

$$\mu(\infty, X \setminus \cup\{S(n_S){:}S \in \mathscr{F}\}) = X \setminus \cup\{cl(S(n_S+2)){:}S \in \mathscr{F}\}.$$

It should be clear that X is EN. □

It is known that paracompact scattered spaces are ultraparacompact (i.e., each open cover is refined by a pairwise–disjoint open covering) [9]. The "only if" of the following result is known with MN as its conclusion [7].

2.2. Proposition. Suppose X is a scattered locally compact space. Then X is hereditarily paracompact iff it is EN.

Proof. The "if" follows from 1.4. Suppose X satisfies the hypothesis. Let Y be the set of all points in X possessing an open EN neighborhood. Then Y is open and, since X has isolated points, $Y \neq \emptyset$. Y is ultraparacompact, so Y is covered by an open pairwise–disjoint family $\mathscr{S}$ consisting of EN sets. Clearly Y is EN. So we are done if $X = Y$. Since X is scattered, there is $x \in X \setminus Y$ with a compact neighborhood K such that $\{x\} = int(K \setminus Y) = K \setminus Y$. From the lemma 2.1, K is EN; hence, $x \in Y$ – a contradiction. □

2.3. Lemma. A Cech complete EN space covered by nowhere dense G_δ- sets is first countable.

Proof. Suppose X satisfies the hypothesis, and $x \in X$. Write $X = \cap_n X_n$, where each X_n is open and dense in βX the Cech–Stone compactification of X. Now $x \in \cap_n G_n$, where each $G_n \subseteq X_n$ is open in βX, and $\mathrm{int}(\cap_n G_n) = \emptyset$. Define $C_0 = G_0$ and let C_{n+1} be open in βX such that $C_{n+1} \cap X = \mu(x, C_n \cap G_n \cap X)$, and hence, $\mathrm{cl}(C_{n+1}) \subseteq C_n$. Since βX is compact, $\{C_n : n \in \omega\}$ is a neighborhood base at the closed set $\cap_n C_n$. Since $\mathrm{int}(\cap_n C_n) = \emptyset$, 1.5 shows $\{x\} = \cap_n C_n$. □

2.4. Proposition. A compact space X is metrizable iff it is EN and its non–isolated points are covered by nowhere dense G_δ – sets.

Proof. The "only if" is obvious, so we prove the "if". Suppose X satisfies the hypothesis, and fix, by 2.3, a finite neighborhood base $\{N_{x,n} : n \in \omega\}$ at each $x \in X$. It suffices to prove that X has a countable base. We build a sequence $\{\mathscr{B}_k : k \in \omega\}$ of finite open covers of X such that $\cup_{k \in \omega} \mathscr{B}_k$ is a base. Let $\mathscr{B}_0$ be a finite open covering of X, and let $\phi_0 : \mathscr{B}_0 \to X$ be a choice function; i.e., each $\phi_0(B) \in B$. Now suppose $k \in \omega$ and $\forall j < k$, we have defined the coverings $\mathscr{B}_j$ and choice functions $\phi_j : \mathscr{B}_j \to X$. We define $\mathscr{B}_k$ and ϕ_k as follows: Find a countable open covering $\mathscr{A}$ of X such that each $\mathrm{cl}(a) \subseteq B(A) \in \mathscr{B}$. Now, $\forall A \in \mathscr{A}$ find a countable open cover $\mathscr{S}(A)$ of $\mathrm{cl}(A) \setminus \mu(\phi(B(A)), N_{\phi(B(A)), m(A)})$, where $m(A) = \min\{n \in \omega : n \geq k, N_{\phi(B(A)), n} \subseteq B\}$, and each $\forall S \in (A)$, $S = \mu(x, G)$ for some $x \in B(A)$ and some open $G \subseteq B(A)$. Let

$$\mathscr{B}_k = \{\mu(\phi(B(A)).N_{\phi(B(A)), m(A)}) : A \in \mathscr{A}\} \cup \{S \cap B(A) : A \in \mathscr{A}, S \in \mathscr{S}(A)\}.$$

and let $\phi_k : \mathscr{B}_k \to X$ be any choice function.

$\forall x \in X$ there is $\mathscr{C} = \{C_n : n \in \omega\} \subseteq \cup_{k\in\omega} \mathscr{B}_k$ such that $\forall n \in \omega$ $\exists x_n \in C_n$, $C_{n+1} \subseteq \mu(x_n, C_n)$, and $x \in \cap\mathscr{C}$. $\cap\mathscr{C}$ is closed, and since X is compact, $\mathscr{C}$ is a neighborhood base at $\cap\mathscr{C}$. From lemma 1.5, either $\{x\} = \cap\mathscr{C}$ or $x = x_n$ for all but finitely many n; however, in this case, $x = \phi(C_n)$ for all but finitely many n. Hence, $\mathscr{C}$ is a neighborhood base at x. □

A space X is called *orderable* provided there is a linear ordering $\leq$ of X whose induced order topology is the topology of X. $C \subseteq X$ is said to be *convex* if $a, b \in C$, $a \leq x \leq b$ implies $x \in C$.

2.5. Corollary. A compact orderable EN space is metrizable.

Proof. A hereditarily paracompact orderable space is first countable. So the result follows from 1.4 and 2.4. □

There is a simple proof **[3]** showing that each orderable space is MN. It is also known that each compact orderable space has a dense subspace which is EN (because it has a base which is a tree **[10]**). The next proposition reproves the latter fact while giving a new, though longer, proof of the former result. First we have a lemma.

2.6. Lemma. Suppose $(X, \leq)$ is a linear ordered set with the induced order topology, and $a, b \in X$, $a < b$. Then there is a continuous order-preserving function $f:[a,b] \to [0,1]$ such that $f(a) = 0$ and $f(b) = 1$.

Proof. Use any standard proof of Urysohn's lemma with $\{a\}$ and $\{b\}$ as the closed sets, and with a chain of open convex sets defining f. □

2.7. Notation. Now suppose $(X \leq)$ in lemma 2.8 is compact, and $f(x) = r$. Then $f^{-1}(r)$ is a (possibly degenerate) closed interval to be denoted by $[u(f,r), v(f,r)]$. Suppose $v(f,r) < q$. If $r \neq 1$, then $\exists$ a largest $\epsilon > 0$, $r + \epsilon < 1$, such that $f^{-1}[r, r+\epsilon) \subseteq [u(f,r), q)$.

Define $\rho(f,x,q) = f^{-1}[r,r+\epsilon)$ if $r\neq 1$; $\rho(f,x,q)=\phi$ otherwise. Suppose $p<u(f,r)$. If $r\neq 0$, then $\exists$ a largest $\delta > 0$, $r-\delta > 0$, $f^{-1}(r-\delta,r] \subseteq (p,v(f,r)]$. Define $\lambda(f,x,p)=f^{-1}(r-\delta,r]$, if $r\neq 0$; $\lambda(f,x,p)=0$ otherwise. Given an OC (for open convex) set C with end–points p and q containing $[u(f,r),v(f,r)]$, define $\mu(f,x,C)=\lambda(f,x,p)\cup\rho(f,x,q)$. Observe that if $x,y\in[a,b]$, $x\neq y$, if $f^{-1}f(x) \subseteq C_0$, $f^{-1}f(y) \subseteq C_1$, and if $\mu(f,x,C_0)\cup\mu(f,y,C_1)\neq\phi$, then there is $i\in\{0,1\}$ such that $\mu(f,y,C_i) \subseteq C_{1-i}$.

2.8. Proposition. Suppose X is a compact orderable space. Then there is a monotone operator μ on X satisfying:

1. Each $\mu(x,G)$ is open and convex.
2. EN 5 on a dense subspace of X.

Proof. We build μ inductively. Given $x\in X=[a,b]$, $f:X \to [0,1]$ as in lemma 2.6, $f(x)=r$, and given an OC set C with end–points p and q containing $[u(f,r),v(f,r)]$, define $\mu(x,C) = \mu(f,x,C)$. Now, for each $r\in[0,1]$ such that $|f^{-1}(r)| \neq 1$, let $f^{-1}(r)=[c,d]$. Suppose $y\in[c,d]$, $g:[c,d] \to [0,1]$ as in claim 1, $g(y)=s$, and suppose H is an open set containing $[u(g,s),v(g,s)]$. If $s\in(0,1)$ or if $r\in\{0,1\}$ and $s=1-r$, define $\mu(y,H)=\mu(g,y,H)$; otherwise, if $r=0$ (respectively, if $r=1$) find a largest $q>v(g,s)$ a smallest $p<u(g,s)$) such that $[v(g,s),q] \subseteq C$ ($[p,u(g,s)] \subseteq C$). Define $\mu(y,H)=\mu(g,y,H)\cup\rho(f,y,q)$ (respectively, $\mu(y,H)=\mu(g,y,H)\cup\lambda(f,y,q)$).

We are, of course, using a tree $\mathcal{T}$ to define μ. The proceeding paragraph tells how to proceed at non–limit ordinal stages. At limit ordinal stages we take each intersection I of a branch of $\mathcal{T}$ as members of $\mathcal{T}$. However, note that μ is already defined at limit ordinal stages – given an open convex set C which contains some I, then C contains some predecessor J of I. Thus $\forall z\in I$, $\mu(z,C)$ since $z\in J$.

Now suppose $D=\{x\in X:x$ is not an end–point of any element of $\mathcal{T}\}$. Then D is dense in X, and the observation in δ shows that EN 5 holds for $\mu|D$. □

REFERENCES

1. Borges, C.J. 1966, On stratifiable spaces, Pacific J. Math. **17**:1–16.

2. Davis, S. 1978, Spaces with linearly ordered local bases, Topology Proc. **3**:37–51.

3. Gruenhage, G. 1984, Generalized metric spaces, Handbook of Set–theoretic Topology, North Holland: 423–502.

4. Gruenhage, G. & Zenor, P. 1977, Notes on protometrizable spaces, Houston J. Math. **3**:47–53.

5. Michael, E. 1959, Yet another note on paracompactness, Proc. AMS **10**:309–314.

6. Nyikos, P. 1980, Order–theoretic base axioms, Surveys in General Topology, Academic Press: 367–398.

7. Nyikos, P. & Purisch, S. 1989, Monotone normality and paracompactness in scattered spaces, Annals on the New York Acad. Sci.: 124–137.

8. Purisch, S. & Williams, S. & Zhou, H., Continuous images of compact orderable spaces. To appear.

9. Telgarsky, R. 1968, Total paracompactness and paracompact dispersed spaces, Bull. Acad. Polon. Sci. Ser. Sci. Math. Astron. Phys. **16**:567–572.

10. Williams, S.W. 119–132, Spaces with dense orderable subspaces, Topology and Order Structures I, Math. Centre Tracts. **142**:27–49.

Distal Minimal Flows

Ta-Sun Wu Case Western Reserve University, Cleveland, Ohio

In memory of Doug C. McMahon

Let (X,T) be a distal minimal flow (see definition on page 3). In the present note, we shall pay specific attention to those flows whose phase groups T, which act on the phase spaces X, are elementary nilpotent Lie groups. By an elementary nilpotent group we mean a group which is isomorphic with a closed subgroup of a simply connected real nilpotent analytic group. An elementary nilpotent group also can be characterized as a torsion free compactly generated nilpotent Lie group. We shall study the structure of a distal minimal flow (X,T) where X is a finite dimensional compact Hausdorff space. In particular we would like to learn more about the case where X is locally connected (then X is a topological manifold). For simplicity, we say that a flow (X,T) is free (or effective) if T acts freely (or effectively) on X. It is well known that an almost connected locally compact group admits an effective distal action on compact spaces if and only if it is an extension of a solvable group of type R by a compact group. (We shall not use these results in this article). The interested reader may consult [MZ] for a proof of above statements and related topics). Hence much of the present work concerns solvable analytic groups.

Now we sketch the main content of this note. In section 1, we show if (M,T) is a distal flow with phase space M a topological manifold and if F is any subgroup of T, then the F–orbit closure of any element m in M is again a manifold. This fact is useful since sometimes it is convenient to construct a distal flow on a manifold for a larger group (cf. section 2). We give a necessary condition for a topological group to admit effective distal flows.

In section 2, we show how to construct free distal flows on manifolds when the phase group is an elementary nilpotent group.

The suspension of a flow is a useful tool to construct flows. In section 3, we give some basic properties of this construction.

Let (X,T) be a distal flow, which is also known as a distal transformation group. The co–dimension of (X,T) is defined as dim X – dim T. When co–dimension of (X,T) is small, it puts certain constraints on T. In section 4, we give some preliminary results in this direction.

I am very grateful to the referee and also P. Misra for many helpful suggestions in the preparation of the present article.

1: Preliminaries: We first establish our notations and summarize some basic results. We will often use X to denote the flow (X,T) as well as the phase space. The phase space of a minimal transformation group (minimal set) is always assumed to be compact and Hausdorff. In many cases the topological group T is assumed to be an analytic group (i.e., a connected Lie group). A subgroup S of an analytic group T is analytic if it is a continuous homomorphic image of an analytic group. Hence an analytic subgroup might not be a closed subgroup. It is known that they are closed in the following situations:

(1) T is simple and S is a dense normal analytic subgroup, then S = T.

(2) T is a simply connected solvable group then also for a dense normal analytic subgroup S, S = T. In fact, by a theorem of Malcev, the closure of analytic subgroup S is SA where A is a compact connected abelian subgroup of T. Because T is simply connected, it does not have any non–trivial compact elements. Therefore $\bar{S} = S$.

Let ϕ be a homomorphism from (X,T) onto (Y,T), i.e., ϕ is a continuous map from X onto Y such that $\phi(x)t = \phi(xt)$ for $x \in X$, $t \in T$. If that is the case then we shall say that (Y,T) is a *factor* of (X,T). We use the notation $(Y,T) <_{\phi} (X,T)$ or $(X,T)_{\phi} > (Y,T)$ to denote this. Let $P(\phi) = \{(x,x'): \phi(x) = \phi(x')$ and there is a net $t_\lambda \in T$ with $\lim xt_\lambda = \lim x't_\lambda\}$. $P(\phi)$ is called the relative (to ϕ) *proximal* relation. A point x in (X,T) is *ϕ–distal if* $P(\phi)[x] = \{x\}$, i.e. $(x,x') \in P(\phi)$ iff $x = x'$. If every point in X is ϕ–distal then we say ϕ is a distal homomorphism. In this case, we also say that (X,T) is a *distal extension* of (Y,T) (with respect to ϕ). In the special case where (Y,T) is a trivial flow, i.e., when Y is a singleton, (X,T) is called a distal flow or distal transformation group. Similarly, we have the definition of *relative regional proximal* relation $Q(\phi)$. Then $Q(\phi) = \{(x,x') \in X{\times}X: \phi(x) = \phi(x')$ and there exist nets x_λ, x'_λ in X, t_λ in T such that $\phi(x_\lambda) = \phi(x'_\lambda)$, $\lim x_\lambda = x$, $\lim x'_\lambda = x'$ and $\lim x_\lambda t_\lambda = \lim x'_\lambda t_\lambda\}$. If $Q(\phi)$ is trivial (i.e., $Q(\phi)[x] = \{x\}$ for all $x \in X$), we say that (X,T) is an almost periodic (relative to ϕ) extension of (Y,T).

Recall that a transformation group (X,T) is a *minimal transformation group* if and only if every orbit is dense in the phase space X, i.e. $\overline{xT} = X$ for all $x \in X$. If the phase space X of the transformation group (X,T) consists of exactly one point, then we say the transformation group is a *trivial transformation group.*

A bi–transformation group (G,X,T) *is a triple where* (X,T) *and* (G,X) *are transformation groups with the property* $(gx)t = g(xt)$ *for* $g \in G$, $x \in X$, $t \in T$, *i.e. the action of* G *on* X *commutes with the action of* T *on* X

We need the following results.

1.1 The addition theorem ([R]): *Let (X,T) be a minimal distal transformation group* and *let ϕ be a homomorphism from (X,T) onto (Y,T), i.e., $(Y,T) <_{\phi} (X,T)$. Then if "dim" denotes the covering dimension, $\dim \phi^{-1}(y)$ is constant for all $y \in Y$ and*

$$\dim Y + \dim \phi^{-1}(y) = \dim X$$

with the convention that $n + \infty = \infty$ ($n = \infty$ or an integer).

1.2 The manifold structure theorem ([R]). *Let* (X,T) *be a minimal distal transformation group and* X *be finite–dimensional with finitely many arcwise–connected components* (*or* X *is locally* tower connected). Then

(*i*) *If* $(Y,T) < (X,T)$, *then* Y *is a topological manifold.*

(*ii*) *Let* (X_0,T) *be the trivial flow. Let* (X_i,T) *be a factor of* (X,T) *which is a maximal almost periodic extension of* (X_{i-1},T). *Let* r *be the least integer such that* $(X_r,T) = (X,T)$. *Then* $r \leq Max(1, dim\ X)$ *and* r *is called the order of* (X,T) *or the height of the Furstenberg tower. The sequence* $(X_0,T) < (X_1,T) <\ldots< (X_r,T) = (X,T)$, *which is essentially unique modulo isomorphism, is called the Furstenberg tower of the canonical almost periodic tower of* (X,T). *For each* (X_i,T) *defined above there exists a minimal distal flow* (Y_i,T), *a compact Lie group* G_i *and a closed subgroup* H_i, *such that* (G_i,Y_i,T) *is a bi–transformation group. Furthermore we have:*

(*a*) $\bigcap_{g\in G_i} g^{-1}H_i g = \{e\}$, *the identity of* G_i.

(*b*) (X_i,T) *is isomorphic with the orbit space of* $(H_i\backslash Y_i,T)$, $Y_i >_{\rho_i} X_i$; (X_{i-1},T) *is isomorphic with orbit space* $(G_i\backslash Y_i,T)$, $Y_i >_{\sigma_i} X_{i-1}$. *Let* π_i *be the homomorphism from* (X_i,T) *onto* (X_{i-1},T) *which gives the almost periodic extension. Then we have the following commutative diagram*

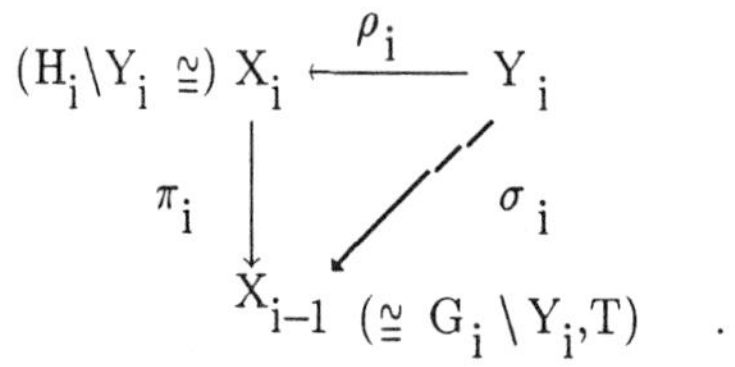

And $(Y_i,X_i,X_{i-1},G_i,H_i,\pi_i,\rho_i,\sigma_i)$ *is a fiber bundle for* $1 \leq i \leq r$. *The* Y_i'*s are manifolds* $(1 \leq i \leq r)$. *Dim* $X_{i+1} >$ *dim* X_i *unless dim* $X = 0$ (*in that case* X *is finite*).

(*c*) G_i/H_i *is connected for* $i \geq 2$ *and* G_1/H_1 *is connected if and only if* X *is connected.*

(note: a similar statement and proof can be found in ([B]).

Now, we are going to show that given any subgroup F of T and any point x in X in the above situation (1.2), the orbit closure $\overline{xF}$ is a manifold. First, we need a lemma.

1.3 Lemma: *Let (G, W, T) be a bi-transformation group where G is a compact Lie group acting freely and W is a manifold and (W, T) is a minimal distal transformation group. Suppose K is a compact subgroup of G. Let $(X, T) = (K\backslash W, T)$, X is the K-orbit space.* Let (Y,T) = (G\W,T). *Let α: $W \to X$, β: $W \to Y$ be the projections. Let w be an element in W and let $x = \alpha(w)$, $y = \beta(w)$. Suppose F is a subgroup of T such that Y'* = $\overline{yF}$ *is a manifold, then $X' = \overline{xF}$ is a manifold.*

Proof: Let W' = $\overline{wF}$. Let G' = {g ∈ G: gw' ∈ W' for some w' ∈ W'}. Note (W',F) is minimal since it is orbit closure of a distal flow. So G' is a closed subgroup of G. Since {W,Y,β,G} is a principal fiber space we have that {W',Y',β',G'} is a principal fiber space where β' is the restriction of β to W'. Then by Theorem 12 of [MOS], it is a principal fiber bundle. Since Y' is a manifold by assumption and G' is a Lie group, W' is a manifold. By Theorem 1.2(i), X' is a manifold.

1.4 Proposition: *Let (X, T) be a distal minimal flow with X a topological manifold. Let F be a subgroup of T and x any point in X. Then the $(\overline{xF}, F)$ is a distal minimal flow and $\overline{xF}$ is a manifold.*

Proof: The restriction of Furstenberg tower of (X,T) to $\overline{xF}$ gives ($\overline{xF}$,F) a tower of almost periodic extensions. (Note: This may not be the Furstenberg tower for ($\overline{xF}$,F)). Now, apply the lemma 1.3 to this tower, we see that $\overline{xF}$ is a manifold with finitely many components. ($\overline{xF}$,F) is hence a distal minimal flow since (X,T) is distal.

1.5 Proposition: *Let ϕ: $(X,H) \to (Y,H)$ be a homomorphism from a distal minimal flow*

(X,H) onto (Y,H). Let $Z = \phi^{-1}(y_0)$ for some $Y_0 \in Y$ and D be the isotropy subgroup of H at y_0. If (Y,H) is transitive, then (Z,D) is a minimal distal flow.

Proof: First, we show that (Z,D) is minimal. Let $x_0 \in \phi^{-1}(y_0)$. Let x be any point in $Z = \phi^{-1}(y_0)$. Since (Y,H) is transitive, $Y = y_0H$, and D is co–compact in H. So there is a compact subset K of H such that $H = DK$. Now, $x = \lim x_0h_\lambda = \lim x_0d_\lambda k_\lambda$ with $d_\lambda \in D$ and $k_\lambda \in K$. We may assume that $\lim k_\lambda = k$. Hence $x = \lim x_0d_\lambda k$. Apply ϕ to x, $\phi(x) = \lim \phi(x_0d_\lambda)k$. So $y_0 = y_0k$, and $k \in D$. Hence x is in the D–orbit closure of x_0. Therefore (Z,D) is minimal.

Since (X,H) is a distal flow, (Z,D) is a distal flow.

As we mentioned in the introduction, an almost connected locally compact group admits an effective distal flow if and only if it is an extension of a solvable group of type R by a compact group ([MZ]). Following is another condition for a topological group. Let T be a topological group. Let t_0 be an element of T. We say that t_0 is unstable if there exists an element t such that for every infinite sequence n_i of integers tending to ∞, we have $\lim t^{n_i} t_0 t^{-n_i} = e$ (identity of T).

1.6 Proposition: *Let (X,T) be a distal minimal flow with T a topological group and X a metric space. If t_0 is an unstable element, then t_0 acts trivially on X.*

Proof: Let $X_0 < X_1 < \cdots$ be a Furstenberg tower associated with (X,T). Suppose i is the least integer such that t_0 acts non–trivially on X_i. Let $x \in X_i$. Let D be the subgroup of T generated by t. Then $(\overline{xD}, D)$ is a distal minimal flow. Hence there exists a sequence of $n_j \to \infty$ such that $\lim xt^{n_j} = x$. Then $\lim xt^{n_j}t_0^{-1} = xt_0^{-1}$. Observe that xt^{n_j} and $xt^{n_j}t_0^{-1}$ are in the same fiber for $X_i \to X_{i-1}$. Because $\lim(xt^{n_j}t_0^{-1}, xt^{n_j})t_0t^{-n_j} = (x,x)$ and X_i is almost periodic over X_{i-1}, $\lim(xt^{n_j}t_0^{-1}, xt^{n_j}) = (x,x)$. Therefore $xt_0 = x$ and t_0 acts trivially on X_i. This is a contradiction. Since

$(X,T) = \varprojlim(X_i,T)$ (inverse limit), hence t_o acts trivially on X. The proof is now complete.

1.7 Example: Let $G = \mathbb{R} \times \mathbb{R}^+$, the affine group of the real line. The group operation is defined by the rule

$$(a,\alpha)(b,\beta) = (a + \alpha b, \alpha\beta).$$

Then every element $(a,1)$ with $a \neq 0$ is unstable. So G does not admit an effective distal minimal action. Let T be the subgroup of G formed by the elements $\{(a,\alpha^n): a \in \mathbb{R}, n \in \mathbb{Z}\}$ with α any constant different from 1. Again every element $(a,1)$ is unstable with $a \neq 0$ and T does not admit effective distal minimal actions. Observe that T is not connected.

2: Construction of distal flows: In this section, we shall show how to construct distal flows on manifolds with certain nilpotent groups as phase groups. We start with a special case first. Let N be the analytic nilpotent group of $n \times n$ upper triangular matrices over the real number field with diagonal entries 1.

Let Γ be the subgroup of N which consists of all the integral matrices in N, i.e. all the matrices in N with integers as its entries. Then Γ is a uniform subgroup of N (i.e. the right coset space $\Gamma\backslash N$ is a compact space). Let d be the diagonal matrix in $GL(n,\mathbb{R})$, $d = \mathrm{dia}(\alpha,\alpha^2,\cdots,\alpha^n)$ where α is an irrational number. Because N is a normal subgroup of the group of upper triangular matrices in $GL(n,R)$, $dNd^{-1} = N$. Therefore $d\Gamma d^{-1}$ is also a uniform subgroup of N and $\Gamma\backslash N \times d\Gamma d^{-1}\backslash N$ is a compact manifold. Let $\Gamma' = d\Gamma d^{-1}$. Let $(\Gamma\backslash N \times \Gamma'\backslash N, N)$ be the transformation group where the action of N on $\Gamma\backslash N \times \Gamma'\backslash N$ is the right multiplication on the coset space, i.e.,

$$(\Gamma n_1,\Gamma' n_2)\, n = (\Gamma n_1 n,\Gamma' n_2 n)$$

Because N is a nilpotent group, it is well known that every closed subgroup of N is subinvariant, hence Γ is subinvariant in N. This implies that $(\Gamma\backslash N,N)$ is a distal flow. Similarly $(\Gamma'\backslash N,N)$ is also a distal flow, hence $(\Gamma\backslash N \times \Gamma'\backslash N,N)$ as a product transformation group with distal factors is a distal flow.

Now, we show that N acts freely on $\Gamma\backslash N \times \Gamma'\backslash N$. Suppose $(\Gamma n_1,\Gamma' n_2)n = (\Gamma n_1,\Gamma' n_2)$. We are going to show that n must be the identity element. Since $\Gamma n_1 n = \Gamma n_1$ and $\Gamma' n_2 n = \Gamma' n_2$, this implies that $n \in n_1^{-1}\Gamma n_1$, $n \in n_2^{-1}\Gamma' n_2 = n_2^{-1} d\Gamma d^{-1} n_2$ and $n \in n_1^{-1}\Gamma n_1 \cap n_2^{-1} d\Gamma d^{-1} n_2$. Let $n = n_1^{-1}\gamma n_1 = n_2^{-1} d\gamma' d^{-1} n_2$ for some γ and γ' in Γ. Then $n_2 n_1^{-1}\gamma n_1 n_2^{-1} = d\gamma' d^{-1}$. Observe the (i,i+1) entry of $n_2 n_1^{-1}\gamma n_1 n_2^{-1}$ is same as the (i,i+1) entry of γ, i.e.

$$(n_2 n_1^{-1}\gamma n_1 n_2^{-1})_{(i,i+1)} = \gamma_{(i,i+1)}.$$

But the (i,i+1) entry of $d\gamma' d^{-1}$ is $\alpha^{-1}\gamma'_{(i,i+1)}$. Since all the entries of γ and γ' are integers, this implies that $\gamma_{(i,i+1)} = \gamma'_{(i,i+1)} = 0$ for $1 \le i \le n$. Now, suppose $\gamma_{(i,j)} = \gamma'_{(i,j)} = 0$ for $1 \le i \le n$, $i + 1 \le j \le k - 1$. Then $(n_2 n_1^{-1}\gamma n_1 n_2^{-1})_{(i,k)} = \gamma_{(i,k)}$. (One way to see this is to use the facts: (1) the inner automorphisms are unipotent, and (2) the matrices under consideration is part of a lower central series of N). On the other hand, by a simple computation we have $(d\gamma' d^{-1})_{(i,k)} = \alpha^{-k}\gamma'_{(i,k)}$. Because γ and γ' have integral entries and α^{-k} is irrational by choice, therefore we have

$$\gamma_{(i,k)} = \gamma'_{(i,k)} = 0 \quad (i < k)$$

Now, we can conclude that n must be the identity matrix and N acts on $\Gamma\backslash N \times \Gamma'\backslash N$ freely. We state the above discussion as a proposition.

2.1 Proposition: *Let T be an elementary nilpotent Lie group. Then there exists a nilmanifold M such that (M,T) is a free distal flow. Any orbit closure is a topological manifold.*

2.2. Now, we shall give a very special example of a distal flow. Let T be the discrete nilpotent group generated by the following matrices in $GL(3,\mathbb{Z})$.

$$A = \begin{bmatrix} 1 & 1 & 0 \\ 0 & 1 & 0 \\ 0 & 0 & 1 \end{bmatrix}, \quad B = \begin{bmatrix} 1 & 0 & 0 \\ 0 & 1 & 1 \\ 0 & 0 & 1 \end{bmatrix}, \quad C = \begin{bmatrix} 1 & 0 & 1 \\ 0 & 1 & 0 \\ 0 & 0 & 1 \end{bmatrix}.$$

Let $X = S' \times S'$, where S' denotes the circle group. Let α and β be any two elements in the circle group with the property that each of them generates a dense subgroup of S' and also the subgroup genrated by α and β has torsion free rank 2.

Now we define the transformation group (X,T) by the following rule.

$$(x,y)A = (\alpha x, xy)$$
$$(x,y)B = (\beta x, y).$$

Since $C = ABA^{-1}B^{-1}$, we have

$$(x,y)C = (x,y)ABA^{-1}B^{-1}$$
$$= (x, \beta y).$$

Let $\langle A\rangle$ denote the subgroup generated by A. Then $(X, \langle A\rangle)$ is the skew product and it is known that $(X, \langle A\rangle)$ is a distal minimal flow. Hence (X,T) is a minimal flow. Let ϕ be the projection from X to the first factor of the product $S' \times S'$. Then $\phi: (X,T) \to (S',T)$ is a homomorphism where the transformation group (S',T) is defined by the rule: $xA = \alpha x$, $xB = \beta x$ and $xC = x$. It is easy to see that (X,T) is a group extension of (S',T), hence (X,T) is a distal flow.

Now suppose that T does not act on X freely. Then there exists an element (x_o,y_o) in X and an element $A^mB^nC^\ell$ in T which leaves (x_o,y_o) fixed. Then $x_oA^mB^nC^\ell = x_o$. Since C acts on S' trivially, so $x_oA^mB^n = x_o$. By the choice of α and β, we have $m = n = o$. Therefore $(x_o,y_o)C^\ell = (x_o,y_o)$. But $(x_o,y_o)C^\ell = (x_o,\beta^\ell y_o) = (x_o,y_o)$, hence $\ell = 0$. We have shown that (X,T) is a free distal flow.

Now, we consider solvable analytic groups. Every solvable analytic group can be obtained from its nilradical by successive semi–direct product of one parameter groups. This is so because if T is an analytic solvable group and if N is the nilradical of T, since T/N is an abelian group, we can find one parameter subgroups $P_1,P_2,\ldots P_\ell$ of T such that $T = (N\cdot P_1)\cdot P_2)\cdots P_\ell$. In this case we have the following construction of flows based upon semi–direct product of groups.

Let T be a locally compact group. Assume $T = H \times_\eta F$, a semi–direct product with the property that $\eta(F)$ has a compact closure Φ in the group of automorphisms of H (automorphism group Aut H is topologized by Birkhoff topology). Let (X,H) and (Y,F) be flows. Define a flow $(X \times Y \times \Phi, T)$ by the following rule.

$$(x,y,\alpha)\cdot(h,f) = (xh^\alpha,yf,\alpha\eta(f)).$$

Here h^α denotes the image of h under the automorphism $\alpha \in \text{Aut } H$. It is straight forward to check that above definition gives a flow. For simplicity, we call the flow $(X \times Y \times \Phi, T)$ the *semi–direct product* of (X,H) and (Y,F) with respect to η.

2.3 Proposition: *Let (X,H) and (Y,F) be distal free flows. Let $T = H\times_\eta F$, be a locally compact group which is a semi–direct product. Assume that $\eta(F)$ has compact closure in Aut H. Then the semi–direct product of (X,H) and (Y,F) with respect to η is a free* *distal flow.*

Proof: (I) We show the action of T is free. Suppose $(x,y,\alpha)(h,f) = (x,y,\alpha)$. Then $(xh^{\alpha},yf,\alpha\eta(f)) = (x,y,\alpha)$. Since (X,H) and (Y,F) both are free, h^{α} and f must be the identities of H and F respectively, and therefore α is the identity in Aut H. Hence T acts freely on $X \times Y \times \Phi$.

(II) We show the action of T is distal. Suppose that

$$\lim(x,y,\alpha)(h_\lambda,f_\lambda) = \lim(x',y',\alpha')(h_\lambda,f_\lambda).$$

Then we have

$$\lim xh_\lambda{}^{\alpha} = \lim x'h_\lambda{}^{\alpha'} \tag{1}$$

$$\lim yf_\lambda = \lim y'f_\lambda \tag{2}$$

$$\lim \alpha\eta(f_\lambda) = \lim \alpha'\eta(f_\lambda) \tag{3}$$

From (3), we have $\alpha = \alpha'$. Since (X,H) and (Y,F) both are distal, $x = x'$ and $y = y'$. Hence the action is distal. The proof is now complete.

We remark here that one may use the Ellis enveloping semigroups instead of Aut H to construct flows. We shall not go into the details of this aspect here.

3: Suspension of a flow: First we recall the definition of the suspension of a flow. Let (X,H) be a flow with H a locally compact group. Let H be a closed uniform subgroup of a locally compact group F. Define the transformation group $(X \times F,F)$ by the rule

$$(x,f)f' = (x,ff')$$

(notice that $X \times F$ in general is not compact, though X is compact by definition).

Define the relation in $X \times F$ by the rule: $(x,f) \sim (x',f')$ if and only if there exists an element $h \in H$ such that $(xh,h^{-1}f) = (x',f')$.

It is straightforward to check that the above relation is an F–invariant equivalence relation. Furthermore, $X \times F/\sim$ is a compact Hausdorff space. We shall use the notation $(X \times_H F,F)$ to denote the transformation group $(X \times F/\sim,F)$ and shall call $(X\times_H F,F)$ the *suspension of* (X,H) *with respect to* F. We use $<x,f>$ to denote the equivalence class of (x,f).

3.1 Proposition: *Let* $(X \times_H F,F)$ *be the suspension of* (X,H) *with respect to* F. *Then we have the following conditions.*

(1) *If* (X,H) *is a minimal* flow, *then* $(X\times_H F,F)$ *is also a minimal flow.*

(2) *If* (X,H) *is free, then* $(X\times_H F,F)$ *is free.*

(3) *Let* ϕ *be a homomorphism from* (X,H) *onto* (Y,H). *If* ϕ *is distal, then* Φ: $(X\times_H F,F) \to (Y\times_H F,F)$ is also distal where $\Phi < x,f> = <\phi(x),f>$. *If* ϕ *is an almost periodic extension, then* Φ *is an almost periodic extension.*

Proof: (1) and (2) follow easily from the definitions. We prove (3). Let $<x,f>$ and $<x',f'>$ be any two points in $X\times_H F$ such that $\Phi(<x,f>) = \Phi(<x',f'>)$. Then $<\phi(x),f> = <\phi(x'),f'>$. By definition, there exists an element h in H such that $(\phi(x),f) = (\phi(x')h,h^{-1}f') = (\phi(x'h),h^{-1}f')$. Let $x'' = x'h$. Then $\phi(x'') = \phi(x)$ and $<x'',f> = <x'h,f> = <x',h^{-1}f> = <x',f'>$. Hence we may express $<x',f'>$ by $<x'',f>$ with $\phi(x'') = \phi(x)$.

Now, assume that ϵ is a distal homomorphism. Suppose $<x,f>$ and $<x',f>$ are Φ–proximal. Since H is a uniform subgroup of F, there exists a net h_λ in H such that $\lim<x,f>h_\lambda = \lim <x',f>h_\lambda = <x_o,f_o>$. Then $\lim<x,fh_\lambda> = \lim<xh_\lambda,h_\lambda^{-1}fh_\lambda>$. Because H is uniform in F, taking a subnet if necessary, we have

$$\lim (xh_\lambda,h_\lambda^{-1}fh_\lambda) = (\lim xh_\lambda,\lim h_\lambda^{-1}fh_\lambda)$$

$$= (\lim xh_\lambda, h_o f_o)$$
$$= \lim (x'h_\lambda, h_o f_o)$$

where h_o is some element in H. Since ϕ is distal, we have $x = x'$, a fortiori, $\langle x,f\rangle = \langle x',f\rangle$. Therefore $\bar{\phi}$ is a distal homomorphism. The proof of almost periodicity of $\bar{\phi}$ when ϕ is almost periodic follows the same argument, hence we skip the detail.

Let $(*,H)$ denote the trivial flow, i.e. the phase space consists of a single point. Let H be a uniform subgroup of F. Then $(* \times_H F,F)$ is equivalent to the coset flow $(H\backslash F,F)$. Thus we have the following corollary.

3.2. Corollary: *If* (X,H) *is a distal flow, then* $(X \times_H F,F)$ *is distal over* $(H\backslash F,F)$. *Similarly, if* (X,H) *is an almost periodic flow, then* $(X \times_H F,F)$ *is almost periodic over* $(H\backslash F,F)$.

In view of the above corollary, it is interesting to know when $(H\backslash F,F)$ is a distal flow. Let F be any noncompact semi-simple analytic group. Let H be any uniform subgroup of F. It is well known that $(H\backslash F,F)$ cannot be a distal flow unless $H = F$.

3.3. Proposition: *Let* H *be a uniform subgroup of a locally compact group* F. *If* H *is a sub-invariant subgroup of* F, *then* $(H\backslash F,F)$ *is a distal flow.*

Proof: By definition, there exists a finite sequence of closed subgroups H_i of F, $1 \leq i \leq n$, such that: (1) $H_1 = H$, and $H_n = F$, and (2) H_i is a normal subgroup of H_{i+1} for $1 \leq i \leq n-1$. Since $H_i\backslash H_{i+1}$ is a compact group, $(H_i\backslash H_{i+1},H_{i+1})$ is an almost periodic minimal flow. This implies that $(H\backslash F,F)$ can be obtained through a sequence of group extensions, therefore $(H\backslash F,F)$ is a distal flow.

Let H be a uniform subgroup of a simply connected solvable analytic group F. In general H is not sub–invariant in F because $(H\backslash F,F)$ is not always a distal flow. However, when F is nilpotent analytic group, we have the following well known proposition.

3.4. Proposition: Let F be a simply connected nilpotent analytic group. Then every closed subgroup of F is sub–invariant in F.

Let X be a compact monothetic group. Let u be a monothetic group generator of X. Let $(X,\mathbb{Z})$ be the discrete flow where the action is simply the multiplication by the generator u, i.e., $xn = xu^n$. Since $\mathbb{Z}$ is uniform in the group $\mathbb{R}$ of additive real numbers, we have the suspension $(X\times_{\mathbb{Z}}\mathbb{R},\mathbb{R})$. Then $X\times_{\mathbb{Z}}\mathbb{R}$ is the solenoidal group. However the situation is drastically different when F is not an abelian group as the following examples show.

3.5. Example: Let F be a noncompact semi–simple analytic group. Let F = KAN, the Iwasawa decomposition with K a compact subgroup, A an abelian group and N a nilpotent subgroup. Then AN is a solvable subgroup. Let (X,AN) be a flow. Then we have the suspension $(X\times_{AN}F,F)$. Even when (X,AN) is a distal flow, $(X\times_{AN}F,F)$ is not a distal flow.

3.6. Example: Let F be the universal covering group of the group of rigid motions of the plane. Then F is the semi–direct product $\mathbb{R}^2 \times \mathbb{R}$. Let (a,b,t) be an element in F. We can express (a,b,t) by the following matrix.

$$\begin{bmatrix} \cos 2\pi t, & -\sin 2\pi t & a \\ \sin 2\pi t & \cos 2\pi t & b \\ 0 & 0 & 1 \end{bmatrix}.$$

Let H be the subgroup of F which consists of all the elements of translations, i.e., (a,b,t) with a,b,t integers. Then H is isomorphic with $\mathbb{Z}^3$. Since $\mathbb{Z}^3$ is a maximal almost periodic group, we can map $\mathbb{Z}^3$ faithfully as a dense subgroup of a compact group X. Now we can form the suspension of (X,H). We show that $(X\times_H F,F)$ is not almost periodic. Since H acts on X freely, by Proposition 3.1, F acts on $X\times_H F$ freely. If $(X\times_H F,F)$ is almost periodic, then F itself has to be a maximal almost periodic group. Since F is solvable and non abelian, F is not a maximal almost periodic group. Therefore $(X\times_H F,F)$ is not an almost periodic group.

Let H be a uniform subgroup of a locally compact group F. If ϕ is an almost periodic homomorphism from (X,H) onto (Y,H), then Φ: $(X\times_H F,F) \to (Y\times_H F,F)$ is almost periodic by Corollary 3.2. In comparison with the construction of solenoid from a monothetic group, it is desirable to improve the Corollary 3.2 for certain special cases. We give one of such instances in the following. First, let (X,H) be a group extension of (Y,H), i.e, there exists a compact group G such that (G,X,H) is a bi–transformation group and Y is the orbit space $G\backslash X$. Let E be a subgroup of H which acts on Y trivially, i.e, yh = y for all $y \in Y$ and $h \in E$. Let H be a uniform subgroup of F. Let Z be a central subgroup of F with the property $Z \cap H = E$ and E uniform in Z. Let β be the homomorphism from F onto F/Z. Since the kernel of β meets H is exactly the subgroup E, we may view H/E as a subgroup of F/Z and we can form the suspension $Y\times_{H/E}F/Z \cong Y\times_{\beta(H)}\beta(F)$ by the rule $\theta<x,f> = <\phi(x),\beta(f)>$ where ϕ is the canonical map: $X \to Y = G\backslash X$. We have the following statement.

3.6. Proposition: *θ is an almost periodic homomorphism from* $(X\times_H F,F) \to (Y\times_{\beta(H)}\beta(F),F)$.

Proof: Given $<x,f> \in X\times_H F$. Then $\theta^{-1}\theta(<x,f>) = \{<x',f'>: \theta(x) = \theta(x')$ and $\beta(f) = \beta(f')\}$. Because Z is central in F, so we may view Z as a group of automorphisms of the transformation group. Thus $(X\times_H F,F)$ is a group extension of $(Y\times_{\beta(H)}\beta(F),F)$ and it is an almost periodic extension. The proof is now complete.

We have assumed that ϕ: (X,H) → (Y,H) is a group extension in Proposition 3.6. This is purely for technical reasons. In fact, we only need to assume that ϕ is almost periodic, because if ϕ: (X,H) → (Y,H) is almost periodic, then there exists a flow (X',H) which is a group extension of (Y,H). And we have the following commutative diagram

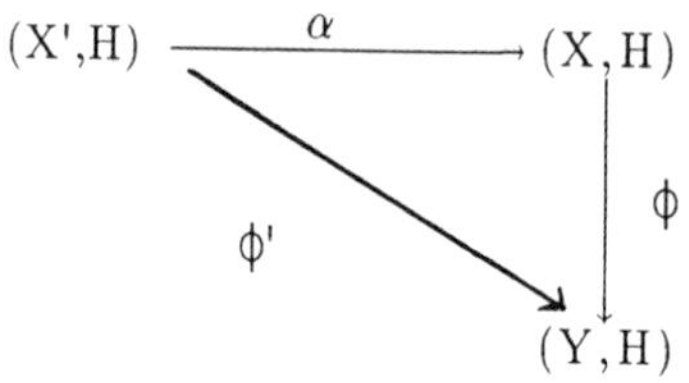

Define θ': $X'\times_H F \to Y\times_{\beta(H)}\beta F$ by the rule $\theta'(\langle x',f\rangle) = (\langle\phi'(x'),\beta(f)\rangle)$. By Proposition 3.6, θ' is almost periodic. Then θ: $X\times_H F \to Y\times_{\beta(H)}\beta(F)$ is almost periodic.

Now we give another proof of proposition 3.6 using the concept of regional proximal relations. Suppose $\langle x_\lambda,f_\lambda\rangle$, $\langle x_\lambda'f_\lambda'\rangle$ be two nets in $X\times_H F$ such that $\theta(\langle x_\lambda,f_\lambda\rangle) = \theta(\langle x_\lambda',f_\lambda')$. We may assume that $f_\lambda' = f_\lambda z_\lambda$ with $z_\lambda \in Z$. Since H is uniform in F, F–regional proximality is the same as H–regional proximality. Let h_λ be a net in H such that

$$\lim \langle x_\lambda,f_\lambda h_\lambda\rangle = \lim \langle x_\lambda',f_\lambda z_\lambda h_\lambda\rangle.$$

Then we have

$$\lim \langle x_\lambda h_\lambda,h_\lambda^{-1}f_\lambda h_\lambda\rangle = \lim \langle x_\lambda'h_\lambda,h_\lambda^{-1}f_\lambda h_\lambda z_\lambda\rangle,$$

because (X,H) is almost periodic over (Y,H). This implies that $\lim x_\lambda = x = \lim x_\lambda'$

and $\lim <x_\lambda, f_\lambda> = \lim <x', f_\lambda'>$. Hence θ is almost periodic and we have the second proof of proposition 3.6.

3.7. Proposition: *Let* (X,T) *be a distal flow. Suppose* $X_o < X_1 < \cdots < X_n$ *is a tower of almost periodic extensions, where* $(X_o,T) = (*,T)$ *and* $(X_n,T) = (X,T)$. *Let* T *be a uniform subgroup of* F. *Suppose there exists a sequence of closed normal subgroups* $Z_o = F > Z_1 > \cdots > Z_n = (e)$ *such that* (1) $Z_i \cap T = E_i$, (2) Z_i/Z_{i-1} *is central in* F/Z_{i-1}, and (3) $E_{i-1}Z_i/Z_i$ is uniform in Z_{i-1}/Z_i. Then $X_o \times_{E_oZ_1/Z_1} F/Z_1 < X_1 \times_{E_1Z_2/Z_2} F/Z_2 < \cdots < X \times_T F$ *is a tower of almost periodic extensions.*

Proof: The proof follows immediately from Proposition 3.6.

4: Co–dimension of a flow: Let (X,T) be a flow. The codimension of (X,T) is defined as $\dim X - \dim T$ where we use covering dimensions for X and T. When the codimension of (X,T) is small, it puts certain constraints on the flow. For instance, if (X,T) is a free distal minimal flow of codimension 1 and T is a nilpotent analytic group, then T must be abelian and the action must be equicontinuous ([IM]). Here we have the following Theorem.

4.1. Theorem: *Let* T *be a connected simply connected solvable analytic group acting freely on a manifold* X. *Consider the distal minimal flow* (X,T). *Suppose* $\dim X = \dim T + 1$. *Then there exists an abelian subgroup* A *of* T *such that* A *is uniform in* T. *The nilradical* N *of* T *is abelian. In particular, if* T *itself is nilpotent, then* T *is abelian.*

Proof: Let $(X_o,T) < (X_1,T) < \ldots < (X_\ell,T) < (X_{\ell+1}, T) = (X,T)$ be the canonical tower of (X,T) (See Theorem 1.2). Let y be a point in $Y = X_\ell$. Let T_y be the isotropy subgroup of T at y. By the dimension theorem (Theorem 1.1), $\dim Y + \dim \pi^{-1}(y) =$

dim T + 1, here π is the canonical map: $X \to Y = X_\ell$. If T is not transitive on Y, then dim Y > dim orbit of $y = \dim T - \dim T_y$. This implies $\dim T_y + 1 > \dim \pi^{-1}(y)$, $\dim T_y = \dim (T_y)_o \geq \dim \pi^{-1}(y)$, $(T_y)_o$ denotes the identity component of T_y. Since $(T_y)_o$ is simply connected and acts freely and $\pi^{-1}(y)$ is compact, dim $(T_y)_o$ can not be the same as dim $\pi^{-1}(y)$. Therefore, (Y,T) has to be transitive and hence (Y,T) is a coset flow. In otherwords, $Y \cong T_y\backslash T$ and T_y is uniform. Now T_y acts equicontinuously on $\pi^{-1}(y)$. Let $(G_{\ell+1}, Y_{\ell+1}, T)$ be the associated principal bundle (relative to π: $X \to Y$) where $G_{\ell+1}$ is a compact Lie group. Because T_y is a solvable Lie group acting freely and equicontinuously on $Y_{\ell+1}$, there is a subgroup T_y' of finite index in T_y such that T_y' is abelian. (We provide here some explanation about this fact. We may identify $G_{\ell+1}$ with a fiber, then T_y may be identified as a subgroup of $G_{\ell+1}$. Through these identifications, the closure of T_y is a compact solvable Lie group. It is well known that a compact connected solvable analytic group is actually a compact torus. Therefore T_y has a subgroup T'_y of finite index which is an abelian group). Let $A = T_y'$. Then A is uniform in T. Let Z^m be a discrete uniform subgroup of A. It is clear that Z^m is uniform in T. By a theorem of Mostow, $Z^m \cap N$ is uniform in N where N is the nilradical of T, and hence the nilradical N of T is abelian. In particular, when T is nilpotent, T = N, and T is abelian. So the proof is complete.

4.2. Example: We shall now construct a distal flow (X,T) such that: (1) T is simply connected solvable analytic group, (2) (X,T) is a distal minimal and the action is free, and (3) dim X = dim T + 1. In view of Theorem 4.1, we know that if such an (X,T) exists then the nilradical of T must be abelian and T must contain a uniform abelian subgroup. Furthermore, the order must be 2 and (X_1,T) must be a coset space, where X_1 is as in Theorem 2.1.

Let $G = \mathbb{R}^3 \times \mathbb{R}$, a semi-direct product where the additive group of real numbers acts on the vector group $\mathbb{R}^3$ over the real numbers by the rule: $\mathbb{R}^3 \times \mathbb{R} \to \mathbb{R}^3$, $((x,y,z),r) \to$

$(x \cos 2\pi r - y \sin 2\pi r,\ x \sin 2\pi r + y \cos 2\pi r,\ z)$. Let $a = (1,0,\alpha,0)$, $b = (0,1,\beta,0)$, $c = (0,0,1,0)$, $d = (0,0,\gamma,1)$ where $\alpha,\beta,\gamma,1$ are rationally independent. Let Γ be the group generated by these four elements. Observe that $(0,0,0,1)$ is central in G so Γ is contained in the abelian groups $\mathbb{Z} \times \mathbb{Z} \times \mathbb{R} \times \mathbb{Z} \subset \mathbb{R}^3 \times \mathbb{Z}$, and $\mathbb{R} \times \mathbb{R} \times \mathbb{R} \times (0)$ is the nilradical N of G. Let $T = (\mathbb{R} \times \mathbb{R} \times (0)) \times \mathbb{R}$ and consider the flow $(\Gamma\backslash G, T)$. This flow is a subflow of the coset flow $(\Gamma\backslash G,G)$. Because the maximal almost periodic factor must be abelian (G is solvable), we must factor out $\overline{[G,G]\Gamma}$ to obtain the almost periodic factor; hence the maximal almost periodic factor is one dimension.

We need to show $(\Gamma\backslash G,T)$ is distal and the action is free. It is distal because Γ is subinvariant and co–compact. To see it is free, suppose $g \in G$, $t \in T$ such that $\Gamma gt = \Gamma g$. This means $t \in g^{-1}\Gamma g$ for some g in G. Observe that any element in Γ has the form $(m,n,\ell + m\alpha + n\beta + k\gamma,k)$, where m,n,ℓ,k are integers and $\ell + m\alpha + n\beta + k\gamma$ is not zero unless ℓ,m,n,k all are zero. Also note that $(0)\times(0)\times R\times(0)$ is central. This means any inner automorphism will leave the third coordinate fixed. On the other hand, every element in T has its third coordinate always 0. Therefore $t \in g^{-1}\Gamma g$ if and only if it is identity. Hence $(\Gamma\backslash G,T)$ has free action.

Let T be a simply connected nilpotent analytic group. Let (X,T) be a distal minimal flow with free action. If $\dim X = \dim T + 1$, then we know that T must be abelian and the action must be equicontinuous. It is natural to ask if $\dim X = \dim T + 2$, can we find such a flow which is not equicontinuous? The following example answers this question.

4.3. Example: Let F be the group

$$\left\{\begin{bmatrix} 1 & x & z \\ 0 & 1 & y \\ 0 & 0 & 1 \end{bmatrix} : x,y,z \text{ are real numbers}\right\}.$$

For simplicity, we shall denote the matrix by [x,y,z]. Let A = [1,0,0], B = [0,1,0] and C = [0,0,1]. Let H be the subgroup generated by A,B,C. Let (X,H) be the flow defined in 2.2. Then $(X\times_H F, F)$ is a free distal minimal flow of co-dimension 2.

REFERENCES

1. [B] I.U. Bronstein, Extensions of minimal transformation groups. Translation from the Russian. Germantown, Maryland: Sijthoff and Noordhoff, VIII, 319 p., (1979).

2. [DM] D. De Riggi and N. Markley, Shear distality and equicontinuity, Pacific J. Math. 70 (1977), 337–345.

3. [DM2] D. De Riggi and N. Markley, The structure of codimension one distal flows with non–trivial isotropy, Rocky Mountain J. Math. 9 (1979), 601–616.

4. [Hahn] F. Hahn "Some embeddings recurrence properties and the Birkhoff–Markov theorem for transformation groups", Duke Math. J. (1960) 513–525.

5. [IM] E. Ihrig and D. McMahon, On distal flows of finite codimension, Indiana Univ. J. of Math. 33 (1984) 345–351.

6. [K] H. Keynes, Topological dynamics in coset transformation groups, Bull. AMS 72 (1966) p. 133–135.

7. [MWW]D. C. McMahon, Jaap Van der Woude, Ta–Sun Wu: Connectedness related to almost periodicity of compositions of flow homomorphisms, Pacific J. of Math. Vol. 127 (1987).

8. [M] C.C. Moore, Distal affine transformation groups, Amer. J. Math, Vol. 90, (1968) p. 733–751.

9. [MZ] C.C. Moore and R.J. Zimmer, Groups admitting ergodic actions with generalized discrete spectrum, Inventiones Math. 51 (1979) 171–188.

10. [MOS] P. Mostert, Sections in principal fibre spaces, Duke Math. J. 23 (1956), 57–71.

11. [R] M. Rees, On the structure of minimal distal transformation groups with topological manifolds as phase space. (University of Warwick).

12. [R'] M. Rees, On the fibres of a minimal distal extension of a transformation group, Bull. London Math. Soc. Vol. 28 (1978) pp. 97–104.

Index

D

E

F